Additive Manufacturing of Polymer Composites

Additive Manufacturing of Polymer Composites

Editor

Matthew Blacklock

 Basel • Beijing • Wuhan • Barcelona • Belgrade • Novi Sad • Cluj • Manchester

Editor
Matthew Blacklock
Mechanical and Construction
Engineering
Northumbria University
Newcastle upon Tyne
United Kingdom

Editorial Office
MDPI
St. Alban-Anlage 66
4052 Basel, Switzerland

This is a reprint of articles from the Special Issue published online in the open access journal *Polymers* (ISSN 2073-4360) (available at: www.mdpi.com/journal/polymers/special_issues/Addit_Manuf_Polym_Compos).

For citation purposes, cite each article independently as indicated on the article page online and as indicated below:

Lastname, A.A.; Lastname, B.B. Article Title. *Journal Name* **Year**, *Volume Number*, Page Range.

ISBN 978-3-7258-0468-9 (Hbk)
ISBN 978-3-7258-0467-2 (PDF)
doi.org/10.3390/books978-3-7258-0467-2

© 2024 by the authors. Articles in this book are Open Access and distributed under the Creative Commons Attribution (CC BY) license. The book as a whole is distributed by MDPI under the terms and conditions of the Creative Commons Attribution-NonCommercial-NoDerivs (CC BY-NC-ND) license.

Contents

Janis Baronins, Maksim Antonov, Vitalijs Abramovskis, Aija Rautmane, Vjaceslavs Lapkovskis and Ivans Bockovs et al.
The Effect of Zinc Oxide on DLP Hybrid Composite Manufacturability and Mechanical-Chemical Resistance
Reprinted from: *Polymers* **2023**, *15*, 4679, doi:10.3390/polym15244679 . . . . **1**

Timothy Russell and David A. Jack
Tensile and Compression Strength Prediction and Validation in 3D-Printed Short-Fiber-Reinforced Polymers
Reprinted from: *Polymers* **2023**, *15*, 3605, doi:10.3390/polym15173605 . . . . **20**

Pablo Zapico, Pablo Rodríguez-González, Pablo Robles-Valero, Ana Isabel Fernández-Abia and Joaquín Barreiro
Influence of Post-Processing on the Properties of Multi-Material Parts Obtained by Material Projection AM
Reprinted from: *Polymers* **2023**, *15*, 2089, doi:10.3390/polym15092089 . . . . **38**

Mustafa Saleh, Saqib Anwar, Abdulrahman M Al-Ahmari and Abdullah Yahia AlFaify
Prediction of Mechanical Properties for Carbon fiber/PLA Composite Lattice Structures Using Mathematical and ANFIS Models
Reprinted from: *Polymers* **2023**, *15*, 1720, doi:10.3390/polym15071720 . . . . **51**

Yingwei Hou and Ajit Panesar
Effect of Manufacture-Induced Interfaces on the Tensile Properties of 3D Printed Polyamide and Short Carbon Fibre-Reinforced Polyamide Composites
Reprinted from: *Polymers* **2023**, *15*, 773, doi:10.3390/polym15030773 . . . . **78**

Xiaojie Zhang, Jianhua Xiao, Jinkuk Kim and Lan Cao
A Comparative Analysis of Chemical, Plasma and In Situ Modification of Graphene Nanoplateletes for Improved Performance of Fused Filament Fabricated Thermoplastic Polyurethane Composites Parts
Reprinted from: *Polymers* **2022**, *14*, 5182, doi:10.3390/polym14235182 . . . . **97**

Sharmad Joshi, Enrique Gazmin, Jayden Glover, Nathan Weeks, Fazeel Khan and Scott Iacono et al.
Improved Electrical Signal of Non-Poled 3D Printed Zinc Oxide-Polyvinylidene Fluoride Nanocomposites
Reprinted from: *Polymers* **2022**, *14*, 4312, doi:10.3390/polym14204312 . . . . **110**

Igor Polyakov, Gleb Vaganov, Andrey Didenko, Elena Ivan'kova, Elena Popova and Yuliya Nashchekina et al.
Development and Processing of New Composite Materials Based on High-Performance Semicrystalline Polyimide for Fused Filament Fabrication (FFF) and Their Biocompatibility
Reprinted from: *Polymers* **2022**, *14*, 3803, doi:10.3390/polym14183803 . . . . **129**

Chethan Savandaiah, Stefan Sieberer, Bernhard Plank, Julia Maurer, Georg Steinbichler and Janak Sapkota
Influence of Rapid Consolidation on Co-Extruded Additively Manufactured Composites
Reprinted from: *Polymers* **2022**, *14*, 1838, doi:10.3390/polym14091838 . . . . **146**

Jiongyu Ren, Rebecca Murray, Cynthia S. Wong, Jilong Qin, Michael Chen and Makrina Totsika et al.
Development of 3D Printed Biodegradable Mesh with Antimicrobial Properties for Pelvic Organ Prolapse
Reprinted from: *Polymers* **2022**, *14*, 763, doi:10.3390/polym14040763 . . . . **160**

Ved S. Vakharia, Lily Kuentz, Anton Salem, Michael C. Halbig, Jonathan A. Salem and Mrityunjay Singh
Additive Manufacturing and Characterization of Metal Particulate Reinforced Polylactic Acid (PLA) Polymer Composites
Reprinted from: *Polymers* **2021**, *13*, 3545, doi:10.3390/polym13203545 . . . . **182**

The Effect of Zinc Oxide on DLP Hybrid Composite Manufacturability and Mechanical-Chemical Resistance

Janis Baronins [1,2,*], Maksim Antonov [3], Vitalijs Abramovskis [1], Aija Rautmane [1,2], Vjaceslavs Lapkovskis [1], Ivans Bockovs [4], Saurav Goel [5,6], Vijay Kumar Thakur [7] and Andrei Shishkin [1]

1. Laboratory of Ecological Solutions and Sustainable Development of Materials, Faculty of Materials Science and Applied Chemistry, Institute of General Chemical Engineering, Riga Technical University, Pulka 3, K-3, LV-1007 Riga, Latvia; vitalijs.abramovskis@edu.rtu.lv (V.A.); aija.rautmane@rtu.lv (A.R.); vjaceslavs.lapkovskis@rtu.lv (V.L.); andrejs.siskins@rtu.lv (A.S.)
2. Latvian Maritime Academy of Riga Technical University, Riga Technical University, Flotes Str. 12 K-1, LV-1016 Riga, Latvia
3. Department of Mechanical and Industrial Engineering, Tallinn University of Technology, Ehitajate Tee 5, 19086 Tallinn, Estonia; maksim.antonov@taltech.ee
4. Faculty of Materials Science and Applied Chemistry, Institute of Polymer Materials, Riga Technical University, 3/7 Paula Valdena Street, LV-1048 Riga, Latvia; ivans.bockovs@rtu.lv
5. School of Engineering, London South Bank University, London SE1 0AA, UK; goels@lsbu.ac.uk
6. Department of Mechanical Engineering, University of Petroleum and Energy Studies, Dehradun 248007, India
7. Biorefining and Advanced Materials Research Center, Scotland's Rural College (SRUC), Kings Buildings, West Mains Road, Edinburgh EH9 3JG, UK; vijay.thakur@sruc.ac.uk
* Correspondence: janis.baronins@rtu.lv

Abstract: The widespread use of epoxy resin (ER) in industry, owing to its excellent properties, aligns with the global shift toward greener resources and energy-efficient solutions, where utilizing metal oxides in 3D printed polymer parts can offer extended functionalities across various industries. ZnO concentrations in polyurethane acrylate composites impacted adhesion and thickness of DLP samples, with 1 wt.% achieving a thickness of 3.99 ± 0.16 mm, closest to the target thickness of 4 mm, while 0.5 wt.% ZnO samples exhibited the lowest deviation in average thickness (±0.03 mm). Tensile stress in digital light processed (DLP) composites with ZnO remained consistent, ranging from 23.29 MPa (1 wt.%) to 25.93 MPa (0.5 wt.%), with an increase in ZnO concentration causing a reduction in tensile stress to 24.04 MPa and a decrease in the elastic modulus to 2001 MPa at 2 wt.% ZnO. The produced DLP samples, with their good corrosion resistance in alkaline environments, are well-suited for applications as protective coatings on tank walls. Customized DLP techniques can enable their effective use as structural or functional elements, such as in Portland cement concrete walls, floors and ceilings for enhanced durability and performance.

Keywords: DLP; additive manufacturing; ZnO; photocured resin; tensile test; corrosion; acidic environment; alkaline environment

1. Introduction

In various production processes, epoxy resin (ER) is widely utilized by industries due to its superior adhesion, ability to eliminate air bubbles, and excellent electrical and thermal insulating properties [1–3]. The rigid molecular structure of the ER matrix provides high thermal stability and mechanical strength [4]. As global initiatives focus on sustainable practices, the European Union Green Deal and worldwide objectives for zero-emission industries emphasize the importance of using manufacturing waste and recycled materials in composites [5].

Additive manufacturing, particularly 3D printing, is a versatile process that can produce a wide range of materials. These include metal oxides such as ZnO, SiO_2, and TiO_2, which can be combined with polymers to create new and innovative products [6–8].

Various techniques such as intense pulsed light (IPL), digital light processing (DLP), fused deposition modeling (FDM), fused filament fabrication (FFF), and stereolithography (SLA) can be used to manufacture these products. It's worth noting that while FDM and FFF share a common additive manufacturing approach, distinctions arise with FDM being a trademarked term by Stratasys, and FFF being a more generic, open-source concept that enables wider material and equipment compatibility [9]. Table 1 summarizes key data for composite materials used in 3D printing, showcasing the potential of metal oxides in different applications.

The combination of ER monomers with prepolymers and photosensitive additives enables rapid liquid photocuring into a complex shaped solid product using techniques such as SLA and DLP [10,11]. Ultraviolet (UV) cured coatings and printable materials, known for their environmental safety in adapted working conditions and lower energy consumption, as compared to thermally cured ER analogs, have gained attention [12,13]. UV-curable coatings find applications in shipbuilding, providing corrosion protection and mechanical durability properly grinded surface [14]. However, challenges exist, such as the limitations of UV-cured thin films in barrier (e.g., gas, and solvents) properties and mechanical strength [15,16].

Polyurethane acrylate (PUA) oligomers, a petrochemical product, play a significant role in UV-cured surfaces [17,18]. Relatively soft UV-cured ER surfaces exhibit low scratch resistance [19,20]. One of the solutions is to increase the polymer matrix hardness. The incorporation of harder additives (e.g., ZnO with typical Vickers hardness from 2 to 4 GPa) [21] into PUA (with hardness from 0.12 to 0.23 GPa) [22] matrix is the simplest approach. It should be noted that Vickers hardness tests are commonly used for ceramics to assess their resistance to indentation, while Shore hardness tests are preferred for polymers to measure their flexibility and elasticity. Comparing the hardness of polymers and ceramics is complicated due to their inherent differences in material properties; polymers are typically more compliant and deformable, while ceramics are significantly more rigid and brittle, making direct comparisons challenging as hardness values are influenced by the materials' distinct responses to applied stress and deformation.

Despite the historically performed tuning of polyurethane mechanical properties with different nanoparticles, such as nano clay [23], carbon nanotubes [24], metal oxide, hydroxyapatite, and graphene (including its derivatives) [25,26], the market offers limited number of solid fillers adopted for SLA and DLP needs. Industry commonly uses rigid inorganic fillers such as kaolin and silica, but the challenge lies in managing surface roughness at high concentrations [27]. Epoxy composites' heterogeneity influences physio-mechanical properties [28], and the use of ceramic micro and nanoparticles as fillers aids in reducing shrinkage and warping errors in SLA and DLP products [29,30].

DLP polymers with inorganic fillers, particularly zinc oxide (ZnO), have attracted attention for their improved thermal stability [31], mechanical strength, reduced vapor permeability [32], and enhanced optical and electrical properties [33]. ZnO, classified as an n-type metal oxide [34], exhibits excellent radiation resistance, electrical and optical properties optical (e.g., absorption at local UV spectrum ranges from around 350 to 362 nm) [35], and finds applications in various industries. Industrial production of gas [36] sensors, humidity sensors [37], catalysts [38], and antibacterial materials [39] are some of typical ZnO application examples.

Dispersion of ZnO in PUA oligomers require high energy. Industry and researchers mostly apply in-situ polymerization [40] or blending [41] processes for manufacturing polyurethane-ZnO composites. Manufacturers currently produce liquid PUA oligomers with blended rigid particles. Several researchers have reported the ZnO mechanical stirring and ultrasonic mixing mostly at room temperature. Different approaches have resulted in different required mixing durations and stability of homogeneous dispersions [42–45]. Unfortunately, the poor compatibility and interfacial interaction between the ZnO particles and liquid PUA oligomers promotes the aggregation and sedimentation of ZnO particles [46]. Conventionally produced PUA exhibits similar corrosion rates in aqueous acidic

(5% HCl) and alkaline (5% NaOH) solutions [47]. However, much less corrosion research data is available on additive manufactured PUA products.

Table 1. Comprehensive summary of composite materials for 3D printing: metal oxides (ZnO, SiO$_2$, and TiO$_2$) containing polymer compositions and properties of products.

No.	Filling Substance	Particle Size	Loading	Additive Manufacturing Method	Properties/Applications	Ref.
1	ZnO	0.7 μm	44–52 vol.%	IPL	The compressive strength ranges from 5.08 to 11.09 MPa at temperatures of 900 to 1500 °C.	[48]
2	ZnO	100 nm	38 wt.%	DLP	The compressive strength ranges from 1.26 to 6.82 MPa for materials with a Gyroid structure and Schwartz P structure.	[49]
3	ZnO	<130 nm	10 wt.%	FDM	New devices are continuously emerging for pertinent applications in fields such as environmental science, energy, and catalysis.	[50]
4	ZnO	Highly concentrated ZnO ink	50 vol.%	Robotic deposition equipment	ZnO optoelectronic devices operate at THz frequencies and can be seamlessly integrated with various optical components such as waveguides and resonators.	[51]
5	SiO$_2$	100 nm	2 wt.%	IPL	The applicability of inkjet 3D printing in the electronics industry is promising with ink characteristics such as a density of 1.05 g·mL^{-1} and a viscosity of 9.53 mPa·s, enabling precise and controlled deposition of conductive materials for circuit fabrication.	[52]
6	SiO$_2$	5–15 nm	0.5–4 wt.%	FFF	Tensile stress ranges from 31 to 35 MPa, with a corresponding tensile modulus of elasticity of 138–148 MPa. Additionally, it has a flexural strength of 40–47 MPa and a flexural modulus of elasticity spanning 786–927 MPa. The impact resistance falls within the range of 3.72–4.01 kJ·m^{-2}, and the microhardness measures between 12.44 and 13.34 HV.	[53]
7	SiO$_2$	The diameter of the fiber is 6.5 μm. 20 nm powder	10 vol.% (fiber) 3.68–11.76 wt.% powder	Direct ink writing	The composite material exhibits a dielectric constant of 1.2 and a dielectric loss tangent of 1.5×10^{-2}. Its bending strength ranges from 11.2 ± 1.1 to 14.15 ± 1.3 MPa, while the apparent porosity falls within the range of 24.36% to 24.48%.	[54]
8	TiO$_2$	10 nm	0–2.5%	SLA	The material demonstrates a tensile stress between 17 and 25 MPa, an impact resistance of 17.5 to 25 kJ·m^{-2}, a hardness of 80 HV, and an elongation at break of 8 to 8.5%.	[55]
9	TiO$_2$	50–300 μm	10–20%	FDM	The grain size distribution plays a crucial role in the frequency-dependent variations of the dielectric constant and loss factor in this ceramic composite. These characteristics are essential for its performance in dielectric applications, including its use in capacitors for A/D converters, filtration capacitors, and dielectric resonant antennas.	[56]

The present article reports the trial research results and discussion on possible DLP PUA mechanical properties and chemical durability enhancement with the help of ZnO at maximum applied concentrations. The test results will be useful for successful 3D parts printing made of polymer based-hybrid composites with extended functionality and better mechanical performance.

2. Materials and Methods

2.1. Applied Materials

The producer is not disclosing the exact composition of the ER Anycubic 3D Printing UV Sensitive Resin, Basic, Clear (Anycubic, Shenzhen, China) for some commercial reasons. The available materials safety data sheet for this product indicates the approximate concentrations of polyurethane acrylate (30–60 wt.%, CAS N° 82116-59-4, epoxy with solvents); isooctyl acrylate (10–40 wt.%, CAS N° 29590-42-9, monomer) and photo initiator (phosphine oxide, 2–5 wt.%). According to the information on producer's webpage, the liquid form of the ER exhibits density of 1.1 g·cm^{-3}, and viscosity of 552 mPa·s. The photo-cured (at the wavelength of 405 ± 8 nm) [57] translucent ER exhibits solid density of 1.284 g·cm^{-3}, yield tensile stress up to 23.4 MPa, and elongation up to 14.2% [58].

As-received 99.9 wt.% pure nanometer spherically (episodically) shaped [59] particles containing zinc oxide (ZnO) powder with particle sizes up to 5 μm (Sigma Aldrich, Saint Louis, state of Missouri, USA, product N° 205532, CAS N° 1314-13-2) was used as the filler for production of reinforced photo-cured ER samples [60].

2.2. Methods for Samples Manufacturing

The planetary mixer Hauschild Speed Mixer DAC 150.1 FVZ-K (Hamm, Germany) with rotational speed of 1150 rpm was set for 5 min to mix each selected composition with the total weight of 140 g (measured by the laboratory scales KERN EMB-S). The resulting suspension was transported to the printing laboratory and used for DLP within 1 h after mixing. In a series of trials, various DLP attempts were made with different ZnO concentrations to ascertain a dependable printable mixture. Ultimately, a decision was made to produce suspensions with ZnO concentrations of 0.5, 1, 1.5, and 2 wt.%. Additionally, reference samples were printed without additives and subsequently subjected to tensile tests to determine only tensile stress at fracture.

The 3D model for the test specimen was drawn with the help of the AutoCad software (version S.51.0.0 AutoCad 2022) according to the specification for the Type IV described in the standard ASTM D638-14 [61]. The obtained computer aided design (CAD) was virtually sliced with the help of the Photon Workshop V2 1.23 RC8 software, as demonstrated in Table 2.

Table 2. Slice settings for manufacturing specimens with the help of the Anycubic Photon Mono 3D printer. Printing time of one sample was approximately 15 min (demonstrated by DLP printer).

Slice Setting Parameter	Value	Unit of Measure
Layer thickness	50	μm
Normal exposure time	2	s
Off time	0.5	s
Bottom exposure time	40	s
Bottom layers	6	layers
Z axis lift distance (after printing of each layer)	6	mm
Z axis lift speed	5	mm·s^{-1}
Z axis retract speed	6	mm·s^{-1}

Settings were chosen for the selected DLP type 3D printer Anycubic Photon Mono (405 nm wavelength, resolution of 1620 × 2560 pixels and pixel size of 51 μm) with detailed specifications demonstrated on the manufacturer webpage [62]. The bottom exposure time of 60 s provided the build plate-resin and initial resin-resin layers adhesion at the beginning of each 3D printing process. The 3D printer was inserted in the insulated box to reduce temperature fluctuation and to stabilize it at +20 °C.

Produced specimens were removed from the build plate, prewashed in isopropanol (99.8%) and immersed in 5 L of isopropanol container and washed with the help of the Anycubic Wash&Cure Machine 2.0 for 10 min. The obtained specimens were additionally cured with UV light (405 nm wavelength, 40 W power) for 10 min inside the same device set in the cure mode. The ZnO containing specimens exhibited the photochromism effect under the applied intensive UV radiation [63].

2.3. Methods for Visual and Mechanical Characterization of Materials and Samples

The granulometric composition analysis of the ZnO powder involved the utilization of the laser analyzer Fritsch Analysette 22, manufactured by FRITSCH GmbH in Germany. This analysis was conducted following the dispersion of the powder in isopropanol. Surfaces of produced and tested samples were observed with the help of digital microscope Keyence VHX-2000 (Keyence Ltd., Osaka, Japan) and scanning electron microscope (SEM) TESCAN VEGA (high vacuum mode was selected) and OLYMPUS SZX10 (Olympus Corporation, Tokyo, Japan). Tensile tests of manufactured specimens were performed with the help of tensile testing machine Zwick/Roell Z150 (ZwickRoell GmbH & Co. KG, Ulm, Germany). The applied loading speed was 5 mm·min^{-1}. The reference specimens (0 wt.%), fixed in the Zwick/Roell Z020 (ZwickRoell GmbH & Co. KG) tensile testing machine, were subjected to the same loading parameters. The recorded data underwent analysis utilizing the TestXpert software (ZwickRoell GmbH & Co. KG, Ulm, Germany, version II). This software facilitated the determination of the modulus of elasticity by extracting it from the slope of the stress-strain curve within the linear region. The dimensions of specimens were measured with the help of the micrometer screw gauge (± 0.005 mm).

2.4. Chemical Corrosion Tests

Two different laboratory-scaled chemical corrosion tests were performed to simulate the effect of typical chemically active liquid substances stored in tanker tanks. Produced specimens were immersed in aqueous acetic acid (10%, pH = 5) solution and 1 M NaOH (pH = 12) solution to estimate the specimen's resistance to acidic and alkaline media, respectively. The universal indicator was used for the measurement of solutions pH. All boxes were sealed, and each corrosion test was left for 7 days at room temperature. Afterward, the samples were extracted from their containers, dried at a constant temperature of 30 °C for one hour until they reached a stable mass, and subsequently subjected to measurement. The loss of mass was measured with the help of the scales KERN EMB-S (KERN & SOHN GmbH, Balingen, Germany) before and after the immersion test.

3. Results
3.1. Visual and Mechanical Properties of Materials and Samples
3.1.1. ZnO Powder Granulometric Analysis Result

The simplified result of granulometric analysis shows that the median particle size (D_{50}) in tested ZnO powder is about 177 nm, as demonstrated in Figure 1. The similar particle size distributions (D_{50} = 117 nm) result has been represented by Meng, F.; et.al. for the same ZnO powder product [64].

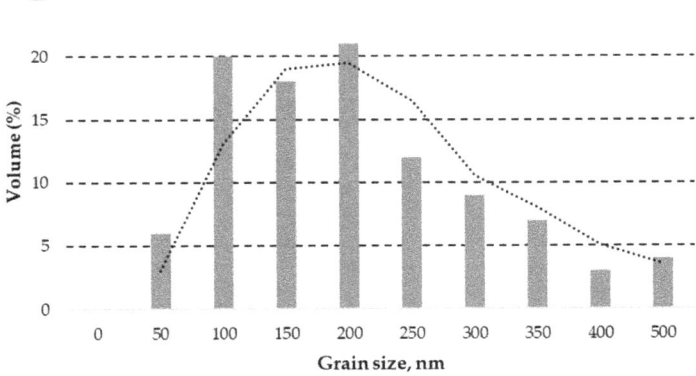

Figure 1. A simplified granulometric analysis of ZnO powder obtained through laser granulometry with two period moving average trendline indicated (dashed line).

Nevertheless, many particles and their agglomerates with particle sizes above 5000 nm (5 μm) were detected by SEM, as demonstrated in Figure 2a,b. ZnO powder manufacturers recommend employing various deagglomeration methods to achieve target particle sizes for additive manufacturing. These methods, including ultrasonic dispersion, mechanical stirring, or high-shear mixing, can effectively break down agglomerates and ensure the uniform dispersion of particles, ultimately enhancing the quality of the 3D printing feedstock.

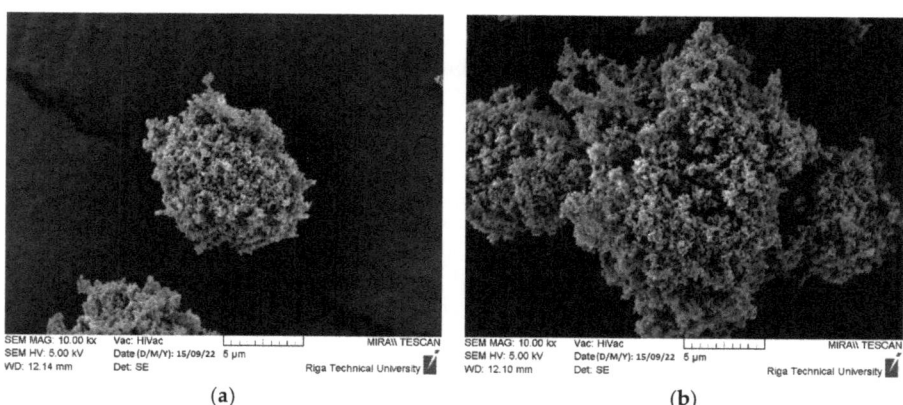

(a) (b)

Figure 2. Detected ZnO particles (**a**) and agglomerates (**b**) with particle sizes above 5 μm.

3.1.2. Visual Characteristics and Tensile Stress of Manufactured Samples

Some of DLP samples sometimes exhibited various defects such as partial delamination from the same area on the build plate, as shown in the Figure 3a,b. The origin of this event can be attributed to the slightly lower cured ER adhesion force due to the insufficient surface roughness on the working area of the build plate. Hence, in specific regions where the adhesion force between the sample and the transparent film surpasses the adhesion between the build plate and the cured sample, or between cured layers, it can significantly impact the overall quality of the DLP specimen. In rare cases, interlayer adhesion problems were observed during first DLP trials with selected target concentrations. The effect was mitigated by thorough premixing of the suspension before pouring it into the resin vat. Only best quality DLP samples were taken for further testing.

The sufficiently low ZnO aggregation and sedimentation process in the liquid ER provided good quality solid particle distribution in 3D printed objects. However, the largest objects (or thinnest layers) would require longest duration which may lead to significantly heterogenous cured polymer structure. The use of silane agents could potentially improve the maintenance of high Zn nanoparticle dispersion in the DLP polymer matrix [65,66].

The measurements of thickness and width at the center for each specimen are presented in Table 3.

An increase in the concentration of ZnO beyond 0.5 wt.% results in a significant escalation in thickness deviation, ranging from 4.3 times (at 2 wt.%) to a substantial 11.3-fold increase (at 1.5 wt.%). However, the 3D printing of specimens from ER with 1 wt.% ZnO leads to thickness values that are closest to the target values, as compared to other specimens. The average values of all specimens show negative deviation from target thickness value (4 mm). Therefore, possible increase in target thickness (or total height along Z axis) should be considered in CAD design to reach closer desired values.

The mean center width values across all samples exhibit a range from 5.93 mm (0 wt.%) to 6.49 mm (1 wt.%). An increase in ZnO concentration above 0.5 wt.% results in a proportional rise in the average deviation in center width, ranging from 1.5 times (1 wt.%) to 5.5 times (2 wt.%). The average values of all ZnO reinforced specimens show positive deviation from target width at the center value (6 mm). Hence, it is advisable to contemplate a potential reduction in the target center width value or a shorter exposure time during CAD design and DLP printing processes. This approach can bring the final 3D printed object closer to the desired specifications, eliminating the need for additional post-printing material removal.

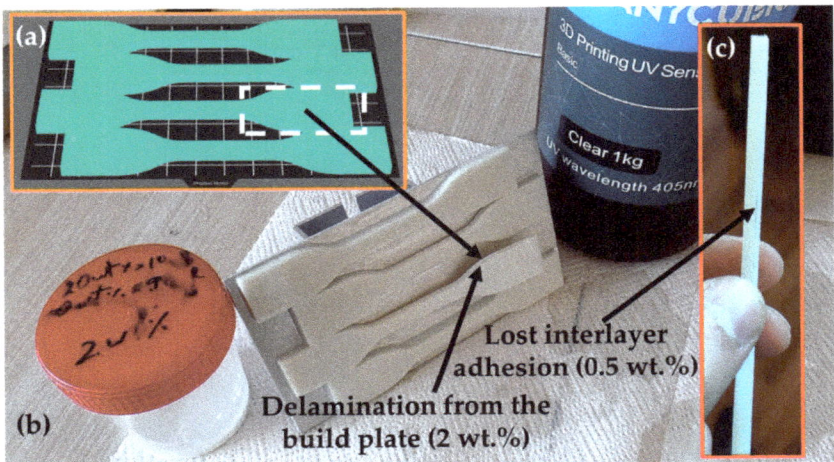

Figure 3. The orientation and layout of CAD objects on the build plate with indicated most failing area (a); DLP specimens with defect caused to one of the specimens by the delamination from the build plate (b); and interlayer adhesion defect (c) caused by transparent film surpassing the adhesion between the build plate and the cured sample.

Table 3. The effect of ZnO concentrations on average thicknesses and widths at the centers of 3D printed specimens and average deviation from target thickness (4 mm) and target width (6 mm) values. A minimum of five samples were measured for each material.

	ZnO Concentration				
	0 wt.%	0.5 wt.%	1 wt.%	1.5 wt.%	2 wt.%
Average thickness, mm	4.01	3.84	3.99	3.72	3.83
(deviation, mm)	(±0.07)	(±0.03)	(±0.16)	(±0.34)	(±0.12)
Average deviation from the target thickness, mm	(~0.000)	−0.160	−0.003	−0.280	−0.170
Average width at the center, mm	5.93	6.33	6.49	6.42	6.47
(deviation, mm)	(±0.03)	(±0.02)	(±0.03)	(±0.07)	(±0.11)
Average deviation from the target width at the center, mm	−0.070	+0.330	+0.490	+0.420	+0.470

Fluctuations in sizes and printed parts warping (see Figure 3b) are important errors typically caused by the shrinkage of acrylate-based ERs [67]. Updated recommendations and instructions should be formulated in accordance with research findings and practical experience across various manufacturing conditions. The polymer bond formation between polymerizing material and slightly denser photocured solid polymer and cooling after exothermic chemical reaction caused thermal expansion are two main factors causing the DLP product shrinkage. The remaining internal stress promotes the warping effect in combination with adhesion forces to the transparent film during every retraction process [68]. Reduction in errors and improved mechanical properties can be achieved by careful post processing by heating and photocuring of DLP parts [69].

The build plate side of each sample displays the replication of visually observable stripes and defects, as illustrated in Figure 4. These defects are attributed to the presence of several ZnO particles with sizes ranging from 1000 to 5000 nm (from 1 to 5 μm), exhibiting a homogeneous distribution.

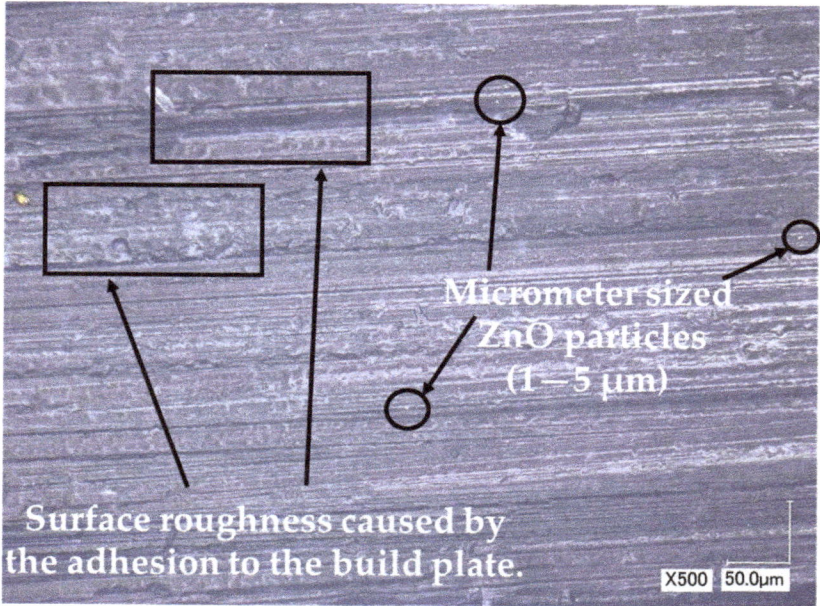

Figure 4. The mimicked rough surface caused from the adhesion to the profiled (textured) build plate and micrometer sized ZnO particles (possible agglomerates) observed by the optical microscope on the surface of 0.5 wt.% ZnO containing sample.

All DLP composites tested with varying ZnO concentrations exhibit comparable tensile stress at yield, ranging from 23.29 MPa (1 wt.%) to 25.93 MPa (0.5 wt.%), with corresponding elongation/deformation values within the range of 0.0089 to 0.011 mm·mm^{-1}, as outlined in Table 4. In contrast, the results of reference samples only provide information on tensile stress at fracture. Notably, it is observed that with higher ZnO concentrations, there is a decrease in tensile stress at fracture. The tensile test measurement curves unequivocally illustrate the increased brittleness of the DLP composite as ZnO concentration is raised, as visually depicted in Figure 5.

Also, the higher ZnO concentration is the lower the elastic modulus, as shown in Figure 6. This value experiences a reduction of approximately 1.4 times at 0.5 wt.% (reaching 2880 MPa) and 2 wt.% (decreasing to 2001 MPa).

Table 4. The average tensile stress at yield and fracture loads of ZnO containing DLP composites. A minimum of five samples were measured for each material.

	Sample				
	0 wt.%	0.5 wt.%	1 wt.%	1.5 wt.%	2 wt.%
Tensile stress at yield, σ_y (MPa)	-	25.93 (±0.44)	23.29 (±0.18)	25.19 (±1.25)	23.41 (±3.21)
Elongation/deformation at yield, ε (mm·mm^{-1})	-	0.0091 (±0.0014)	0.0089 (±0.0012)	0.0110 (±0.0001)	0.0091 (±0.0036)
Tensile stress at fracture, σ_{UTS} (MPa)	43.1 (±4.06)	38.76 (±0.03)	35.40 (±0.03)	29.93 (±0.03)	24.04 (±0.03)
Elongation/deformation at fracture, ε_{UTS} (mm·mm^{-1})	-	0.0156 (±0.0001)	0.0172 (±0.0007)	0.0146 (±0.0011)	0.0112 (±0.0051)

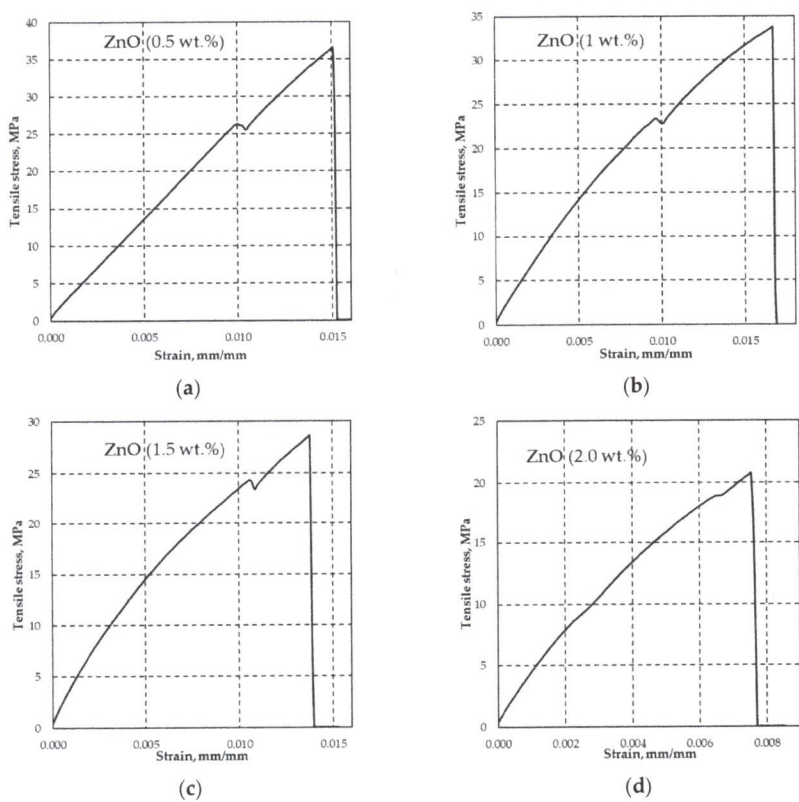

Figure 5. Examples of effect of ZnO concentrations of 0.5 wt.% (**a**); 1 wt.% (**b**); 1.5 wt.% (**c**); and 2 wt.% (**d**) on tensile stress at yield and fracture strains. The increase in ZnO concentrations leads to more brittle composite behavior under applied destructive tensile force.

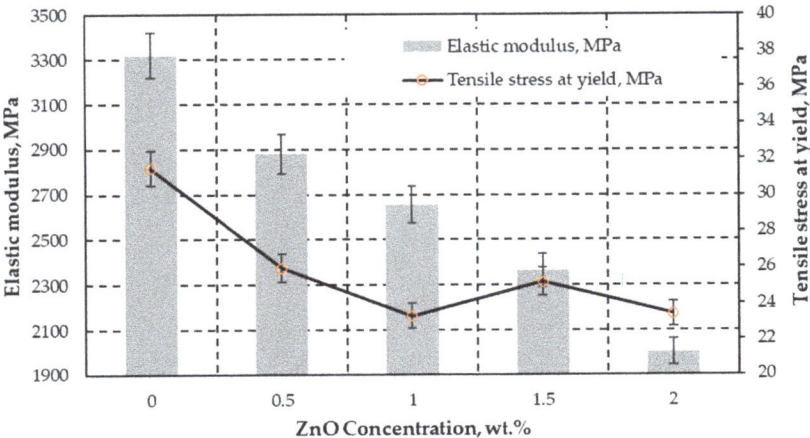

Figure 6. The graph displays the correlation between the elastic modulus (E) in megapascals (MPa) on the left Y-axis and the tensile stress at the yield point (σ_y) in MPa on the right Y-axis. These measurements are analyzed with respect to different ZnO concentrations. Reference sample (0 wt.%) exhibit fracture at yield.

The shape, size, dispersion, and interfacial interaction between ZnO filler and DLP polymer matrix influence the mechanical properties of the product [32,65,66]. The 1 wt.% of tetrapod shaped ZnO whiskers provide similar tensile stress at break (36 MPa) in thin film PUA matrix [70], as compared to present test result (35.4 MPa). The needle shaped ZnO structure would be interesting alternative due to potential DLP matrix reinforcement in case of avoiding the agglomeration and sedimentation without significant impact on surface roughness [27].

3.1.3. DLP ZnO Composites Corrosion Resistance in Acetic Acid Solution (pH = 5)

The ZnO reacts with acetic acid and generates zinc acetate salts according to chemical Equation (1):

$$2CH_3COOH + ZnO = Zn(CH_3COO)_2 + H_2O \qquad (1)$$

The zinc acetate generated crystal hydrates of $Zn(CH_3COO)_2(H_2O)_2$ in the presence of water [71]. The samples with lowest ZnO concentration (0.5 wt.%) exhibited almost no visually observable deterioration signs after immersion in acetic acid, as shown in Figure 7a. The increase in the weight by 2.31% (see Figure 8a) indicates the embedment of almost all generated corrosion products (zinc acetic crystals) in the pores of plastic composites with 0.5 wt.% ZnO. The corrosion of samples with 1, 1.5 and 2 wt.% ZnO leads to detachment of deteriorated plastic particles from the surfaces of exposed materials, as demonstrated in Figure 7b–d, Figures 8b and 9a,b.

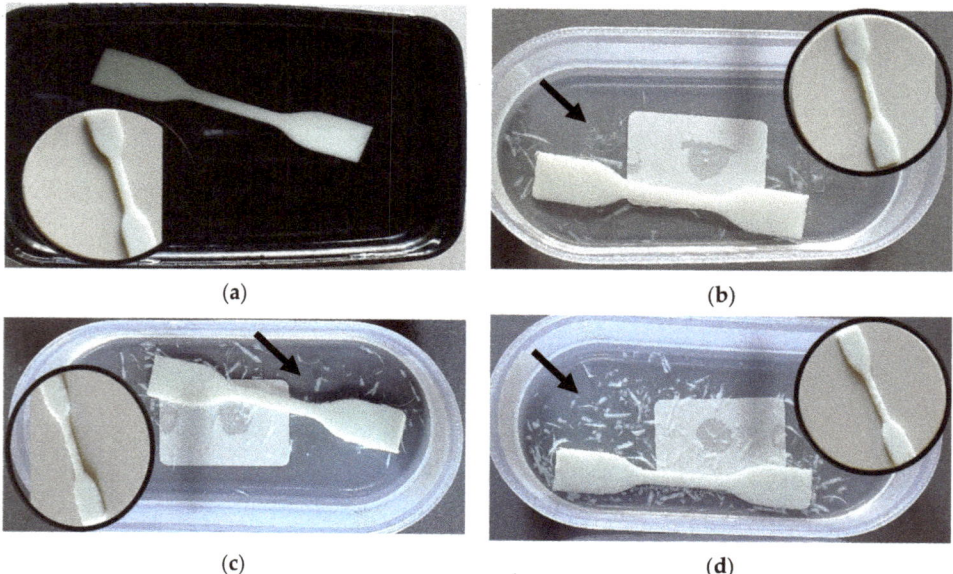

Figure 7. Degradation results after 7 days exposure of polymer composites with ZnO concentrations of 0.5 wt.% (**a**); 1 wt.% (**b**); 1.5 wt.% (**c**); and 2 wt.% (**d**) in acetic acid (pH = 5) solution. Black arrows indicate the degradation products caused by ZnO reaction with acetic acid and destruction of polymer structure by the formation of relatively large crystalline salt (zinc acetate) grains.

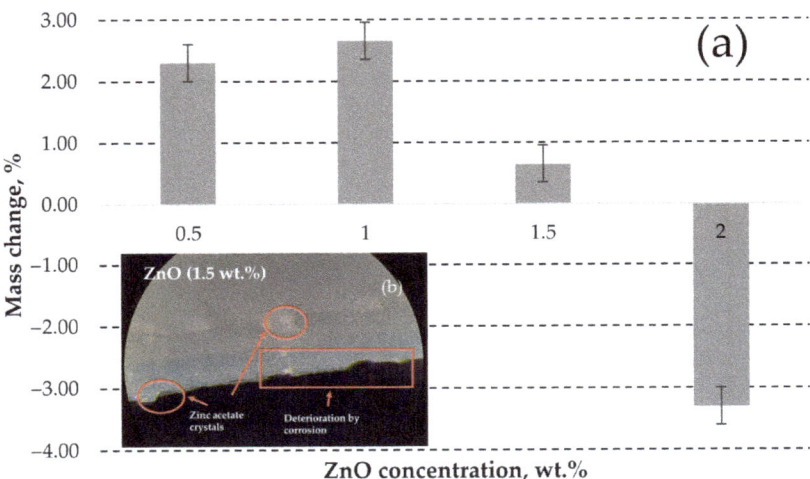

Figure 8. The change of ZnO containing polymer composites mass after exposure in acetic acid for 7 days (**a**); and optical image (magnification ×126) of the corroded 1.5 wt.% sample (**b**) with indicated corrosion products and deterioration defects. Positive values indicate that the gain in mass by corrosion products is higher than the loss of mass by composite deterioration.

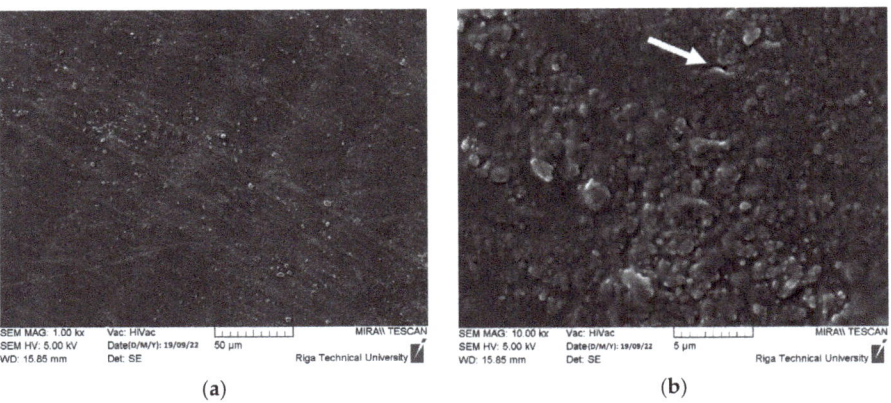

Figure 9. Degradation results after seven days of exposure of polymer composites with ZnO concentration of 1 wt.% in acetic acid (pH = 5) solution captured by SEM with 1000× (**a**) and 10,000× (**b**) resolutions. White arrow indicates the mechanical degradation of cured ER by swelling of ZnO and acetic acid reaction products.

The deterioration of 1 and 1.5 wt.% ZnO containing composites also leads to increase in weight of samples (see Figure 8), swelling reaction products increases the surface roughness and cause the formation of new pores on the surface of DLP products, as demonstrated in Figure 9b. Swelling by reaction products occurs when chemical reactions within a material lead to the formation of new compounds with larger molecular structures, causing the material to expand or increase in volume. Therefore, corrosive solution can access more ZnO particles under the corroded surface. This effect indicates the exposure of higher ZnO particles to reactive environment and embedment in the corroded surface pores. Despite the solubility of 43 g·100 mL$_{water}^{-1}$ [72], the zinc acetate preservation occurred due to specimen immersion in the standing aqueous solution. Therefore, corrosion products significantly limited the penetration of the corrosive substance to fresh ZnO sites inside the DLP composites. Additionally, the free water can remain mechanically trapped in the porous structure of the deteriorated DLP composites after removing the samples from aqueous solutions and drying.

However, a significant weight loss was observed in the case of the sample containing 2 wt.% ZnO, attributed to the intensive detachment of deteriorated material, as illustrated in Figure 7d.

3.1.4. DLP ZnO Composites Corrosion Resistance in Sodium Hydroxide Solution (pH = 12)

The ZnO reacts with aqueous sodium hydroxide solution and generates sodium tetrahydroxozincate according to chemical Equation (2):

$$2NaOH + ZnO + H_2O = Na_2[Zn(OH)_4] \qquad (2)$$

The sodium tetrahydroxozincate also forms crystal hydrates ($Na_2[Zn(OH)_4] \cdot 2H_2O$) in the presence of aqueous solution. The corrosion products and mechanically trapped water inside the defects of DLP printed ZnO composites leads to increase in weight of corroded samples by up to 2.6%, as demonstrated in Figure 10a.

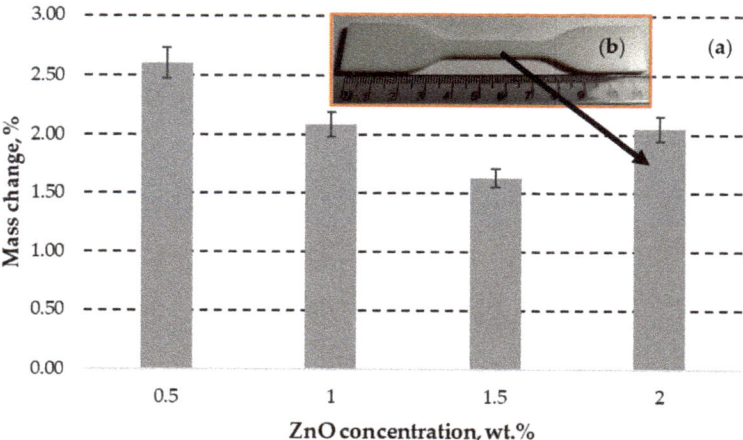

Figure 10. The increase in ZnO containing polymer composites mass after exposure in the aqueous 1 M NaOH solution (pH = 11) for 7 days (**a**); and photo of the corroded 2 wt.% ZnO containing sample without significant impact on surface morphology (**b**).

Remarkably, that increase in ZnO concentration up to 1 and 1.5 wt.% leads to slightly lower increase in mass change, as compared to 0.5 wt.% specimen. This effect can be attributed to the more efficient block of defects by corrosion products in standing (stagnated) aqueous NaOH solution, as observed by SEM (see Figure 11a,b). Reaction leads to formation of smoother grains as compared to fresh ZnO powder. However, the specimen with 2 wt.% ZnO exhibits similar mass change (2.05%) as compared to the specimen with 1 wt.% ZnO (2.09%). This effect indicates the limit of blocking effect by ZnO at concentrations between 1.5 and 2 wt.% in tested conditions. However, the NaOH and reaction products lead almost no effect on visually observable shape and morphology of any sample, as demonstrated in Figure 10b.

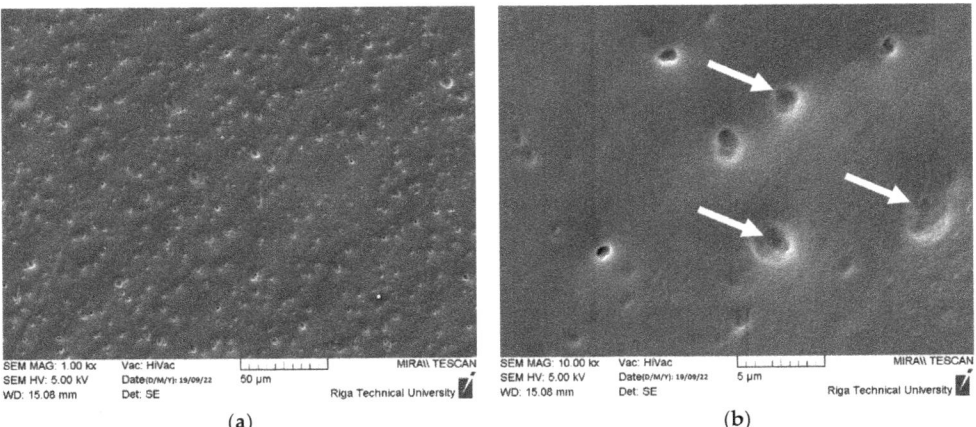

Figure 11. Degradation results after 7 days exposure of polymer composites with ZnO concentration of 2 wt.% in 1 M NaOH solution (pH = 11) solution captured by SEM with 1000× (**a**) and 10,000× (**b**) resolutions. The reaction leads to the formation of smooth particles (indicated by arrows) with passivation effect.

The ester groups of acrylic polymer and acrylic styrene copolymer hydrolyses into carboxylate salt of sodium [73]. However, in case of tested PUA composites, the surface of polymer material passivates the chemical corrosion. Further detailed studies are necessary to provide a better understanding of corrosion mechanisms.

4. Discussion

The DLP method gets attention due to its operational simplicity and for quickly obtainable outputs. Unfortunately, limited supplementing (raw) materials availability on the market [74] and manufacturing precision (spatially controlled solidification) [75] leads to demand for intensive R&D process to obtain instructions for profitable and reliable materials and methods. These factors contribute to the production speed, product durability, and the miniaturization of objects for DLP printing, particularly for small-scale and highly detailed items [76].

The observed dimensional variations (thickness and width, see Table 3) in a 3D-printed structure using DLP printing with ZnO reinforcement in PUA are influenced by multiple factors affecting the final product's dimensions. The polymerization and crosslinking of PUA may cause Z-axis shrinkage, while the addition of ZnO nanoparticles can induce expansion along the X and Y axes [77]. Uneven curing, post-curing conditions, ZnO distribution, and layer orientation contribute to differential dimensional changes. Mitigating photocuring shrinkage is essential for enhancing material performance. Various methods, including hyperbranched polymers and thiol-ene photopolymerization, have proven effective [78]. All of these aspects should be researched in detail in further studies.

Obtained DLP PUA with ZnO additives provide stable and durable protective solutions for both indoor [79] and outdoor [80] conditions. However, the chemical durability under different corrosive conditions typical for chemical transportation and storage tanks, household chemical containers, and many other exposed compartments should be studied in detail. The demonstrated strong corrosion resistance when subjected to a tested sodium hydroxide solution with a pH of 12 not only expands the potential applications of this composite to include the transportation of alkaline solutions and dry substances but also paves the way for its utilization in groundbreaking construction materials characterized by a high pH. One notable example is the development of innovative building materials, such as a 3D printed concrete composite based on a Portland cement binder [81].

An innovative approach involves hybrid concrete-DLP 3D printers, allowing for the incorporation of ER and polymer elements (e.g., polymer films and fibers) [82,83] during the concrete deposition process. These printers could come in various sizes to accommodate polymer reinforced lightweight concrete based [84] large-scale cylindrical tanks [85] and other construction. ZnO reinforcement in ERs can sustain alkaline corrosion during concrete hardening during curing process and periodical wetting during service period. Thus, both concrete properties can be improved, and new functions can be provided.

The sedimentation of the ZnO powder become observable inside the container about 10 h after mixing with the planetary mixer, as demonstrated in Figure 12a. The maximum ZnO concentration for the successful 3D printing was preliminarily detected by studying the range of concentrations from 0.5 to 10 wt.% and it was found that materials with content of 2 wt.% or lower are enabling to produce multilayered structure without delamination from the build plate and stronger adhesion to the transparent film, as demonstrated in Figure 12b,c.

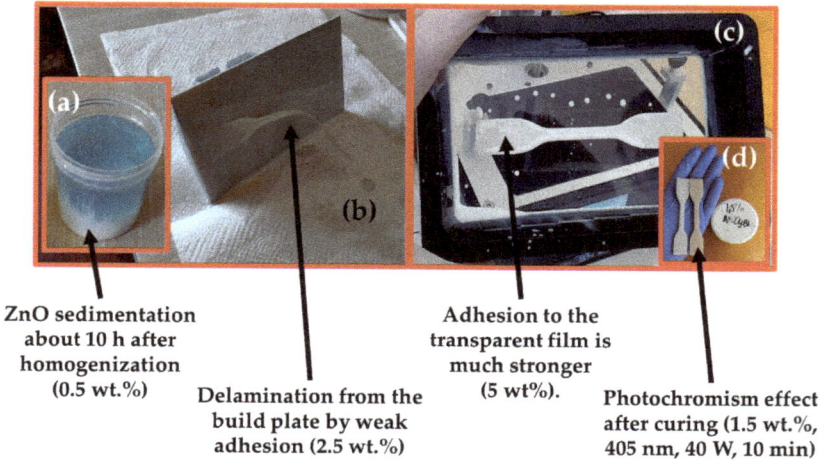

Figure 12. The sedimentation (white colored substance) of the ZnO in raw ER (**a**); first layer delamination from the build plate (**b**); DLP failure by too strong first layer adhesion to the transparent film (**c**); and photochromism effect after high power (40 W) UV curing (**d**).

Successfully achieving product printing with a relatively low concentration of ZnO powder (from 0.5 to 2 wt.%) highlights the imperative for thorough multifactorial studies. Subsequent investigations are poised to elevate concentrations of ZnO and other fillers up to 52 wt.%, as depicted in Table 1. This approach not only promises enhanced properties for the DLP product but also opens avenues for novel functional attributes.

The attachment of active ER premixing system would reduce the sedimentation effect on 3D printing performance. Further studies need to test different active mixing techniques.

When ZnO-reinforced polymers exhibit the photochromism effect, it can significantly enhance their aesthetic properties. Photochromic materials, such as those containing ZnO nanoparticles, can change color or optical characteristics in response to UV light, as demonstrated in Figure 12d. For example, this effect can be valuable for shielding polymers from excessive UV exposure during the disinfection process of tanker walls.

In-depth studies on the behavior of ZnO-reinforced PUA composites under shock loading conditions (e.g., pendulum impact test) are also required in future studies. Stukhlyak et al. conducted such research exploring the impact of meso- and macroscale processes on the dynamic fracture mechanisms of cross-linked epoxy composites [28]. This involves cross-linking during the curing of thermosetting plastics, transforming epoxy resins into cross-linked polymers through the polymerization of polyfunctional monomers or oligomers. The cross-bonding of linear and branched macromolecules with reactive groups of an epoxy polymer is a crucial aspect, with cross-links in these polymers being chemical, physical, and topological [86].

Predicting PUA based composite behavior involves accounting for microstructural complexity. In nanocomposites, non-dimensional models show no significant nanoscopic size effect below the transition temperature, but above it, a notable discrepancy emerges between experimental modulus and theory. This suggests negligible effects in the glassy state become relevant in the rubbery domain, necessitating evaluation of mechanical interactions using multi-scale approaches or finite element modeling. This aligns with growing industrial interest for potential applications in sporting equipment, automotive parts, and packaging due to high barrier properties and enhanced flame retardance. Industrial nanocomposites, mainly using rubbery, thermosets, or semi-crystalline matrices, exhibit typical viscoelastic behaviors [87].

When using ZnO powder as a filler in DLP photocured resins, it's essential to consider its ecological impact, including proper disposal and recycling measures to mitigate potential environmental consequences.

5. Conclusions

As-received 99.9 wt.% pure ZnO powder with D_{50} = 177 nm and maximum particle (including agglomerate) sizes up to about 5 μm were mixed in commercially available polyurethane acrylate, isooctyl acrylate, and phosphine oxide mixture with ZnO concentrations from 0.5 up to 2 wt.% (from 0.5 to 10 wt.%, during preliminary studies). Adhesion weakening to the build plate and next layer (stronger adhesion to transparent film) limits the possibility to increase the ZnO concentration under selected DLP settings. The change in ZnO concentration causes thickness deviations in DLP samples. The 1 wt.% samples achieve closest value of thickness (3.99 ± 0.16 mm) to target thickness value (4 mm), while 0.5 wt.% ZnO containing samples exhibit lowest deviation in average thickness (±0.03 mm).

Nonetheless, the deviation in the width of the DLP product consistently escalates, increasing by a factor of 1.5–5.5 times as the concentration rises from 1 wt.% to 2 wt.%. This results in an exceeding of the target value of 6 mm in the horizontal plane for all concentrations, with a minimum overage of 0.33 ± 0.02 mm. To address this deviation, it can usually be corrected by simply reducing the exposure time while keeping the initially set conditional parameters intact.

All tested DLP composites with ZnO exhibits similar tensile stress from 23.29 (1 wt.%) up to 25.93 MPa (0.5 wt.%) at similar elongation/deformation values (0.0089 up to 0.0110), however, the increase in ZnO concentration increase brittleness of the product (tensile stress at fracture reduces to 24.04 MPa and elastic modulus reduces to 2001 MPa at ZnO concentration of 2 wt.%).

The increase in ZnO concentration increases the total area for reaction with acetic acid solution and causes visually observable loss of ER material from DLP products from 1 wt.% ZnO. The reaction products (zinc acetate salts) swell and cause formation of new defects under the surface layer. Therefore, use of ZnO reinforced DLP resin is limited to acidic environments.

All samples exhibit good corrosion resistance to applied aqueous sodium hydroxide solution (pH = 12). The reaction products (mainly generates sodium tetrahydroxozincate) form on the surface of specimens and passivates further reaction when subjected to standing (stagnating) solution. Reaction products cause smooth surface above the ZnO, therefore reduces the possible friction caused defects. Future studies should include the measurement and reporting of roughness values.

Generally, the produced samples are suitable for application in alkaline environment and can be applied as protective coatings over tank walls or used as structural or functional elements (e.g., in Portland cement concrete walls).

Author Contributions: Conceptualization, J.B., V.K.T. and A.S.; methodology, J.B.; software, J.B. and I.B.; validation, A.S., M.A. and V.A.; formal analysis, V.A., S.G. and V.K.T.; investigation, V.A., I.B. and V.L.; resources, J.B. and A.S.; data curation, J.B.; writing—original draft preparation, J.B., A.S., S.G. and V.K.T.; writing—review and editing, M.A., A.R., S.G. and A.S.; visualization, J.B. and V.A.; supervision, J.B.; project administration, A.S. and V.L.; funding acquisition, J.B. and A.S. All authors have read and agreed to the published version of the manuscript.

Funding: This research was funded by the VPP AIPP project Nr. VPP-AIPP-2021/1-0015, "Combined lightweight, high-temperature resistant hybrid composite for combined protection of drones from Direct Energy Weapon", as well as B57 RTU research funds (ZM-2023/10) and grants from the Estonian Research Council (PRG643 and "DuplexCER" MNHA22040). The article was published with financial support from the Riga Technical University Research Support Fund.

Institutional Review Board Statement: Not applicable.

Data Availability Statement: Data are contained within the article.

Acknowledgments: The authors extend their sincere appreciation to the team at the Latvian Maritime Academy of Riga Technical University for their generous provision of access to the DLP printer.

Conflicts of Interest: The authors declare no conflict of interest.

References

1. Becker, O.; Simon, G.P. Epoxy Nanocomposites Based on Layered Silicates and Other Nanostructured Fillers. In *Polymer Nanocomposites*; Elsevier: Amsterdam, The Netherlands, 2006; pp. 29–56.
2. Fuchi, Y.; Yoshida, K.; Kozako, M.; Hikita, M.; Kamei, N. Comparison of Electrical Insulation Properties of Hydrocarbon-Based Thermosetting Resin and Epoxy Resin. In Proceedings of the 2016 IEEE Conference on Electrical Insulation and Dielectric Phenomena (CEIDP), Toronto, ON, Canada, 16–19 October 2016; IEEE: Piscataway, NJ, USA, 2016; pp. 133–136.
3. Tanaka, T. Dielectric Nanocomposites with Insulating Properties. *IEEE Trans. Dielectr. Electr. Insul.* **2005**, *12*, 914–928. [CrossRef]
4. García, J.M.; García, F.C.; Serna, F.; de la Peña, J.L. High-Performance Aromatic Polyamides. *Prog. Polym. Sci.* **2010**, *35*, 623–686. [CrossRef]
5. Maisonneuve, L.; Lamarzelle, O.; Rix, E.; Grau, E.; Cramail, H. Isocyanate-Free Routes to Polyurethanes and Poly(hydroxy urethane)s. *Chem. Rev.* **2015**, *115*, 12407–12439. [CrossRef] [PubMed]
6. Varghese, G.; Moral, M.; Castro-García, M.; López-López, J.J.; Marín-Rueda, J.R.; Yagüe-Alcaraz, V.; Hernández-Afonso, L.; Ruiz-Morales, J.C.; Canales-Vázquez, J. Fabrication and Characterisation of Ceramics via Low-Cost DLP 3D Printing. *Boletín La Soc. Española Cerámica Y Vidr.* **2018**, *57*, 9–18. [CrossRef]
7. Popov, V.; Koptyug, A.; Radulov, I.; Maccari, F.; Muller, G. Prospects of Additive Manufacturing of Rare-Earth and Non-Rare-Earth Permanent Magnets. *Procedia Manuf.* **2018**, *21*, 100–108. [CrossRef]
8. Fidan, I.; Huseynov, O.; Ali, M.A.; Alkunte, S.; Rajeshirke, M.; Gupta, A.; Hasanov, S.; Tantawi, K.; Yasa, E.; Yilmaz, O.; et al. Recent Inventions in Additive Manufacturing: Holistic Review. *Inventions* **2023**, *8*, 103. [CrossRef]
9. Haryńska, A.; Carayon, I.; Kosmela, P.; Szeliski, K.; Łapiński, M.; Pokrywczyńska, M.; Kucińska-Lipka, J.; Janik, H. A Comprehensive Evaluation of Flexible FDM/FFF 3D Printing Filament as a Potential Material in Medical Application. *Eur. Polym. J.* **2020**, *138*, 109958. [CrossRef]
10. Xu, J.; Pang, W.; Shi, W. Synthesis of UV-Curable Organic–Inorganic Hybrid Urethane Acrylates and Properties of Cured Films. *Thin Solid Films* **2006**, *514*, 69–75. [CrossRef]
11. Wang, Y.; Gao, M.; Wang, D.; Sun, L.; Webster, T.J. Nanoscale 3D Bioprinting for Osseous Tissue Manufacturing. *Int. J. Nanomed.* **2020**, *15*, 215–226. [CrossRef]
12. Choi, J.-S.; Seo, J.; Khan, S.B.; Jang, E.S.; Han, H. Effect of Acrylic Acid on the Physical Properties of UV-Cured Poly(urethane acrylate-co-acrylic acid) Films for Metal Coating. *Prog. Org. Coat.* **2011**, *71*, 110–116. [CrossRef]
13. Lü, N.; Lü, X.; Jin, X.; Lü, C. Preparation and Characterization of UV-Curable ZnO/Polymer Nanocomposite Films. *Polym. Int.* **2007**, *56*, 138–143. [CrossRef]
14. Gaidukovs, S.; Medvids, A.; Onufrijevs, P.; Grase, L.; Brunavs, J. Development of UV-Cured Epoxy Resin Based on Marineline. Formulation for Applications as Efficient Repair System of Boat Tanks' Protective Coating. In Proceedings of the Baltic Polymer Symposium, Jurmala, Latvia, 12–14 September 2018; Institute of Physics Publishing (IOP): Jurmala, Latvia, 2018; pp. 12–14.
15. Wang, F.; Hu, J.Q.; Tu, W.P. Study on Microstructure of UV-Curable Polyurethane Acrylate Films. *Prog. Org. Coat.* **2008**, *62*, 245–250. [CrossRef]
16. Bao, F.; Shi, W. Synthesis and Properties of Hyperbranched Polyurethane Acrylate Used for UV Curing Coatings. *Prog. Org. Coat.* **2010**, *68*, 334–339. [CrossRef]
17. Hu, Y.; Zhu, G.; Zhang, J.; Huang, J.; Yu, X.; Shang, Q.; An, R.; Liu, C.; Hu, L.; Zhou, Y. Rubber Seed Oil-Based UV-Curable Polyurethane Acrylate Resins for Digital Light Processing (DLP) 3D Printing. *Molecules* **2021**, *26*, 5455. [CrossRef] [PubMed]

18. Chattopadhyay, D.K.; Raju, K.V.S.N. Structural Engineering of Polyurethane Coatings for High Performance Applications. *Prog. Polym. Sci.* **2007**, *32*, 352–418. [CrossRef]
19. Nakayama, N.; Hayashi, T. Synthesis of Novel UV-Curable Difunctional Thiourethane Methacrylate and Studies on Organic–Inorganic Nanocomposite Hard Coatings for High Refractive Index Plastic Lenses. *Prog. Org. Coat.* **2008**, *62*, 274–284. [CrossRef]
20. Mishra, R.S.; Mishra, A.K.; Raju, K.V.S.N. Synthesis and Property Study of UV-Curable Hyperbranched Polyurethane Acrylate/ZnO Hybrid Coatings. *Eur. Polym. J.* **2009**, *45*, 960–966. [CrossRef]
21. Yoshimura, H.N.; Molisani, A.L.; Narita, N.E.; Manholetti, J.L.A.; Cavenaghi, J.M. Mechanical Properties and Microstructure of Zinc Oxide Varistor Ceramics. *Mater. Sci. Forum* **2006**, *530–531*, 408–413. [CrossRef]
22. Choi, H.M.; Kwon, S.; Jung, Y.-G.; Cho, Y.T. Comparison of Durability for PUA Type Resin Using Wear and Nano-Indentation Test. *J. Korean Soc. Manuf. Process Eng.* **2018**, *17*, 8–15. [CrossRef]
23. Nik Pauzi, N.N.P.; Majid, R.A.; Dzulkifli, M.H.; Yahya, M.Y. Development of Rigid Bio-Based Polyurethane Foam Reinforced with Nanoclay. *Compos. B Eng.* **2014**, *67*, 521–526. [CrossRef]
24. Madhan Kumar, A.; Mizanur Rahman, M.; Gasem, Z.M. A Promising Nanocomposite from CNTs and Nano-Ceria: Nanostructured Fillers in Polyurethane Coatings for Surface Protection. *RSC Adv.* **2015**, *5*, 63537–63544. [CrossRef]
25. Zhu, M.; Li, S.; Sun, Q.; Shi, B. Enhanced Mechanical Property, Chemical Resistance and Abrasion Durability of Waterborne Polyurethane Based Coating by Incorporating Highly Dispersed Polyacrylic Acid Modified Graphene Oxide. *Prog. Org. Coat.* **2022**, *170*, 106949. [CrossRef]
26. Rahman, M.M. Polyurethane/Zinc Oxide (PU/ZnO) Composite—Synthesis, Protective Property and Application. *Polymers* **2020**, *12*, 1535. [CrossRef] [PubMed]
27. Fu, J.; Wang, L.; Yu, H.; Haroon, M.; Haq, F.; Shi, W.; Wu, B.; Wang, L. Research Progress of UV-Curable Polyurethane Acrylate-Based Hardening Coatings. *Prog. Org. Coat.* **2019**, *131*, 82–99. [CrossRef]
28. Stukhlyak, P.D.; Buketov, A.V.; Panin, S.V.; Maruschak, P.O.; Moroz, K.M.; Poltaranin, M.A.; Vukherer, T.; Kornienko, L.A.; Lyukshin, B.A. Structural Fracture Scales in Shock-Loaded Epoxy Composites. *Phys. Mesomech.* **2015**, *18*, 58–74. [CrossRef]
29. Mohamed, O.A.; Masood, S.H.; Bhowmik, J.L. Experimental Investigation of Time-Dependent Mechanical Properties of PC-ABS Prototypes Processed by FDM Additive Manufacturing Process. *Mater. Lett.* **2017**, *193*, 58–62. [CrossRef]
30. Eckel, Z.C.; Zhou, C.; Martin, J.H.; Jacobsen, A.J.; Carter, W.B.; Schaedler, T.A. Additive Manufacturing of Polymer-Derived Ceramics. *Science* **2016**, *351*, 58–62. [CrossRef]
31. de Azeredo, H.M.C. Nanocomposites for Food Packaging Applications. *Food Res. Int.* **2009**, *42*, 1240–1253. [CrossRef]
32. Duncan, T.V. Applications of Nanotechnology in Food Packaging and Food Safety: Barrier Materials, Antimicrobials and Sensors. *J. Colloid Interface Sci.* **2011**, *363*, 1–24. [CrossRef]
33. Tang, E.; Liu, H.; Sun, L.; Zheng, E.; Cheng, G. Fabrication of Zinc Oxide/Poly(Styrene) Grafted Nanocomposite Latex and Its Dispersion. *Eur. Polym. J.* **2007**, *43*, 4210–4218. [CrossRef]
34. Yan, M.F. Zinc Oxide. In *Concise Encyclopedia of Advanced Ceramic Materials*; Pergamon: Oxford, UK, 1991; Volume 1, pp. 523–525.
35. Song, Z.; Kelf, T.A.; Sanchez, W.H.; Roberts, M.S.; Rička, J.; Frenz, M.; Zvyagin, A.V. Characterization of Optical Properties of ZnO Nanoparticles for Quantitative Imaging of Transdermal Transport. *Biomed. Opt. Express* **2011**, *2*, 3321. [CrossRef] [PubMed]
36. Khabazipour, M.; Anbia, M. Removal of Hydrogen Sulfide from Gas Streams Using Porous Materials: A Review. *Ind. Eng. Chem. Res.* **2019**, *58*, 22133–22164. [CrossRef]
37. Ates, T.; Tatar, C.; Yakuphanoglu, F. Preparation of Semiconductor ZnO Powders by Sol–Gel Method: Humidity Sensors. *Sens. Actuators A Phys.* **2013**, *190*, 153–160. [CrossRef]
38. Li, K.; de Rancourt de Mimérand, Y.; Jin, X.; Yi, J.; Guo, J. Metal Oxide (ZnO and TiO_2) and Fe-Based Metal–Organic-Framework Nanoparticles on 3D-Printed Fractal Polymer Surfaces for Photocatalytic Degradation of Organic Pollutants. *ACS Appl. Nano Mater.* **2020**, *3*, 2830–2845. [CrossRef]
39. Bykkam, S.; Narsingam, S.; Ahmadipour, M.; Dayakar, T.; Venkateswara Rao, K.; Shilpa Chakra, C.; Kalakotla, S. Few Layered Graphene Sheet Decorated by ZnO Nanoparticles for Anti-Bacterial Application. *Superlattices Microstruct.* **2015**, *83*, 776–784. [CrossRef]
40. Siyanbola, T.O.; Sasidhar, K.; Rao, B.V.S.K.; Narayan, R.; Olaofe, O.; Akintayo, E.T.; Raju, K.V.S.N. Development of Functional Polyurethane-ZnO Hybrid Nanocomposite Coatings from Thevetia Peruviana Seed Oil. *J. Am. Oil Chem. Soc.* **2015**, *92*, 267–275. [CrossRef]
41. Mhd Haniffa, M.A.C.; Ching, Y.C.; Chuah, C.H.; Ching, K.Y.; Liou, N.S. Synergistic Effect of (3-Aminopropyl)Trimethoxysilane Treated ZnO and Corundum Nanoparticles under UV-Irradiation on UV-Cutoff and IR-Absorption Spectra of Acrylic Polyurethane Based Nanocomposite Coating. *Polym. Degrad. Stab.* **2019**, *159*, 205–216. [CrossRef]
42. Soares, R.R.; Carone, C.; Einloft, S.; Ligabue, R.; Monteiro, W.F. Synthesis and Characterization of Waterborne Polyurethane/ZnO Composites. *Polym. Bull.* **2014**, *71*, 829–838. [CrossRef]
43. Zhang, J.; Li, Y.; Hu, C.; Huang, W.; Su, L. Anti-Corrosive Properties of Waterborne Polyurethane/Poly(o-Toluidine)-ZnO Coatings in NaCl Solution. *J. Adhes. Sci. Technol.* **2019**, *33*, 1047–1065. [CrossRef]
44. Ye, X.; Wang, Z.; Ma, L.; Wang, Q.; Chu, A. Zinc Oxide Array/Polyurethane Nanocomposite Coating: Fabrication, Characterization and Corrosion Resistance. *Surf Coat Technol.* **2019**, *358*, 497–504. [CrossRef]

45. Salazar-Bravo, P.; del Angel-López, D.; Torres-Huerta, A.M.; Domínguez-Crespo, M.A.; Palma-Ramírez, D.; Brachetti-Sibaja, S.B.; Ferrel-Álvarez, A.C. Investigation of ZnO/Waterborne Polyurethane Hybrid Coatings for Corrosion Protection of AISI 1018 Carbon Steel Substrates. *Metall. Mater. Trans. A* **2019**, *50*, 4798–4813. [CrossRef]
46. Posthumus, W.; Magusin, P.C.M.M.; Brokken-Zijp, J.C.M.; Tinnemans, A.H.A.; van der Linde, R. Surface Modification of Oxidic Nanoparticles Using 3-Methacryloxypropyltrimethoxysilane. *J. Colloid Interface Sci.* **2004**, *269*, 109–116. [CrossRef] [PubMed]
47. Mai, N.T.; Anh, B.T.M.; Vuong, N.T. Acid and Alkali Resistance of Acrylic Polyurethane/R-SiO$_2$ Nanocomposite Coating. *Vietnam. J. Chem.* **2020**, *58*, 67–73. [CrossRef]
48. Tubío, C.R.; Guitián, F.; Gil, A. Fabrication of ZnO Periodic Structures by 3D Printing. *J. Eur. Ceram. Soc.* **2016**, *36*, 3409–3415. [CrossRef]
49. Lu, J.; Dong, P.; Zhao, Y.; Zhao, Y.; Zeng, Y. 3D Printing of TPMS Structural ZnO Ceramics with Good Mechanical Properties. *Ceram. Int.* **2021**, *47*, 12897–12905. [CrossRef]
50. Waheed, S.; Rodas, M.; Kaur, H.; Kilah, N.L.; Paull, B.; Maya, F. In-Situ Growth of Metal-Organic Frameworks in a Reactive 3D Printable Material. *Appl. Mater. Today* **2021**, *22*, 100930. [CrossRef]
51. Tubío, C.R.; Nóvoa, J.A.; Martín, J.; Guitián, F.; Salgueiro, J.R.; Gil, A. Broadband Terahertz ZnO Photonic Crystals Fabricated by 3D Printing. *Ceram. Int.* **2019**, *45*, 6223–6227. [CrossRef]
52. Lee, D.K.; Sin, K.S.; Shin, C.; Kim, J.H.; Hwang, K.T.; Kim, U.S.; Nahm, S.; Han, K.S. Fabrication of 3D Structure with Heterogeneous Compositions Using Inkjet Printing Process. *Mater. Today Commun.* **2023**, *35*, 105753. [CrossRef]
53. Vidakis, N.; Petousis, M.; Velidakis, E.; Tzounis, L.; Mountakis, N.; Korlos, A.; Fischer-Griffiths, P.E.; Grammatikos, S. On the Mechanical Response of Silicon Dioxide Nanofiller Concentration on Fused Filament Fabrication 3d Printed Isotactic Polypropylene Nanocomposites. *Polymers* **2021**, *13*, 2029. [CrossRef]
54. Zhao, Z.; Zhou, G.; Yang, Z.; Cao, X.; Jia, D.; Zhou, Y. Direct Ink Writing of Continuous SiO$_2$ Fiber Reinforced Wave-Transparent Ceramics. *J. Adv. Ceram.* **2020**, *9*, 403–412. [CrossRef]
55. Liu, Y.; Chen, J.; Ning, L.; Sun, J.; Liu, L.; Zhao, K. Preparation and Properties of Nano-TiO2-Modified Photosensitive Materials for 3D Printing. *E-Polymers* **2022**, *22*, 686–695. [CrossRef]
56. Veselý, P.; Froš, D.; Hudec, T.; Sedláček, J.; Ctibor, P.; Dušek, K. Dielectric Spectroscopy of PETG/TiO$_2$ Composite Intended for 3D Printing. *Virtual Phys. Prototyp.* **2023**, *18*, e2170253. [CrossRef]
57. Deng, Y.; Li, J.; He, Z.; Hong, J.; Bao, J. Urethane Acrylate-based Photosensitive Resin for Three-dimensional Printing of Stereolithographic Elastomer. *J. Appl. Polym. Sci.* **2020**, *137*, 49294. [CrossRef]
58. Anycubic Translucent UV Resin. Available online: https://www.anycubic.com/products/clear-uv-resin (accessed on 24 July 2022).
59. Kołodziejczak-Radzimska, A.; Jesionowski, T. Zinc Oxide—From Synthesis to Application: A Review. *Materials* **2014**, *7*, 2833–2881. [CrossRef] [PubMed]
60. MERCK Zinc Oxide. Available online: https://www.sigmaaldrich.com/LV/en/product/sigald/205532 (accessed on 24 July 2022).
61. ASTM International. ASTM D638-14. Standard Test Method for Tensile Properties of Plastics. In *Standards and Publications*; ASTM International: West Conshohocken, PA, USA, 2014; p. 17.
62. Anycubic ANYCUBIC Photon Mono. Available online: https://www.anycubic.com/products/photon-mono-resin-3d-printer (accessed on 25 July 2022).
63. Ito, H.; Yoshioka, D.; Hamada, M.; Okamoto, T.; Kobori, Y.; Kobayashi, Y. Photochromism of Colloidal ZnO Nanocrystal Powders under Ambient Conditions. *Photochem. Photobiol. Sci.* **2022**, *11*, 1781–1791. [CrossRef] [PubMed]
64. Meng, F.; King, M.D.; Hassan, Y.A.; Ugaz, V.M. Localized Fluorescent Complexation Enables Rapid Monitoring of Airborne Nanoparticles. *Environ. Sci. Nano* **2014**, *1*, 358. [CrossRef]
65. Wang, Y.; Shi, J.; He, Z.B.; Bai, H.W. Preparation and Mechanical Properties of T-ZnOw/PS Composites. *Chin. J. Polym. Sci. Engl. Ed.* **2009**, *27*, 173–181. [CrossRef]
66. Zhou, J.P.; Qiu, K.Q.; Fu, W.L. The Surface Modification of ZnOw and Its Effect on the Mechanical Properties of Filled Polypropylene Composites. *J. Compos. Mater.* **2005**, *39*, 1931–1941. [CrossRef]
67. Manapat, J.Z.; Mangadlao, J.D.; Tiu, B.D.B.; Tritchler, G.C.; Advincula, R.C. High-Strength Stereolithographic 3D Printed Nanocomposites: Graphene Oxide Metastability. *ACS Appl. Mater. Interfaces* **2017**, *9*, 10085–10093. [CrossRef]
68. Prakash, K.S.; Nancharaih, T.; Rao, V.V.S. Additive Manufacturing Techniques in Manufacturing -An Overview. *Mater. Today Proc.* **2018**, *5*, 3873–3882. [CrossRef]
69. Ngo, T.D.; Kashani, A.; Imbalzano, G.; Nguyen, K.T.Q.; Hui, D. Additive Manufacturing (3D Printing): A Review of Materials, Methods, Applications and Challenges. *Compos. B Eng.* **2018**, *143*, 172–196. [CrossRef]
70. Kim, D.; Jang, M.; Seo, J.; Nam, K.H.; Han, H.; Khan, S.B. UV-Cured Poly(urethane acrylate) Composite Films Containing Surface-Modified Tetrapod ZnO Whiskers. *Compos. Sci. Technol.* **2013**, *75*, 84–92. [CrossRef]
71. van Niekerk, J.N.; Schoening, F.R.L.; Talbot, J.H. The Crystal Structure of Zinc Acetate Dihydrate, Zn(CH$_3$COO)$_2$·2H$_2$O. *Acta Crystallogr.* **1953**, *6*, 720–723. [CrossRef]
72. Tyner Chainer, T. Zinc Acetate Dihydrate. In *ACS Reagent Chemicals*; American Chemical Society: Washington, DC, USA, 2017.
73. Vuong, N.T.; Hiep, N.A. The Alkaline Hydrolysis Degradation of a Water-Borne Styrene Acrylic Coating. *Vietnam. J. Chem.* **2016**, *54*, 249.

74. Chartrain, N.A.; Williams, C.B.; Whittington, A.R. A Review on Fabricating Tissue Scaffolds Using Vat Photopolymerization. *Acta Biomater.* **2018**, *74*, 90–111. [CrossRef]
75. Bae, J.H.; Won, J.C.; Lim, W.b.; Min, J.G.; Lee, J.H.; Kwon, C.R.; Lee, G.H.; Huh, P. Synthesis and Characteristics of Eco-Friendly 3D Printing Material Based on Waterborne Polyurethane. *Polymers* **2020**, *13*, 44. [CrossRef] [PubMed]
76. Bae, J.H.; Won, J.C.; Lim, W.b.; Lee, J.H.; Min, J.G.; Kim, S.W.; Kim, J.H.; Huh, P. Highly Flexible and Photo-Activating Acryl-Polyurethane for 3D Steric Architectures. *Polymers* **2021**, *13*, 844. [CrossRef] [PubMed]
77. Li, S.; Cui, Y.; Li, J. Thiol-terminated Hyperbranched Polymer for DLP 3D Printing: Performance Evaluation of a Low Shrinkage Photosensitive Resin. *J. Appl. Polym. Sci.* **2021**, *138*, 50525. [CrossRef]
78. Chen, L.; Wu, Q.; Wei, G.; Liu, R.; Li, Z. Highly Stable Thiol–Ene Systems: From Their Structure–Property Relationship to DLP 3D Printing. *J. Mater. Chem. C Mater.* **2018**, *6*, 11561–11568. [CrossRef]
79. Nguyen, T.V.; Do, T.V.; Ha, M.H.; Le, H.K.; Le, T.T.; Linh Nguyen, T.N.; Dam, X.T.; Lu, L.T.; Tran, D.L.; Vu, Q.T.; et al. Crosslinking Process, Mechanical and Antibacterial Properties of UV-Curable Acrylate/Fe_3O_4-Ag Nanocomposite Coating. *Prog. Org. Coat.* **2020**, *139*, 105325. [CrossRef]
80. Vuong, N.T.; Linh, N.T. The Accelerated Weathering Aging of a Water-Borne Styrene Acrylic Coating. *Vietnam. J. Chem.* **2016**, *54*, 139.
81. Voicu, G.; Tiuca, G.A.; Badanoiu, A.I.; Holban, A.M. Nano and Mesoscopic SiO_2 and ZnO Powders to Modulate Hydration, Hardening and Antibacterial Properties of Portland Cements. *J. Build. Eng.* **2022**, *57*, 104862. [CrossRef]
82. Guo, S.Y.; Zhang, X.; Chen, J.Z.; Mou, B.; Shang, H.S.; Wang, P.; Zhang, L.; Ren, J. Mechanical and Interface Bonding Properties of Epoxy Resin Reinforced Portland Cement Repairing Mortar. *Constr. Build. Mater.* **2020**, *264*, 120715. [CrossRef]
83. Silva, D.A.; Betioli, A.M.; Gleize, P.J.P.; Roman, H.R.; Gómez, L.A.; Ribeiro, J.L.D. Degradation of Recycled PET Fibers in Portland Cement-Based Materials. *Cem. Concr. Res.* **2005**, *35*, 1741–1746. [CrossRef]
84. Liu, J.; Lv, C. Properties of 3D-Printed Polymer Fiber-Reinforced Mortars: A Review. *Polymers* **2022**, *14*, 1315. [CrossRef]
85. Thevendran, V.; Thambiratnam, D.P. Cylindrical Concrete Water Tanks: Analysis and Design. In *Numerical Techniques for Engineering Analysis and Design*; Springer: Dordrecht, The Netherlands, 1987; pp. 163–170.
86. Irzhak, V.; Rozenberg, A.; Enikolopyan, N. Cross-Linked Polymers. In *Cross-Linked Polymers. Synthesis, Structure, Properties*; Nauka: Moscow, Russia, 1979; pp. 105–157. (In Russian)
87. Chazeau, L.; Gauthier, C.; Vigier, G.; Cavaillé, J.Y. Relationships between Microstructural Aspects and Mechanical Properties in Polymer Based Nanocomposites. In *Handbook of Organic-Inorganic Hybrid Materials and Nanocomposites*; HAL Open Science: Villeurbanne Cedex, France, 2003.

Disclaimer/Publisher's Note: The statements, opinions and data contained in all publications are solely those of the individual author(s) and contributor(s) and not of MDPI and/or the editor(s). MDPI and/or the editor(s) disclaim responsibility for any injury to people or property resulting from any ideas, methods, instructions or products referred to in the content.

Article

Tensile and Compression Strength Prediction and Validation in 3D-Printed Short-Fiber-Reinforced Polymers

Timothy Russell and David A. Jack *

Department of Mechanical Engineering, Baylor University, Waco, TX 76798, USA;
timothy_russell@alumni.baylor.edu
* Correspondence: david_jack@baylor.edu; Tel.: +1-254-710-3347

Abstract: In the current study, a methodology is validated for predicting the internal spatially varying strength properties in a single 3D-printed bead composed of 13%, by weight, carbon-fiber-filled acrylonitrile butadiene styrene. The presented method allows for the characterization of the spatially varying microstructural behavior yielding a local anisotropic stiffness and strength that can be integrated in a finite element framework for a bulk estimate of the effective stiffness and strength. The modeling framework is presented with a focus on composite structures made from large area additive manufacturing (LAAM). LAAM is an extrusion-based process yielding components on the order of meters, with a typical raster size of 10 mm. The presented modeling methods are applicable to other short-fiber-reinforced polymer processing methods as well. The results provided indicate the modeling framework yields results for the effective strength and stiffness that align with experimental characterization to within ~1% and ~10% for the longitudinal compressive and tensile strength, respectively, and to within ~3% and ~50% for the longitudinal compressive and tensile stiffness, respectively.

Keywords: Jeffery model; closure; fiber orientation; short-fiber composites; structural properties; large area additive manufacturing

Citation: Russell, T.; Jack, D.A. Tensile and Compression Strength Prediction and Validation in 3D-Printed Short-Fiber-Reinforced Polymers. *Polymers* 2023, *15*, 3605. https://doi.org/10.3390/polym15173605

Academic Editor: Chenggao Li

Received: 16 July 2023
Revised: 2 August 2023
Accepted: 24 August 2023
Published: 30 August 2023

Copyright: © 2023 by the authors. Licensee MDPI, Basel, Switzerland. This article is an open access article distributed under the terms and conditions of the Creative Commons Attribution (CC BY) license (https://creativecommons.org/licenses/by/4.0/).

1. Introduction

The development of better manufacturing techniques promotes the development of new engineering materials, and vice versa. This concept is well embodied in the area of 3D-printed short-fiber-reinforced polymers (SFRPs). These materials offer the attractive combination of good mechanical properties compared to virgin polymers and a relatively easy processing. In addition, they help enable large area additive manufacturing (LAAM), which often uses carbon fiber to provide crucial thermal and mechanical property advantages (see, e.g., [1]). However, 3D-printed SFRPs introduce added difficulty in engineering design since their anisotropic properties are more complicated to model. Additive manufacturing allows one to customize tool paths to tailor that anisotropy to one's advantage (for example, one can orient the way a part is built so that it has greater stiffness and strength in a load-bearing direction), but good methods for being able to predict these anisotropic properties must be established so that one can generate designs with high confidence.

The anisotropic properties of 3D-printed SFRPs are a function of the internal, spatially varying fiber orientation state. Thus, predicting the properties of SFRP parts oftentimes involves the problem of predicting their internal fiber orientation state first, then predicting the properties as functions of that orientation state. Research on predicting the spatially varying fiber orientation state in a fluid flow has its roots going back about a hundred years ago to George B. Jeffery [2]. The works of Charles Tucker III and his students in the past few decades have contributed greatly to the current understanding of this topic as well. Tucker and his student Suresh Advani brought orientation tensors to light as a general but very compact way to describe orientation states, and these have since become mainstream, allowing for the prediction of material properties as functions of those orientation tensors [3]. Folgar and Tucker [4] introduced a fiber interaction parameter to allow for the modeling of dense suspensions, and in the past decade, multiple models have been developed to advance the initial work of Folgar and Tucker to implement fiber orientation kinematics for industrial flow processes (see, e.g., [5–7]), each of which uses the orientation tensor approach to characterize the fiber alignment state. This topic is discussed in detail in Section 4 while introducing the mathematical framework. Modern modeling techniques now give the ability to predict the final properties of SFRP parts based on the processing conditions used to build said parts. We wish to bring this capability to light in the additive manufacturing industry so that SFRP parts can be evaluated prior to printing, saving time and money and improving engineers' confidence level in their designs.

The next two sections of this paper discuss experimental specimen preparation and testing, respectively. A miniature LAAM system was constructed for printing specimens. Section 4 then covers some of the mathematical models used to make predictions. Finally, Section 5 compares the experimental findings with the predicted structural behavior. The novelty of the present work is the experimental validation of a methodology, also demonstrated in [8], which fully integrates fiber orientation kinetics with structural property predictions for additively manufactured composites made from the LAAM process. This work is extended to compare against experimental observations and highlights the need to properly account for both the proper fiber orientation kinetics slowness parameter and the void content within the deposited material, the latter of which is able to be neglected in other industrial processes that study fiber orientation.

2. Specimen Preparation

2.1. Baylor's Large Area Additive Manufacturing System

For fabricating test specimens, a miniature LAAM system was constructed at Baylor University. This system has been utilized in a variety of studies by various researchers (see e.g., [9,10]). The LAAM system has a print area of approximately 1.2 m by 1.2 m and the extruder has approximately 15 cm of vertical travel space. The bed can also be manually unfastened from the perforated tubing it is attached to, lowered, and refastened for added vertical space. The materials for the main structure of the LAAM system include cold-rolled steel railings, a 6.4 mm aluminum plate for the print bed, and 76 mm steel tubing for the frame. The LAAM system also includes a heated print bed composed of multiple zones.

The extruder on the Baylor LAAM system is a Strangpresse Extruder Model 19, which has a pellet-fed hopper at the top which was a custom-made part fabricated at Baylor University. The Strangpresse extruder shown Figure 1 is about 1 m tall with the hopper. The Strangpresse human–machine interface (HMI) allows an operator to adjust the temperatures in three zones along the length of the extruder as well as the extruder screw revolutions per minute (RPM) to control the deposition rate. The gantry system used for the motion control of the extruder is a Magnum II system from Precision Plasma LLC. The Magnum II system was made from an assembly kit for computer numerical control (CNC) plasma cutters, but its ability to move in three dimensions makes it suitable for custom LAAM systems as well, if the plasma cutter head is replaced with a suitable extruder. The gantry system is controlled using the Mach software Mach3 by Newfangled Solutions (Livermore Falls, ME, USA) managed on a local computer. All the experimental specimens for this study were fabricated using G-codes manually written by the author and ported to Mach3. The individual pellets, shown in Figure 2a, are fed into the hopper and then the filament is extruded, as shown in Figure 2b, and then deposited on to the print bed as shown in Figure 2c. During each of these various steps, the polymer is subjected to shearing and elongation, thus increasing the potential for fiber damage. The beads in the present study were composed of 13%, by weight, carbon-fiber-reinforced acrylonitrile butadiene styrene (CF-ABS).

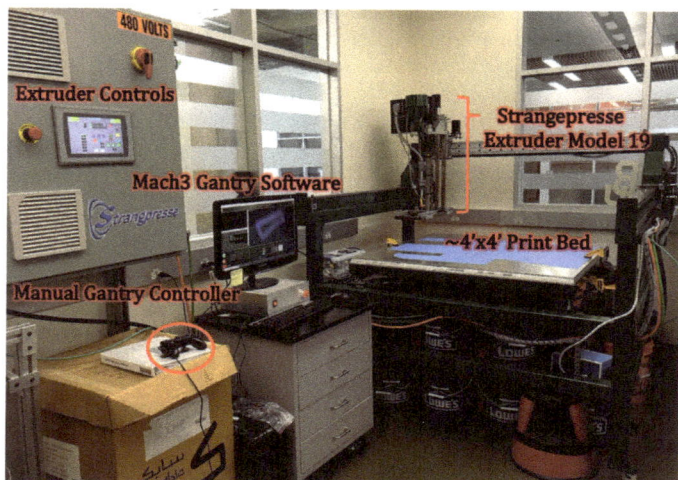

Figure 1. Baylor's large area additive manufacturing (LAAM) system.

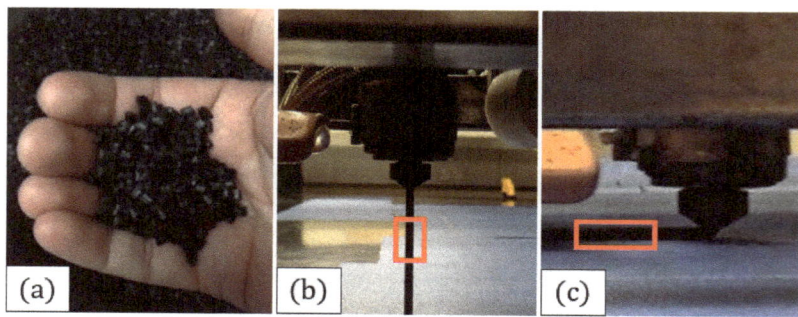

Figure 2. Three states of printed material: (**a**) a preprocessed pellet, (**b**) extrudate, and (**c**) a deposited bead.

2.2. Reduction of Additively Manufactured Bead to a Test Specimen

For the fiber length distribution and aspect ratio characterization, single beads of acrylonitrile butadiene styrene (ABS) pellets with 13% carbon fiber, by weight, were printed at 250 °C on a 95 °C bed. Single beads were selected so as to keep the present focus on the individual deposited bead. Future studies proposed by the authors will focus on multiple deposited beads to study issues presented due to the presence of an interface. Specimens were gathered at the three different points in the printing process indicated in Figure 2, the preprocessed pellet, predeposition extrudate, and the deposited bead. Obtaining the fiber aspect ratio at each stage enables one to identify the most damaging part of the printing process and quantify its effect on the material property predictions. It was found that the volume fraction of fibers was consistent across all three stages of the process and over multiple samples processed over various batches. Thus, it was reasonably assumed that the fibers were well dispersed in the end product, but this did not suggest that there would not exist a small number of local variations in the fiber packing density as demonstrated in the companion study [11]. The present paper is focused on local variations in orientation and for simplicity, assumes that the volume fraction of fibers is homogeneous.

For the strength specimens, the 13% CF-ABS was dried for 4 h at 82 °C prior to printing and stored in an environmental chamber with a −40 °C dew point. For printing the specimens, the three extruder temperature zones were set to 200 °C, 205 °C, and 210 °C, and the extruder RPM was set to 2250 with a nozzle-to-bed distance of 3 mm.

After printing, the upper surface of the beads were milled to remove the roughness of the surface of the bead. After machining, the specimens were sectioned with a Buehler IsoMet™ Low Speed Precision Cutter (Lake Bluff, IL, USA). The tensile specimens were 165 mm in length with 25 mm long tabs and a nominal thickness of 3 mm. The dimensions of the compression specimens were 81 mm in length with 38 mm long tabs and a nominal thickness of 3 mm. Specimen dimensions were of a single deposited raster and the length dimension was along the raster deposition dimension. After the specimens were cut, they were then cleaned, and the tabs were adhered using a Hysol 3039A adhesive and cured following the manufacturer's recommended curing cycle. Tensile specimens were then heat-treated at 110 °C for 10–15 min to reduce the warpage induced by milling the top surface and adhesive bonding. For the compression samples there was no measurable warpage, so they were not subjected to any heat treatment. After assembly of the tabs, samples were milled on their sides to ensure a constant, rectangular cross section and dimensional consistency between specimens. Figure 3 shows the prepared tensile specimens.

Figure 3. Photo of 13% CF-ABS tensile specimens prior to testing.

3. Experimental Characterization of Fiber-Reinforced Additive Manufactured Composite

The key experimental results of this paper are discussed in this section. Each subsection both describes an experimental setup and presents the results from that experimental setup. The fiber length distribution and aspect ratio study is discussed first. Next, the tensile stiffness and strength results are given. Finally, the compression stiffness and strength results are presented.

3.1. Fiber Length Distribution and Aspect Ratio Characterization

The process of characterizing the fiber length distribution consisted of five major steps. The first involved obtaining specimens. Specimens were gathered from the three points in the printing process shown in Figure 2. That is, a pellet, a 25.4 mm extrudate specimen, and a 25.4 mm printed specimen were gathered, with the printed specimen being of primary importance in this study. The second step was to isolate the fibers from each sample without damaging them. The third step was to capture micrographs of the fibers, stitch them together, and measure the fibers using a custom MATLAB code. The fourth and final step was to plot the fiber length distribution, get the number-average and weight-average fiber aspect ratios, and predict other material properties of interest.

Extracting the individual fibers from the matrix was accomplished by using a thermal digestion of each specimen. The complete test setup is shown in Figure 4, and the test setup where pellets were placed in a TA Instruments Q50 Thermogravimetric Analyzer (TGA) is shown in Figure 4a (New Castle, DE, USA). The burn-off procedure consisted of (1) ramping up the temperature to 600 °C, (2) holding it isothermal for 1 h in nitrogen at 600 °C, and (3) holding it isothermal for 10 min in air at 600 °C. The inert, nitrogen atmosphere prevented the carbon fibers from degrading. However, to get rid of remaining residue after step 2, step 3 was added so that the fibers could completely separate from each other. Adding step 3 is not ideal because the fibers may degrade slightly, skewing the measured fiber length distribution. Therefore, a correction factor was applied to the measurement data to mitigate these effects, as discussed in [12].

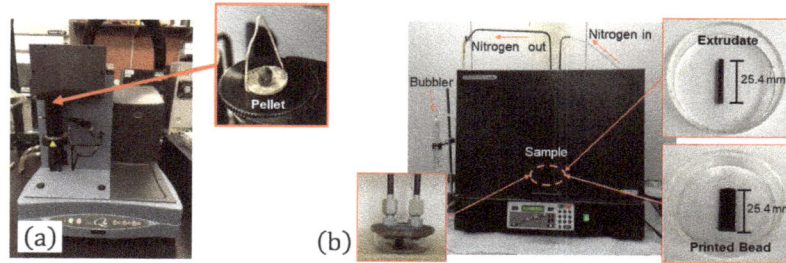

Figure 4. Burn-off testing setups. (**a**) TA Instruments Q50 TGA machine with a 13% CF-ABS pellet. (**b**) Ney Vulcan 3-1750 box furnace and 1-inch extrudate and printed specimens. These specimens were enclosed in a Petri dish with a custom-machined aluminum lid and inserted into the furnace.

To decrease the effect of cut fibers on skewing the fiber length distribution of the extrudate and printed specimens, 25.4 mm specimens were used since they were relatively large compared to the nominal length of a fiber, 100∼500 µm. These specimens were too large to fit in the TGA machine though, so a Ney Vulcan 3-1750 box furnace (Scotia, NY, USA) was used as shown in Figure 4b. The specimens were enclosed in a Petri dish with an aluminum lid. Two pipes attached to the aluminum lid purged the enclosure with nitrogen. A similar burn-off procedure as for the pellet was repeated for the extrudate and printed specimens.

The third step in characterizing the fiber length distribution was to capture images of the fibers under an MZ7 microscope, (Scienscope, New Haven, CT, USA) shown in Figure 5a, and stitch them together. After burn-off testing, the burn-off specimens were dropped into beakers of acetone and water and dispersed using a Branson Digital Sonifier 450 (Branson, MO, USA). The solutions were poured into Petri dishes before the dispersed fibers had time to settle completely. After pouring the dispersed fibrous solution into a Petri dish, the Petri dish was examined under a microscope and several micrographs were captured and stitched together. Figure 5b shows a micrograph, stitched in Adobe Photoshop CC 2017 and the resulting image was used for the fiber length measuring.

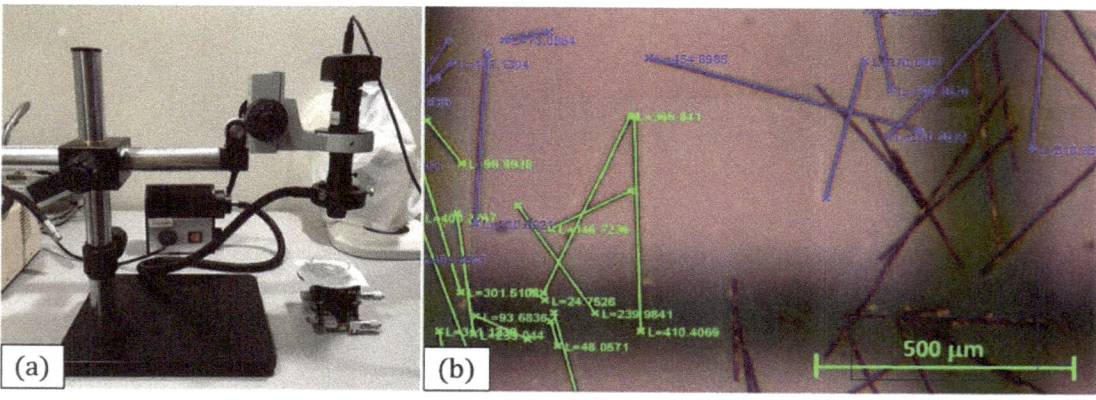

Figure 5. (**a**) An MZ7 microscope, used to capture micrographs of the fibers in a Petri dish. (**b**) Fibers being measured in a stitched micrograph. Green graph paper was placed under the Petri dish to provide a helpful frame of reference. The measurements in blue were measured in a current session, whereas the green measurements were from a previous measurement session.

After stitching, a custom code written in the MATLAB environment was used to measure the fibers. The code allows one to measure a fiber by performing a mouse-click on both ends of the fiber. First, however, a microscopic ruler must be measured to find the proper pixel-to-millimeter scaling factor. Once a fiber is measured, the length of the fiber is automatically scaled and logged. The code allows a user to measure several fibers at once, close out of the measurement session, and return to it at a later time without losing any data.

The fourth step was to plot the probability distribution function (PDF) of the fiber length and calculate other quantities of interest. Figure 6 shows the PDF of the fiber length, with the PDF being cast in units of μm^{-1} for the three manufacturing stages. From the data in Figure 6, the number- and weight-average fiber lengths were calculated, respectively, by

$$L_n = \frac{\sum N_i L_i}{\sum N_i} \quad (1)$$

$$L_w = \frac{\sum N_i L_i^2}{\sum N_i L_i} \quad (2)$$

In the above equations, N_i is the number of fibers in the ith histogram bin, and L_i is the average fiber length within the ith histogram bin. Similarly, the number- and weight-average aspect ratios were approximated, respectively, as

$$a_{rn} = \frac{L_n}{d_n} \quad (3)$$

$$a_{rw} = \frac{L_w}{d_w} \quad (4)$$

where d_n and d_w are taken from [12]. The results for the fiber diameters, lengths, and aspect ratios for both the weight and number averages are given in Table 1 for this study. Note, these data were corrected for fiber degradation due to thermal heating in the presence of oxygen to remove the polymer matrix as detailed in [12].

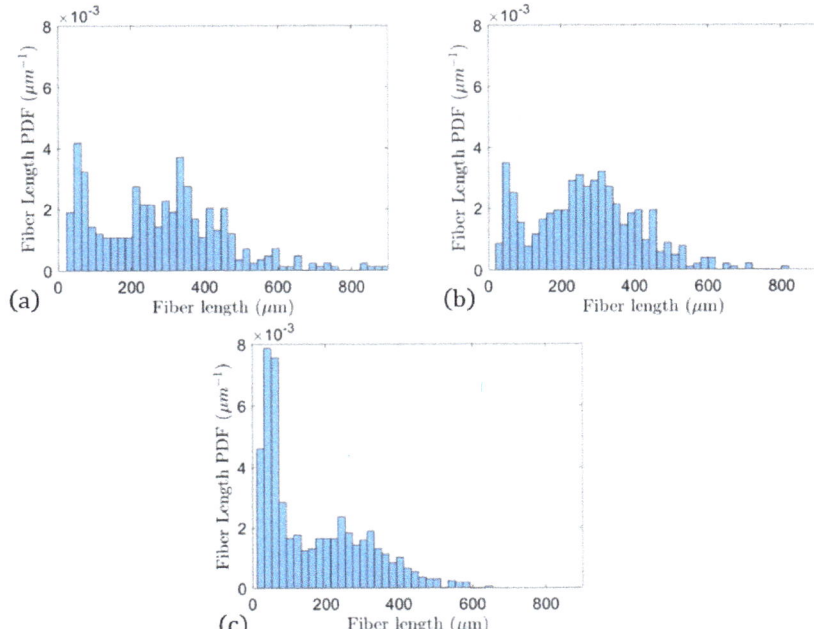

Figure 6. Fiber length distributions from (**a**) the preprocessed pellet, (**b**) the extrudate (predeposition), and (**c**) the printed bead (postdeposition).

Table 1. Fiber length and aspect ratio data, corrected for fiber degradation.

Parameter	Unit	Pellet	Extrudate	3D-Printed Bead
L_n	µm	293	271	169
L_w	µm	424	348	279
d_n	µm	7.83	7.51	7.19
d_w	µm	7.83	7.51	7.20
a_{rn}	µm	37.4	36.1	23.5
a_{rw}	µm	54.1	46.3	38.8

3.2. Stiffness and Strength Characterization

This section discusses the stiffness and strength testing. The results for the tensile properties are given first, followed by the compression results.

3.2.1. Tensile Testing

A Test Resources 810 Series fatigue tester with a F2500-B load cell was used for the tensile testing. Figure 7a shows a mounted specimen with an extensometer for measuring strain. The specimens were tested at a speed of 0.059 in/min with a data collection rate of 20 Hz and at a temperature of approximately 21 °C (70 °F). Three width and three thickness measurements were taken for each of the five specimens used in the tensile study using micrometers in or near the gage region prior to testing. The width and thickness measurements were averaged to obtain the cross-sectional area of each specimen, from which the stress was calculated. A typical tensile stress–strain plot is shown in Figure 7b.

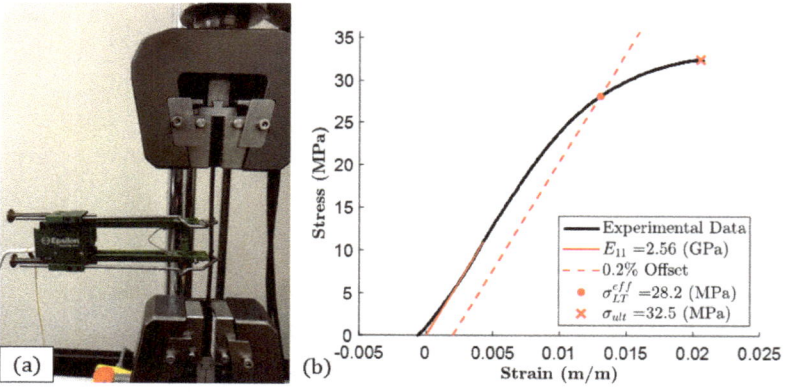

Figure 7. (a) Tensile testing setup with a 1 inch Epsilon 3542-0100-100-HT2 extensometer. (b) Tensile stress–strain plot for 13% CF-ABS.

The effective longitudinal tensile stiffness of the tensile specimens was defined as E_{LT}^{eff}, where the LT subscript indicates the longitudinal tensile component and the eff superscript indicates the bulk stiffness of the locally varying stiffness. The effective longitudinal tensile stiffness was defined over the strain range 0.003–0.005. In addition, the effective longitudinal yield strength, defined as the term σ_{LT}^{eff}, was determined by using an offset as defined in ASTM D638 [13]. In the present study, a 0.2% offset was selected due to the initial nonlinear loading region and to align with the literature for which the models were based upon. That is, a line parallel to the elastic region was extended from (0, 0.002) up to the stress–strain curve, and the stress at the intersection was taken as σ_{LT}^{eff}. A toe region sometimes appeared at the beginning of the stress–strain curves, which can be seen in Figure 7b. Therefore, the data were horizontally shifted to where the E_{LT}^{eff} lines would intersect the origin, prior to applying the 0.2% offset method. A summary of the results of four tests is given in Table 2, where $\bar{x}$, s_{n-1}, and C_V are, respectively, the average, standard deviation, and coefficient of variation for the five specimens studied.

Table 2. Tensile testing statistics for 13% CF-ABS specimens.

Property	Unit	Minimum	Maximum	$\bar{x}$	s_{n-1}	C_V
E_{LT}^{eff}	GPa	2.56	2.88	2.71	0.163	6.04%
σ_{LT}^{eff}	MPa	25.9	28.2	27.3	1.04	3.79%
σ_{LT}^{ult}	MPa	30.5	32.8	32.0	1.04	3.24%

3.2.2. Confined Compression Testing

The same Test Resources 810 Series fatigue tester with the same F2500-B load cell was also used for compression testing. Figure 8a shows a specimen in a modified ASTM D695 compression test fixture (Boeing BSS 7260) [14], which was placed between two compression platens mounted on the testing machine. Three width measurements of the five specimens were taken for each compression specimen, two on either end and one in the middle. The compression specimens had tabs covering most of their length, so only one thickness measurement was taken in the middle of the specimens, and this was done with calipers since the micrometer anvils did not fit between the tabs. The average width and thickness measurements were used to calculate the specimen's cross-sectional areas from which the stress was derived. The length of each of the five specimens was also measured and used to calculate the approximate strain. Before initiating a test, the upper compression platen was lowered in contact with the specimen. Each test was conducted at 0.051 in/min (1.3 mm/min, as called for by ASTM D695) with a data collection rate of 20 Hz and an environmental temperature of about 21–22 °C (70–72 °F).

A typical compression stress–strain curve is shown in Figure 8b. Five compression tests were performed. The toe region of each curve was trimmed, this time by determining the stiffness over the stress range 25–35 MPa and horizontally shifting the data. In addition, the effective longitudinal compression yield strength, σ_{LC}^{eff}, was determined using a 0.2% strain offset. The results of the compression tests are given in Table 3. The ultimate compression strength is not reported because the upper platen would often come into contact with the top of the metal fixture before the ultimate failure of a specimen. In addition, it is noteworthy that both the experimental compression stiffness and strength surpass the experimental tensile stiffness and strength, respectively.

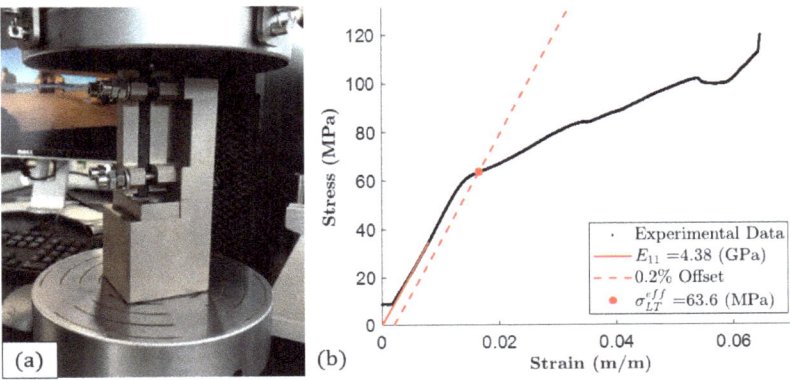

Figure 8. (a) A compression specimen mounted in a modified ASTM D695 compression test fixture (a Boeing BSS 7260) and between two platens. (b) A typical compression stress–strain plot for a 13% CF-ABS specimen.

Table 3. Compression testing statistics for 13% CF-ABS.

Property	Unit	Minimum	Maximum	$\bar{x}$	s_{n-1}	C_V
E_{LC}^{eff}	GPa	3.65	4.42	4.06	0.336	8.28%
σ_{LC}^{eff}	MPa	45.4	64.7	58.6	7.72	13.2%

4. Fiber Flow Modeling for Stiffness and Strength Prediction

4.1. Fiber Orientation Kinematics

Over the past few decades there has been considerable work in the area of fiber motion kinetics and the resulting structural performance of the processed composite. The following section pulls together a summary of the high points along with various components that were used in the present study. Consider a fluid flow domain with the velocity profile $\mathbf{v}(\mathbf{x})$ and velocity gradients $\mathbf{L}(\mathbf{x})$, where $\mathbf{x}$ denotes position. Jeffery's foundational work in expressing fiber motion for a dilute suspension in such a domain describes the time rate of change in orientation of a single, rigid, inertialess ellipsoid [2],

$$\dot{\mathbf{p}} = \mathbf{W} \cdot \mathbf{p} + \xi(\mathbf{D} \cdot \mathbf{p} - \mathbf{D} : \mathbf{ppp}) \tag{5}$$

In the above equation, the material derivative of the unit vector $\mathbf{p}$, which points along the long axis of the fiber, is used. That is, $\dot{\mathbf{p}} = \frac{d\mathbf{p}}{dt} = \frac{\partial \mathbf{p}}{\partial t} + \mathbf{v} \cdot \nabla \mathbf{p}$. In addition, the vorticity tensor is defined as $\mathbf{W} = \frac{1}{2}(\mathbf{L} - \mathbf{L}^T)$, and the rate of deformation tensor is similarly defined as $\mathbf{D} = \frac{1}{2}(\mathbf{L} + \mathbf{L}^T)$. ξ is the fiber geometric term, which allows Equation (5) to be used for other axisymmetric shapes besides ellipsoids as long as an equivalent ellipsoidal aspect ratio, r_e, is properly defined. Expressing the fiber geometric term ξ in terms of r_e gives (see, e.g., [15])

$$\xi = \frac{(r_e^2 - 1)}{(r_e^2 + 1)} \tag{6}$$

Jeffery's equation as expressed in Equation (5) may be used in dilute fiber solutions, but a fiber interaction model must be used for concentrated solutions. The concentrated regime may be defined as $V_f > (d/L)$, where V_f is the fiber volume fraction and d and L are the fiber diameter and length. For the carbon fibers in the CF-ABS used in the present study $d/L = 6.4$ μm/278 μm = 2.3%. The CF-ABS used in the present study had $V_f = 8.11\%$ (this corresponds to a 13% weight fraction) and $8.11\% > 2.3\%$, thus the CF-ABS flow was considered concentrated and therefore a fiber interaction model was used. In addition, the orientation tensor approach popularized by Advani and Tucker [3] was used to describe the orientation state of a population of fibers and save orders of magnitude in computational time. Cast in terms of orientation tensors, there are multiple options for the proper characterization of the fiber interaction kinetics, such as the retarding principal rate model from Tseng et al. [16], the interaction coefficient approach of Folgar and Tucker [4], and the anisotropic rotary diffusion model of Phelps and Tucker [6]. In the present study, we focused on the fiber interaction model of Wang et al. [5],

$$\dot{\mathbf{A}} = \mathbf{W} \cdot \mathbf{A} - \mathbf{A} \cdot \mathbf{W} + \xi \{\mathbf{D} \cdot \mathbf{A} + \mathbf{A} \cdot \mathbf{D} - 2[\mathbf{A_4} + (1-\kappa)(\mathbf{L_4} - \mathbf{M_4} : \mathbf{A_4})] : \mathbf{D}\}$$
$$+ 2\kappa C_I \dot{\gamma}(\mathbf{I} - 3\mathbf{A}) \tag{7}$$

where the second- and fourth-order orientation tensors, as discussed by Advani and Tucker [3], are defined as

$$\mathbf{A} = \oint \mathbf{pp}\psi(\mathbf{p})d\mathbf{p}, \quad \mathbf{A_4} = \oint \mathbf{pppp}\psi(\mathbf{p})d\mathbf{p} \tag{8}$$

In the above, $\psi(\mathbf{p})$ is the orientation probability density function. In addition, κ is the slowness factor as discussed by Wang et al. [5], C_I is the fiber interaction constant introduced by Folgar and Tucker [4], $\dot{\gamma}$ is the scalar magnitude of $\mathbf{D}$, $\mathbf{L_4} = \sum_{i=1}^{3} \lambda_i \mathbf{e}_i \mathbf{e}_i \mathbf{e}_i \mathbf{e}_i$,

$M_4 = \sum_{i=1}^{3} e_i e_i e_i e_i$, and λ_i and e_i are the ith eigenvalue and eigenvector of A. Based on the work of Bay and Tucker [17], we take

$$C_I = 1.84 \times 10^{-2} \exp\left(-0.7148 a_r V_f\right) \quad (9)$$

Note, that a_r is the geometric fiber aspect ratio ($a_r = L/D$) and is related to the equivalent ellipsoidal aspect ratio of a cylindrical fiber as given by Zhang et al. [15].

$$r_e = 0.000035 a_r^3 - 0.00467 a_r^2 + 0.764 a_r + 0.404 \quad (10)$$

Furthermore, based upon earlier work by the present authors (see, e.g., [8]), a value for the slowness parameter of $\kappa = 0.125$ was selected. Finally, the eigenvalue-based orthotropic closure discussed by Wetzel [18] and VerWeyst [19] was used to estimate A_4 due to its accuracy and efficiency. Equation (7) and all the equations in this section which it depends on were custom-coded by the authors in MATLAB, such that all flow kinematics were performed in the MATLAB environment. This was conducted along several streamlines in the flow domain to capture a fine-resolution solution for the fiber orientation state across the flow domain.

4.2. Fiber Micromechanics

The linear elastic nature of a material is often described using Hooke's law to relate the stress, σ_{ij}, to the strain, ϵ_{ij}, through the stiffness, C_{ijkl}, or the compliance, S_{ijkl}, as

$$\sigma_{ij} = C_{ijkl} \epsilon_{kl}, \qquad \epsilon_{ij} = S_{ijkl} \sigma_{kl} \quad (11)$$

Fibers and matrix have different values for their respective stiffness and due to the length scales considered, an effective stiffness is considered for the composite, $\langle C_{ijkl} \rangle$. This effective stiffness, also termed the homogenized stiffness, due to its relationship to the associated underlying unidirectional stiffness tensor of the composite, $\bar{C}_{ijkl}$, and the fiber orientation distribution function, is expressed as

$$\langle C_{ijkl} \rangle = \int_0^{2\pi} \int_0^{\pi} \bar{C}_{ijkl} \psi(\theta, \phi) \sin\theta \, d\theta \, d\phi \quad (12)$$

The underlying unidirectional stiffness tensor of the associated composite $\bar{C}_{ijkl}$ may be found in a variety of methods, the Halpin–Tsai and rule of mixtures often being the most popular due to their ease of implementation. In the present study, the authors used the Tandon and Wang [20] model with the closed-form solution suggested in Tucker and Liang [21] and with the mathematical form expressed in Zhang [22]. The choice of the Tandon and Wang model for the underlying unidirectional stiffness tensor of the associated composite was based upon the results from [21] in comparison to other available unidirectional composite models. Thus, with knowledge of the underlying stiffness tensor $\bar{C}_{ijkl}$ and the fiber orientation state from Equation (7), the effective stiffness tensor at a point could be computed. Advani and Tucker [3] provided a mathematical form of the effective stiffness tensor in terms of the orientation tensors as

$$\begin{aligned} \langle C_{ijkl} \rangle &= B_1 A_{ijkl} + B_2 \left(A_{ij}\delta_{kl} + A_{kl}\delta_{ij}\right) + B_3 \left(A_{ik}\delta_{jl} + A_{il}\delta_{jk} + A_{jl}\delta_{ik} + A_{jk}\delta_{il}\right) \\ &+ B_4 \left(\delta_{ij}\delta_{kl}\right) + B_5 \left(\delta_{ik}\delta_{jl} + \delta_{il}\delta_{jk}\right) \end{aligned} \quad (13)$$

where the coefficients B_i can be cast in terms of the underlying unidirectional stiffness tensor $\bar{C}_{ijkl}$ as presented in [3]. Observe that in Equation (13), the fourth-order orientation tensor A_{ijkl} appears. The same orthotropic closure of VerWeyst [19] and Wetzel [18] used in Equation (7) is also used in Equation (13).

The Tsai–Wu [23] model for the onset of failure was used and is expressed in terms of the second-order and fourth-order strength tensors, respectively, f_i and F_{ij}, as

$$f_m \sigma_m + F_{mn} \sigma_m \sigma_n = 1 \qquad (14)$$

In Equation (14), a contracted notation is used where the index pairs $[11, 22, 33, 23, 13, 12]$ are replaced by $[1, 2, 3, 4, 5, 6]$. For example, the stress tensor in index notation σ_{ij} can be expressed in contracted notation as σ_m where $[\sigma_{11}, \sigma_{22}, \sigma_{33}, \sigma_{23}, \sigma_{13}, \sigma_{12}] \rightarrow [\sigma_1, \sigma_2, \sigma_3, \sigma_4, \sigma_5, \sigma_6]$. In Equation (14), the summation convention continues to be used where $\{m, n\} \in \{1, 2, \ldots, 6\}$. Alternatively, the failure envelope in terms of the contracted strain tensor is

$$g_m \epsilon_m + G_{mn} \epsilon_m \epsilon_n = 1 \qquad (15)$$

where g_m and G_{mn} are, respectively, the contracted second- and fourth-order strength tensors. One can obtain the relationship between the tensors f_m, F_{mn}, g_m, and G_{mn} as (see e.g., [23])

$$g_m = f_o C_{om}, \quad G_{mn} = F_{op} C_{om} C_{pn} \qquad (16)$$

where the summations on the indices o and p are from one to six, and C_{mn} is the contracted form of the stiffness tensor discussed in Equation (13). As discussed in [8], the strength tensor relationship for the failure envelope can be recast in terms of the underlying unidirectional strength tensors, $\bar{f}_m$, $\bar{F}_{mn}$, $\bar{g}_m$, and $\bar{G}_{mn}$, where the bar indicates the unidirectional composite property, using a homogenization similar to that of the stiffness tensor of Equation (12) as

$$\langle g \rangle_m \epsilon_m + \langle G \rangle_{mn} \epsilon_m \epsilon_n = 1 \qquad (17)$$

For a unidirectional-fiber-reinforced composite with all fibers aligned along the x_1 axis, the unidirectional strength tensors $\bar{f}_m$ and $\bar{F}_{mn}$ may be expressed in terms of the tensile and compression strengths in the form

$$\begin{aligned}
\bar{f}_1 &= \frac{1}{\bar{\sigma}_{LT}} - \frac{1}{\bar{\sigma}_{LC}}, \quad \bar{f}_2 = \bar{f}_3 = \frac{1}{\bar{\sigma}_{TT}} - \frac{1}{\bar{\sigma}_{TC}}, \quad \bar{F}_{11} = \frac{1}{\bar{\sigma}_{LT}\bar{\sigma}_{LC}}, \\
\bar{F}_{22} &= \bar{F}_{33} = \frac{1}{\bar{\sigma}_{TT}\bar{\sigma}_{TC}}, \quad \bar{F}_{44} = 2(\bar{F}_{22} - \bar{F}_{23}), \quad \bar{F}_{55} = \bar{F}_{66} = \frac{1}{\tau^2}
\end{aligned} \qquad (18)$$

where the first subscript of the term $\bar{\sigma}$ indicates the longitudinal (L) axis along the fiber or the transverse (T) axis, and the second subscript indicates a tensile (T) load or compression (C) load. For example, $\bar{\sigma}_{TC}$ is the experimental transverse strength found from compression loading. In the present study, we assumed the tensile and compression behavior for the transverse load direction of the underlying unidirectional composite was identical, $\bar{\sigma}_{TT} = \bar{\sigma}_{TC} = \sigma_m$, where σ_m is the experimentally obtained strength of the matrix. We also assumed the corresponding shear strength of the matrix was found as $\tau = \sigma_m/\sqrt{3}$. Using the form suggested in [24], the remaining nonzero terms for $\bar{F}_{mn}$ were expressed as $\bar{F}_{12} = \bar{F}_{13} = -\bar{F}_{11}/4$ and $\bar{F}_{23} = -\bar{F}_{22}$. Then, using Equation (16), the unidirectional strength tensors $\bar{g}_m$ and $\bar{G}_{mn}$ were obtained. As shown in [8], the homogenized fourth-order strength tensor was found using an identical form as Equation (13) in terms of the orientation tensors. The homogenized second-order strength tensor in terms of the second-order orientation tensor and second-order strength tensor of the underlying unidirectional SFRP was found as (see, e.g., [25])

$$\begin{aligned}
\langle g \rangle_1 &= (\bar{g}_1 - \bar{g}_2) A_{11} + \bar{g}_2, \quad \langle g \rangle_2 = (\bar{g}_1 - \bar{g}_2) A_{22} + \bar{g}_2, \quad \langle g \rangle_3 = (\bar{g}_1 - \bar{g}_2) A_{33} + \bar{g}_2 \\
\langle g \rangle_4 &= (\bar{g}_1 - \bar{g}_2) A_{23}, \quad \langle g \rangle_5 = (\bar{g}_1 - \bar{g}_2) A_{13}, \quad \langle g \rangle_6 = (\bar{g}_1 - \bar{g}_2) A_{12}
\end{aligned} \qquad (19)$$

In the present study, the composite was considered at the edge of failure when the equality of Equation (17) was satisfied. For simulation purposes, a loading state was imposed on the composite and the following expression was evaluated

$$\varphi(\epsilon) = \langle g \rangle_m \epsilon_m + \langle G \rangle_{mn} \epsilon_m \epsilon_n \qquad (20)$$

where ϵ is the strain state from the imposed load on the structure. The SFRP is not predicted to fail as long as $\varphi(\epsilon) < 1$. When $\varphi(\epsilon)$ reaches one, the SFRP is predicted to fail.

The longitudinal tensile strength $\bar{\sigma}_{LT}$ for the unidirectional composite was found using the modified rule of mixtures suggested by Van Hattum and Bernardo [25] as

$$\bar{\sigma}_{LT}(L/d) = V_f \frac{\tau L}{d} + \sigma'_m \left(1 - V_f\right) \tag{21}$$

where the above expression is only valid when the fiber length, L, is less than the critical fiber length, L_c. The fibers in the current study had a typical fiber length of 200–300 µm in the final processed composite structure, whereas the critical fiber length assumed in this study, $L_c \simeq 0.9$ mm (taken from [25]), was much greater than the largest fiber length identified in Figure 6. The term σ'_m is the stress of the matrix at the failure strain of the fiber and is experimentally obtained once the strain at failure for the fiber is known.

The longitudinal compression strength was cast in the form suggested by Hayashi and Koyama [26] as expressed by Bajracharya et al. [27]

$$\bar{\sigma}_{LC} = \chi E_f \epsilon_m^* V_f + \left(1 - V_f\right) \sigma_m^* \tag{22}$$

where E_f and E_m are the moduli of, respectively, the fiber and the matrix, and $\sigma_m^* = E_m \epsilon_m^*$, where ϵ_m^* is the strain at which the matrix yields. The parameter χ is expressed in terms of the weight-average length of the fibers, L_w, and the critical fiber length, L_c, where $\chi = L_w/(2L_c)$.

It is well known that the presence of porosity in the composite reduces the stiffness and the strength. In the thesis of Nargis [10], the average void fraction identified in her specimen was 13.84%, a value that was used in this paper. Using the estimation of Zhang et al. [28] that the knockdown factor for the stiffness and the tensile strength is the same and is a function of the void volume fraction, the effective stiffness, E^{eff}, and strength, σ^{eff}, were expressed as

$$\frac{E^{eff}}{E_{without\ porosity}} = \frac{\sigma^{eff}}{\sigma_{without\ porosity}} = (1 - V_{voids})^n \tag{23}$$

where in the present study $n = 2$ was used. Equation (23) was used to find the effective stiffness values E_{LT}^{eff}, E_{LC}^{eff}, and E_{LF}^{eff} and the effective strength values σ_{LT}^{eff}, σ_{LC}^{eff}, and σ_{LF}^{eff}.

4.3. Deposition Flow Domain

The flow domain in the present study is depicted in Figure 9 with the dimensions selected based upon the nozzle shown in Figure 2. Solutions for the Navier–Stokes equations were performed using COMSOL Multiphysics. The interior walls of the nozzle were assumed to be no-slip boundary conditions along with the surface of the print bed, whereas the surfaces of the melt flow that are outside of the confined nozzle were considered to have slip boundary conditions. The geometry of the flow domain at the leading edge of the flow was constructed such that the surface was stress-free, and the normal component of the velocity was also zero using the approach from Heller et al. [29] to identify the surface using an iterative approach. Once the velocity field was obtained from the finite element simulation, the spatially varying fiber orientation state, $A_{ij}(\mathbf{x})$, was obtained using Equation (7) with $C_I = 1.9 \times 10^{-3}$, $\kappa \in \{0.05, 0.2\}$, and λ cast as a function of a_{rw} of the 3D-printed bead as given in Table 1. The fiber equations of motion were solved numerically using a custom program created in the MATLAB environment written by the authors. Then, using the material properties for the fiber and the matrix listed in Table 4, Equation (20) was evaluated across the flow domain using a custom script created by the authors that took the spatial location and returned the failure parameter at the specified location.

Table 4. Properties of ABS matrix and carbon fiber used for modeling.

Property	Unit	ABS Matrix	Ref.	Carbon Fiber	Ref.
Elastic modulus	GPa	$E_m = 2.41$	[30]	$E_f = 218$	[25]
Poisson's ratio	-	$\nu_m = 0.35$	[31]	$\nu_f = 0.26$	[25]
Density	kg/m^3	$\rho_m = 1040$	[30]	$\rho_f = 1760$	[32]
Dynamic viscosity	Pa·s	$\mu_m = 3200$	[33]	-	-
Tensile yield strength	MPa	$\sigma_m = 37.9$	[30]	$\sigma_f = 6024$	[25] †
Compression yield strain	$\frac{m}{m}$	$\epsilon_m^* = 0.04$	[34]	-	-
Matrix stress at fiber failure strain	MPa	$\sigma_m' = 24$	[32,35] ‡	-	-
Shear strength	MPa	$\tau = 21.9$	[25] †	-	-
Fiber length	μm	-	-	$L_f = 279$	††
Fiber diameter	μm	-	-	$d_f = 7.20$	††
Geometric aspect ratio	$\frac{m}{m}$	-	-	$a_r = 38.8$	††
Ellipsoidal aspect ratio	$\frac{m}{m}$	-	-	$r_e = 25.0$	[15] †
Critical fiber length	μm	-	-	$L_c = 835$	[25] †
Fiber weight fraction	$\frac{kg}{kg}$	-	-	$w_f = 0.13$	[36]
Fiber volume fraction	$\frac{m^3}{m^3}$	-	-	$V_f = 8.11\%$	[37] †

† These references are for the equations used to calculate the property. ‡ Fiber failure strain taken from [32] and using the stress–strain curve from [35] for ABS, the matrix stress was obtained. †† Refers to experimental results in the present paper.

4.4. Structural Simulation

The orientation state at the end of the flow domain, $A_{ij}(x_1 = 0, x_2)$, shown at the origin on the bottom left of Figure 9, was taken as the steady-state orientation and was used for all structural simulations. This orientation state was then projected along the direction of printing to form a specimen with an orientation that changed along x_2 but was constant along x_1. The tensile and compression specimen model domains are shown in Figure 10. Each domain is of a single deposited bead. The bead subjected to tension or compression is allowed to freely slide in the vertical direction, except in the lower, right corner. That is, the displacement u_2 is zero in the lower, right corner and unconstrained everywhere else. The right edge of the bead is fixed horizontally, i.e., $u_1 = 0$, and equal to some static displacement value in the x_1 direction, δ, on the left edge.

For the applied displacement δ, the spatially varying strain, $\epsilon(\mathbf{x})$, was computed from the finite element results along with the surface stresses. The effective stiffness values for tension or compression, E_{LT}^{eff} and E_{LC}^{eff}, respectively, were obtained by taking the average of the stress divided by the applied strain on the right end of the loaded specimen. Then, using the spatially varying strain values along with Equation (20), the spatially varying value for $\varphi(\epsilon)$ was obtained at every point in the solution domain. If the maximum value for φ was less than one, then the applied displacement δ was increased, whereas if φ at any point in the domain was more than one, the applied load was decreased. This process continued until the maximum spatial value for φ was one, thus suggesting that the member was at the onset of failure. At this point, σ_{LT}^{eff} and σ_{LC}^{eff} were calculated. A representative solution is shown in Figure 11a,b for a displacement resulting in a tensile and compression failure, respectively.

The finite element process described above was extended to study the stiffness and strength response over a range of values for κ. This first study fixed the fiber length and diameter to be that in the deposited bead, yielding a weight-average fiber aspect ratio of 38.8 as given in Table 1. The fibers were assumed to be randomly orientated initially for the flow domain of Figure 9, and once deposited, the structural domain shown in Figure 10 was analyzed for the stiffness and strength estimates for tension and compression. Using the fiber aspect ratio of 38.8 and a weight fraction of 13% (i.e., a volume fraction of 8.11%), the interaction coefficient from Equation (9) evaluated to $C_I = 1.9 \times 10^{-3}$. The range for κ was taken to be 0.05 to 0.2, within the range suggested by Wang et al. [5]. The remaining modeling inputs in this study are given in Table 4.

The results for the stiffness and strength parameters as a function of fiber slowness parameter κ are provided in Figure 12 for both the scenario when the porosity is neglected and the scenario when the porosity is accounted for in the structural simulations. Notice in Figure 12 that the tensile and compression stiffness values are graphically indistinguishable regardless of the porosity. This is by construction in the model, as the stiffness response is assumed to be equivalent in tension and compression. Conversely, the compression strength is measurably more than that of the tensile strength. Observe that the stiffness and strength increase as a function of the increasing value of the slowness parameter. This correlates to the increase in the orientation. As the slowness parameter, κ, increases from zero to one, the rate of alignment increases as well; thus, a higher value of the parameter κ correlates with a higher value of the alignment along the flow direction, and thus a higher value for the stiffness and strength for fibers that are stiffer and stronger than the surrounding matrix. It is also worth noting the significant reduction in the part performance as a result of porosity. This highlights the importance of accounting for porosity and the need to identify a means to reduce porosity generation in the deposited bead during processing.

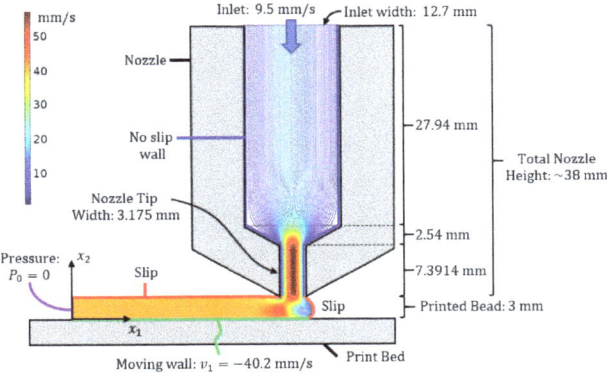

Figure 9. Finite element modeling domain at the tip of the LAAM nozzle including the deposition region.

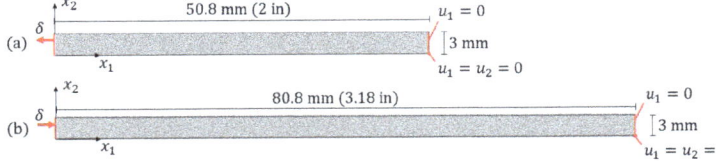

Figure 10. Specimen domains used for the structural analysis for the (**a**) tensile test and (**b**) compression test.

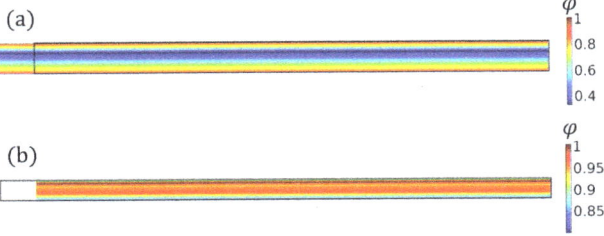

Figure 11. Structural simulation results showing the failure parameter $\varphi(\mathbf{x})$ for the (**a**) tensile test and (**b**) compression test.

The last modeling study focused on the importance of including the reduction in the fiber length in the structural simulations. Taking the three length distributions presented in Table 1 for the case of the pellet, the extrudate, and the deposited bead, the weight-average length is shown to decrease from 424 μm to 348 μm to 279 μm. Performing the flow simulation with the corresponding fiber aspect ratios, followed by the structural simulations led to the stiffness and strength predictions provided in Table 5. Observe that by properly including the reduction in the fiber length, there is a reduction of nearly 12% in the stiffness, a reduction of over 30% in the tensile strength, and a reduction of over 14% in the compression strength.

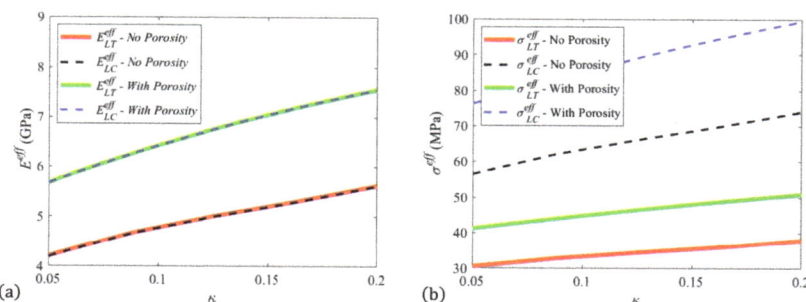

Figure 12. Effective stiffness and strength predictions of the deposited bead for $a_r = 38.8$. (**a**) Effective stiffness and (**b**) effective strength.

Table 5. Tensile and compression stiffness and strength predictions as a function of the weight-average aspect ratios from Table 1 along the flow domain for $\kappa = 0.125$.

Parameter	$E_{LT}^{eff} = E_{LC}^{eff}$ (GPa)	σ_{LT}^{eff} (MPa)	σ_{LC}^{eff} (MPa)
L_w = 424 μm (pellet)	5.58	44.8	75.4
L_w = 348 μm (extrudate)	5.34	39.3	71.2
L_w = 279 μm (bead)	5.00	34.4	66.1

5. Comparison between Model and Experiment

Table 6 summarizes the experimental results of this study and Table 7 summarizes the model predictions for comparison. These results show that the model predictions are on the high side of the experimental results. However, there are several possible culprits for this.

Table 6. Summary of experimental results, with the error taken from the standard deviation of experimental results.

Property	Average Property
E_{LT}^{eff}	2.71 ± 0.16 GPa
E_{LC}^{eff}	4.06 ± 0.34 GPa
σ_{LT}^{eff}	27.3 ± 1.0 MPa
σ_{LC}^{eff}	58.6 ± 7.7 MPa

Table 7. Tensile and compression stiffness and strength predictions of the deposited bead using the RSC model with a random initial orientation state ($A_{ij} = \frac{1}{3}\delta_{ij}$), $a_{rw} = 38.8$, and $\kappa \in \{0.05, 0.20\}$. Equation (23) was used to generate the second row from the first row.

Accounting for Porosity?	$E_{LT}^{eff} = E_{LC}^{eff}$ (GPa)	σ_{LT}^{eff} (MPa)	σ_{LC}^{eff} (MPa)
No	5.67–7.54	41.0–50.8	76.2–99.4
Yes	4.20–5.62	30.4–37.8	56.4–74.0

On the experimental side, the top surfaces of the specimens were machined down to 10–20% of the specimen height to eliminate waviness and to ensure uniform cross sections. While uniform cross sections are desirable, the skin of the specimens contained a high fiber alignment in the load-bearing direction. Thus, removing this skin from the tops of the specimens could have potentially caused a noticeable drop in the mechanical properties.

On the modeling side, Table 7 shows that a relatively large range of predictions can be found depending on the parameters fed into the models. The lower range for E_{LT}^{eff} and E_{LC}^{eff} should be selected in the present study based on the recommendations found in recent fiber interaction model papers to decrease the value of κ. κ has a greater effect than C_I, but there are other parameters that were estimated in the present study that could also influence the accuracy of the models. These include the initial orientation state and the constituent material properties in Table 4. Thus, it is important to zero in on accurate modeling parameters when using the proposed models in order to tighten the prediction window. In addition, the recent work by Nargis and Jack in [11] showed that the porosity was found to be significantly larger than anticipated, coupled with a broad spectrum of void sizes. Thus, accounting for porosity and selecting the lower range for the slowness factor κ, the predicted value for the compressive stiffness, E_{LC}^{eff}, came out to be 4.2 GPa, whereas the measured value was 4.06 GPa; the predicted value for the tensile stiffness, E_{LT}^{eff}, was 4.2 GPa, whereas the measured value was 2.71 GPa; the predicted value for the compressive strength σ_{LC}^{eff}, was 56 MPa, whereas the measured value was 59 MPa; and the predicted value for the tensile strength, σ_{LT}^{eff}, was 30 MPa, whereas the measured value was 27 MPa. Each of these values, with the exception of the tensile stiffness, is well within an acceptable range. This study highlights a limitation of the present modeling that should be addressed in a future study. The present model assumes that the tensile and compressive stiffness are identical, which is not the case based upon the results presented. Thus, a future expansion of the model to allow for discontinuous tensile and compressive behavior will be required prior to full-scale industrial implementation and will be the scope of a future study.

6. Conclusions

Having so many modeling parameters increases the difficulty to fully validate the proposed models. However, the results provided indicate the modeling framework yields results for the effective strength and stiffness that align with experimental characterization to within ∼1% and ∼10% for the longitudinal compressive and tensile strength, respectively, and to within ∼3% and ∼50% for the longitudinal compressive and tensile stiffness, respectively. In addition, a framework has been developed that allows for improvement by model substitution as more accurate modeling inputs are found. The modeling methodology goes from the properties of the individual fibers and surrounding matrix, through the fiber kinematic equations of motion through the final part performance. Results show the sensitivity of the final predicted response to subtle changes in the fiber interaction behavior as well as a reduction in the final part performance through the incorporation of the known porosity of the final processed part.

For future work on the experimental side, improvement in specimen preparation should be sought. For example, nonmachined, multibead tensile and compression specimens could be used to reduce the effect anomalous defects might have, or injection-molded specimens (which have less void content) could be used to reduce the effects of porosity. On the modeling side, in general, efforts to isolate and independently verify each underlying model and model input would be helpful. The fiber orientation could be physically measured as in Nargis' work [10], for example, and directly input into the stiffness and strength models to validate these models independently from a fiber orientation model. In addition, it is well understood that polymers are sensitive to changes in temperature, and performing tests across a temperature range would be of interest. The present model is linearly elastic, and extending the work to viscoelastic models would be of interest. A first pass at a 3D model for stiffness prediction which incorporated Nargis' results was performed in [12], but this area still needs more work. In addition, the effects of porosity need to be accounted for better, both in modeling the fiber orientation state and predicting the final mechanical properties as a function of the orientation state. The flow and orientation kinematics relationship was decoupled in the present study, but coupling them may make a difference in the predicted final orientation state. Another consideration for a future study may be a fiber aspect ratio distribution, rather than a single-value aspect ratio.

Author Contributions: Conceptualization, T.R. and D.A.J.; methodology, T.R. and D.A.J.; software, T.R. and D.A.J.; formal analysis, T.R. and D.A.J.; investigation, T.R.; resources, D.A.J.; data curation, T.R. and D.A.J.; writing—original draft preparation, T.R.; writing—review and editing, T.R. and D.A.J.; visualization, T.R. and D.A.J.; supervision, D.A.J.; project administration, D.A.J.; funding acquisition, D.A.J. All authors have read and agreed to the published version of the manuscript.

Funding: This research received was internally funded through Baylor University.

Institutional Review Board Statement: Not applicable.

Acknowledgments: The authors are grateful for the financial support of Douglas Smith in the purchase of some of the translation hardware for the LAAM system. In addition, the authors thank Douglas Smith and Blake Heller for graciously allowing the use of their finite element flow model. Finally, the authors would also like to thank Luke Battershell from Baylor University for his help in the assembly and software development of the LAAM system and James Orr from Baylor University for his help in the machining of the various components.

Conflicts of Interest: The authors declare no conflict of interest.

Abbreviations

The following abbreviations are used in this manuscript:

ABS	Acrylonitrile butadiene styrene
ASTM	American Society for Testing and Materials
CF-ABS	Carbon-fiber-reinforced acrylonitrile butadiene styrene
CNC	Computer numerical control
HMI	Human–machine interface
LAAM	Large area additive manufacturing
PDF	Probability distribution function
RPM	Revolutions per min
SFRP	Short-fiber-reinforced polymer
TGA	Thermogravimetric analyzer

References

1. Love, L.; Kunc, V.; Rios, O.; Duty, C.; Elliott, A.; Post, B.; Smith, R.; Blue, C. The Importance of Carbon Fiber to Polymer Additive Manufacturing. *J. Mater. Res.* **2014**, *29*, 1893–1898. [CrossRef]
2. Jeffery, G. The Motion of Ellipsoidal Particles Immersed in a Viscous Fluid. *Proc. R. Soc. Lond. A* **1923**, *102*, 161–179.
3. Advani, S.; Tucker, C. The Use of Tensors to Describe and Predict Fiber Orientation in Short Fiber Composites. *J. Rheol.* **1987**, *31*, 751–784. [CrossRef]
4. Folgar, F.; Tucker, C. Orientation Behavior of Fibers in Concentrated Suspensions. *J. Reinf. Plast. Compos.* **1984**, *3*, 98–119. [CrossRef]
5. Wang, J.; O'Gara, J.; Tucker, C. An Objective Model for Slow Orientation Kinetics in Concentrated Fiber Suspensions: Theory and Rheological Evidence. *J. Rheol.* **2008**, *52*, 1179–1200. [CrossRef]
6. Phelps, J.; Tucker, C. An Anisotropic Rotary Diffusion Model for Fiber Orienation in Short- and Long Fiber Thermoplastics. *J. Non-Newton. Fluid Mech.* **2009**, *156*, 165–176. [CrossRef]
7. Sommer, D.; Favaloro, A.; Pipes, R. Coupling anisotropic viscosity and fiber orientation in applications to squeeze flow. *J. Rheol.* **2018**, *62*, 669–679. [CrossRef]
8. Russell, T.; Jack, D. Strength prediction in single beads of large area additive manufactured short-fiber polymers. *Polym. Compos.* **2021**, *42*, 6534–6550. [CrossRef]
9. Heller, B. Non-Isothermal Non-Newtonian Flow and Interlayer Adhesion in Large Area Additive Manufacturing Polymer Composite Deposition. Ph.D. Thesis, Baylor University, Waco, TX, USA, 2019.

10. Nargis, R. Internal Fiber Orientation Measurements and Void Distribution for Large Area Additive Manufactured Parts Using Optical and SEM Imaging Techniques. Master's Thesis, Baylor University, Waco, TX, USA, 2021.
11. Nargis, R.; Jack, D. Fiber Orientation Quantification for Large Area Additively Manufactured Parts Using SEM Imaging. *Polymers* **2023**, *15*, 2871. [CrossRef]
12. Russell, T. Mechanical and Thermal Property Prediction in Single Beads of Large Area Additive Manufactured Short-Fiber Polymer Composites. Ph.D. Thesis, Baylor University, Waco, TX, USA, 2021.
13. *ASTM D638*; Standard Test Method for Tensile Properties of Plastics. ASTM: West Conshohocken, PA, USA, 2022.
14. *ASTM D695*; Standard Test Method for Compressive Properties of Rigid Plastics. ASTM: West Conshohocken, PA, USA, 2015.
15. Zhang, D.; Smith, D.; Jack, D.; Montgomery-Smith, S. Numerical Evaluation of Single Fiber Motion for Short-Fiber-Reinforced Composite Materials Processing. *J. Manuf. Sci. Eng.* **2011**, *133*, 051002. [CrossRef]
16. Tseng, H.-C.; Chang, R.-Y.; Hus, C.-H. Phenomenological Improvements to Predictive Models of Fiber Orientation in Concentrated Suspensions. *J. Rheol.* **2013**, *57*, 1597–1631. [CrossRef]
17. Bay, R.; Tucker, C. Stereological Measurement and Error Estimates for Three-Dimensional Fiber Orientation. *Polym. Eng. Sci.* **1992**, *32*, 240–253. [CrossRef]
18. Wetzel, E. Modeling Flow-Induced Microstructure of Inhomogeneous Liquid-Liquid Mixtures. Ph.D. Thesis, University of Illinois at Urbana Champaign, Champaign, IL, USA, 1999.
19. VerWeyst, B. Numerical Predictions of Flow Induced Fiber Orientation in Three-Dimensional Geometries. Ph.D. Thesis, University of Illinois at Urbana Champaign, Champaign, IL, USA, 1998.
20. Tandon, G.; Weng, G. The Effect of Aspect Ratio of Inclusions on the Elastic Properties of Unidirectionally Aligned Composites. *Polym. Compos.* **1984**, *5*, 327–333. [CrossRef]
21. Tucker, C.; Liang, E. Stiffness Predictions for Unidirectional Short-Fiber Composites: Review and Evaluation. *Compos. Sci. Technol.* **1999**, *59*, 655–671. [CrossRef]
22. Zhang, C. Modeling of Flexible Fiber Motion and Prediction of Material Properties. Master's Thesis, Baylor University, Waco, TX, USA, 2011.
23. Tsai, S.; Wu, E. A General Theory of Strength for Anisotropic Materials. *J. Compos. Mater.* **1971**, *5*, 58–80. [CrossRef]
24. DeTeresa, S.; Larsen, G. *Derived Interaction Parameters for the Tsai-Wu Tensor Polynomial Theory of Strength for Composite Materials*; Technical Report; UCRL-JC-145041; Lawrence Livermore National Lab.: Livermore, CA, USA, 2001.
25. Van Hattum, J.; Bernardo, C. A Model to Predict the Strength of Short Fiber Composites. *Polym. Compos.* **1999**, *20*, 524–533. [CrossRef]
26. Hayashi, T.; Koyama, K. Theory and Experiments of Compressive Strength of Unidirectionally Fiber-reinforced Composite Materials. *Mech. Behav. Mater.* **1972**, *5*, 104–112.
27. Bajracharya, R.; Manalo, A.; Karunasena, W.; Lau, K. Experimental and Theoretical Studies on the Properties of Injection Moulded Glass Fibre Reinforced Mixed Plastics Composites. *Compos. Part A* **2016**, *84*, 393–405. [CrossRef]
28. Zhang, Y.; Rodrigue, D.; Ait-Kadi, A. High Density Polyethylene Foams. III. Tensile Properties. *J. Appl. Polym. Sci.* **2003**, *90*, 2130–2138. [CrossRef]
29. Heller, B.; Smith, D.; Jack, D. Planar Deposition Flow Modeling of Fiber Filled Composites in Large Area Additive Manufacturing. *Addit. Manuf.* **2019**, *25*, 227–238. [CrossRef]
30. Polyhedron Laboratories Inc. *ABS Plastics Testing Services—ABS Testing—ABS Plastic Strength Test*; Technical Report; Polyhedron Laboratories Inc.: Houston, TX, USA, 2021. Available online: https://www.polyhedronlab.com/ (accessed on 1 May 2021).
31. Engineers Edge LLC. *Specifications for Common Plastic Molding Design Material*; Technical Report; Engineers Edge LLC: Monroe, GA, USA, 2021. Available online: https://www.engineersedge.com/ (accessed on 1 May 2021).
32. Minus, M.; Kumar, S. The Processing, Properties, and Structure of Carbon Fibers. *JOM J. Miner. Met. Mater. Soc.* **2005**, *57*, 52–58. [CrossRef]
33. Russell, T.; Heller, B.; Jack, D.; Smith, D. Prediction of the Fiber Orientation State and the Resulting Structural and Thermal Properties of Fiber Reinforced Additive Manufactured Composites Fabricated Using the Big Area Additive Manufacturing Process. *J. Compos. Sci.* **2018**, *2*, 26. [CrossRef]
34. Wang, J.; Xu, D.; Sun, W.; Du, S.; Guo, J.; Xu, G. Effects of Nozzlebed Distance on the Surface Quality and Mechanical Properties of Fused Filament Fabrication Parts. *IOP Conf. Ser. Mater. Sci. Eng.* **2019**, *479*, 012094. [CrossRef]
35. Banjanin, B.; Vladic, G.; Pal, M.; Balos, S.; Dramicanin, M.; Rackov, M.; Knezevic, I. Consistency Analysis of Mechanical Properties of Elements Produced by FDM Additive Manufacturing Technology. *Materia* **2018**, *23*, e12250. [CrossRef]
36. MatWeb LLC. *Avient Stat-Tech AS-13CF/000 Black Acrylonitrile Butadiene Styrene (ABS)*; Technical Report; MatWeb LLC: Blacksburg, VA, USA, 2021. Available online: http://www.matweb.com/ (accessed on 1 May 2021).
37. Barbero, E. *Introduction to Composite Material Design*, 3rd ed.; CRC Press: Boca Raton, FL, USA, 2017.

Disclaimer/Publisher's Note: The statements, opinions and data contained in all publications are solely those of the individual author(s) and contributor(s) and not of MDPI and/or the editor(s). MDPI and/or the editor(s) disclaim responsibility for any injury to people or property resulting from any ideas, methods, instructions or products referred to in the content.

Article

Influence of Post-Processing on the Properties of Multi-Material Parts Obtained by Material Projection AM

Pablo Zapico [1], Pablo Rodríguez-González [2,*], Pablo Robles-Valero [2], Ana Isabel Fernández-Abia [2] and Joaquín Barreiro [2]

1. Department of Construction and Manufacturing Engineering, University of Oviedo, Campus of Gijón, 33204 Gijón, Spain; zapicopablo@uniovi.es
2. Department of Mechanical, Informatics and Aerospace Engineering, University of León—Universidad de León, Campus de Vegazana, 24071 León, Spain; probv@unileon.es (P.R.-V.); aifera@unileon.es (A.I.F.-A.); joaquin.barreiro@unileon.es (J.B.)
* Correspondence: prodrg@unileon.es

Abstract: The great geometric complexity that additive manufacturing allows in parts, together with the possibility of combining several materials in the same part, establishes a new design and manufacturing paradigm. Despite the interest of many leading sectors, the lack of standardization still makes it necessary to carry out characterization work to enjoy these advantages in functional parts. In many of these techniques, the process does not end with the end of the machine cycle, but different post-processing must be carried out to consider the part finished. It has been found that the type of post process applied can have a similar effect on part quality as other further studied process parameters. In this work, the material projection technique was used to manufacture multi-material parts combining resins with different mechanical properties. The influence of different post-processing on the tensile behavior of these parts was analyzed. The results show the detrimental effect of ultrasonic treatment with isopropyl alcohol in the case of the more flexible resin mixtures, being advisable to use ultrasonic with mineral oil or furnace treatment. For more rigid mixtures, the furnace is the best option, although the other post-processing techniques do not significantly deteriorate their performance.

Keywords: additive manufacturing; multi-material; material projection; post-processing

1. Introduction

Additive manufacturing technology (AM) is based on part manufacturing by adding layers of material. Originally called rapid prototyping, it was conceived to produce prototypes quickly and easily from digital models. Given its advantages in terms of simplicity of use, great freedom of design and customization without a significant increase in cost, the possibility of reducing the number of components, etc., it is currently of great interest to many sectors such as medical [1], aerospace [2], energy industry [3], sports [4,5], and those geared towards the manufacture of micrometer-scale components [6].

One of the AM techniques that has received great attention in recent times is the Material Projection (MP) technique [7]. This technique allows for the addition of material layers by selective deposition of light-curing resin droplets that are subsequently cured by an ultraviolet light source. For this purpose, AM machines have a head that incorporates piezoelectric injectors arranged in a matrix to deposit the resin, as well as a blade or roller and several ultraviolet light lamps, which flatten and cure the resin once it has been deposited [8].

The main manufacturers of AM machines based on this technique (3D Systems [9] and Stratasys [10]) have developed equipment capable of mixing several light-curable polymer resins simultaneously at the voxel level (3D pixel) to manufacture multi-material parts [11,12]. This new paradigm undoubtedly marks a milestone in terms of the ability to simplify designs by reducing the number of components, due to the variety of properties that these resins can present. The support material used in this technique consists of a paraffin or hydrophobic material that is deposited next to the photopolymers that must be removed by post-processing out of the machine.

The MP technique has proven useful for manufacturing functional parts, in general [13,14], for manufacturing plastic injection molds for short or one-off series [15,16] and for validating prototype parts for series production [17,18]. In addition to the aforementioned advantages of additive manufacturing, this technique has minimal material waste, lower cost, and greater supply chain efficiency compared to machining or small series injection molding [19]. This has made it of great interest to industries as varied as biomedical [20,21] and aeronautics [22].

Nevertheless, MP does not escape the challenges that remain to be addressed for additive manufacturing, among which can be highlighted both the lack of design and manufacturing guidelines, as well as the lack of specific standardization [23]. This has led several researchers to focus in recent years on studying this technique in depth, mainly focusing on the material properties' characterization [24,25], studying the geometric and surface quality [26–28], analyzing the anisotropy of the mechanical properties of the printed parts [29], and manufacturing structures reinforced with multiple materials [30,31].

The interesting results achieved by these works reinforce the need to address the AM challenges mentioned above. Specifically for MP, there is a key issue related to the development of manufacturing procedures: the post-processing operation for removing support material. This operation is mandatory and common to all parts obtained by this technique. As seen in many of the aforementioned works, researchers have focused on analyzing the properties and the economy of the parts depending on their orientation and position in the machine working volume, while avoiding the implications of the needed post-processing. In this respect, a few works mention the problems related to this post-processing. Liu et al. [32] note the advantage of using integrated lattices over snap-fitted lattices in terms of energy absorption but state the difficulty of removing the support material from the former when the structure is very intricate. Something similar is highlighted by Meisel et al. [33] in their work evaluating different limitations of the process, among which stands out the elimination of support material in the case of thin sections. Eren et al. [34] recognize that the removal of support is extremely difficult when printing delicate and thin cell structures, as this limits the minimum cell size and affects the final mechanical response of the reticular structures. Abayazid et al. [31] note that the use of the post-processing recommended by manufacturers can generate warping and geometric distortions of the printed parts. Meanwhile, He et al. [35] note the extreme force that must be applied and its adverse effects on the final quality of the parts in the case of using certain support materials that are different from the usual ones. Therefore, there is a clear need to analyze in depth the influence of post-processing.

On the other hand, it is worth highlighting the few works that analyze the properties or establish design guidelines for using something that is relatively common in this technique, i.e., the combination of different resins with different properties in the same part. This characteristic undoubtedly makes this technique stand out within additive manufacturing due to the simplification of the design in terms of reducing the number of components, as it allows different materials with different properties to be combined in the same part. With regard to the combination of materials, Boopathy et al. [36] studied the energy absorption capacity of printed parts against impact, observing that the different combinations of materials and the direction of the load played an important role in the behavior of the composite material.

This work focused on analyzing two questions that have been little studied or outright overlooked in other works on MP and that are of vital importance to safely exploit the advantages of this AM technique: the good tensile behavior of components that combine several materials with different mechanical properties, as well as the influence that the post-processing to be applied has on this behavior. For that, in the first part of the work, the tensile mechanical behavior of standardized specimens combining a stiff and tough resin with a more flexible one is analyzed. After verifying the good performance of the multi-material specimens, the second part of the work analyzes the influence of different usual post-processing procedures, i.e., using heat or wet ultrasound (with vegetable oil and with isopropanol), on the tensile behavior of multi-material specimens, identifying the most suitable post-processing for the different resin combinations.

2. Materials and Methodology

2.1. Materials

The AM Material Projection Machine ProJet® MJP 5600 (3DSystems, Rock Hill, SC, USA) was used in this study. The working volume of this machine is $518 \times 381 \times 300$ mm^3, with a maximum resolution in the XYZ axes of $750 \times 750 \times 2000$ DPI. As manufacturing material, this machine uses various resins, including mixtures of these in different proportions. The resins used in this work, both supplied by the 3D System (Rock Hill, SC, USA), are VisiJet CR-CL 200 (translucent acrylic based photopolymer resin) and VisiJet CE-NT (amber acrylic-based photopolymer resin). After MP manufacturing, the VisiJet CR-CL 200 resin is comparatively much stiffer and stronger than VisiJet CE-NT, which shows a more flexible behavior. Table 1 shows the tensile strengths of these resins and of the different resin mixtures configurable in the manufacturing software [9]. These tensile strengths are represented in Figure 1, where it can be seen that the higher the proportion of VisiJet CR-CL 200 in the material mixture, the higher the tensile strength achieved. For the pure resins, the supplier also provides the tensile modulus, 0.27–0.43 MPa and 1400–2100 MPa, and the elongation at break, 160–230% and 14–22%, VisiJet CE-NT and VisiJet CR-CL 200, respectively. Figure 2 shows the AM machine used and the carbides of the different VisiJet materials.

Table 1. Main mechanical properties of resins and mixtures (according to ASTM D638) [9].

Material/Mixture	Tensile Strength (MPa)	Material/Mixture	Tensile Strength (MPa)
VisiJet CE-NT	0.20–0.40	D50	1–3
A40	0.23–0.32	D55	2–3
A50	0.35–0.48	D60	4–5
A60	0.48–0.77	D65	8–10
A70	0.75–1.10	D70	12–16
A80	1.30–1.70	D75	19–27
A90	1.40–1.90	VisiJet CR-CL 200	30–43

As support material, this machine uses VisiJet S500, also supplied by 3D Systems, which is a non-toxic wax. During the MP process, both the resins and the support material are heated to around 60 °C, so that they acquire an appropriate viscosity for projection onto the manufacturing platform. After projection, the resins are cured by ultraviolet light, while the wax is solidified by cooling.

In order to evaluate the tensile mechanical properties of components containing areas made of different resins or resin mixtures, a tensile specimen was designed with a central area which can be made of a different material than that used for the heads. This specimen was designed according to the recommendations of ISO 527-2 [37], with the geometrical parameters represented in Figure 3. As can be seen, the specimen is always manufactured with the strongest material, CR-CL 200, in the head area to ensure the correct gripping of the tensile grips in the tensile test, while in the core, any of the 14 possibilities shown in Table 1 can be selected. Therefore, this specimen not only allows us to determine the tensile behavior of the weakest material used, which is the weakest link in the chain, but also to validate the good behavior of the interface between materials (see Figure 3). All the specimens manufactured were oriented horizontally and with their longitudinal axis perpendicular to the direction of movement of the machine head in order to avoid the main effect of the characteristic anisotropy of this process [38].

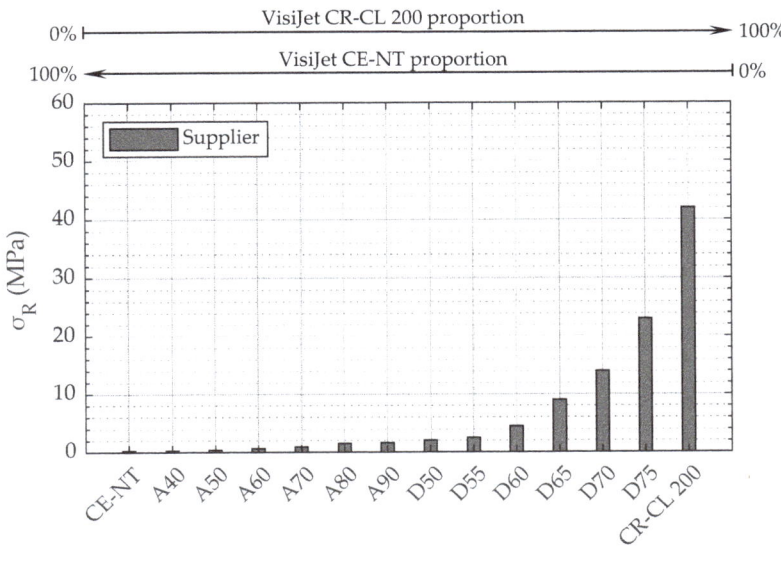

Figure 1. Tensile strength of VisiJet CE-NT, VisiJet CR-CL 200, and the mixtures established in 3D Sprint software [9].

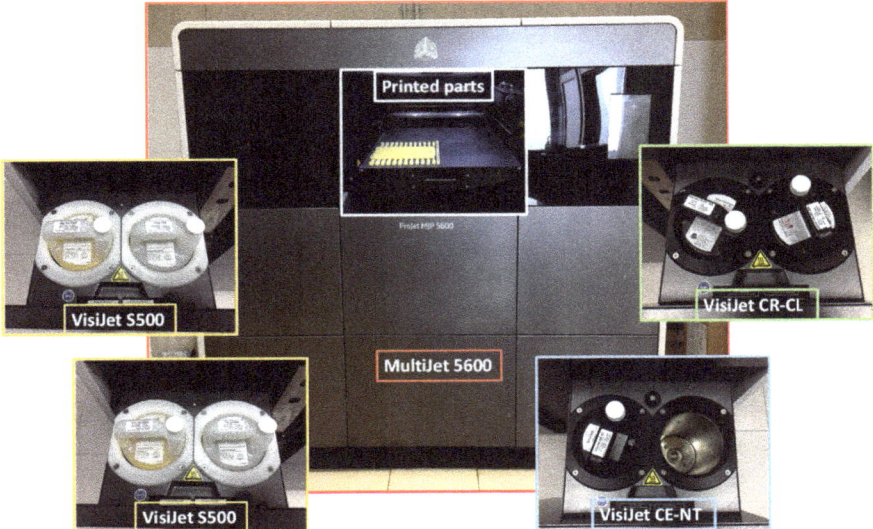

Figure 2. Projet MJP 5600 machine with a set of specimens and VisiJet materials.

Before the parts produced with this machine can be used, it is necessary to remove the VisiJet S500 material in which they are embedded immediately after the MP process. For this purpose, both a DIGITHEAT-TFT furnace (JP Selecta, Barcelona, Spain) and a Digital Pro ultrasonic tank (GT Sonic, Meizhou, China) were used, using either isopropyl alcohol (99.9% purity) or vegetable oil as the fluid in the latter. In some cases which are described in the Methodology section below, the roughness of the central area of the specimens was measured using a Surftest SJ-500 roughness tester (Mitutoyo, Kawasakishi,

Japan). The different manufactured specimens were tensile tested using a ME-402 universal machine (Servosis, Madrid, Spain) equipped with a 5 kN load cell. In order to determine the actual dimensions of the center section of the specimens, all of them were measured with a Mitutoyo universal calliper before being tested. During the tests, the length of the central area of the specimen, l_1, was taken as the reference length for the calculation of the length deformations.

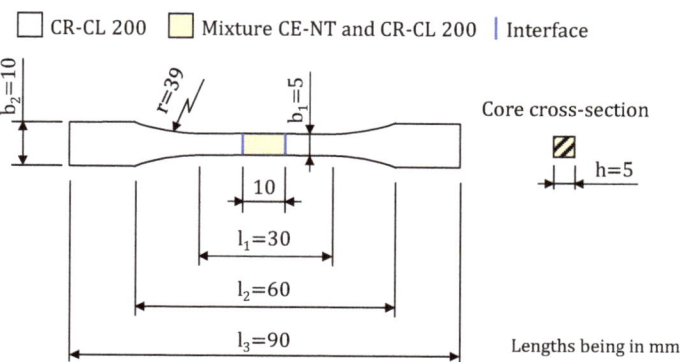

Figure 3. Schematic of the 1BA multi-material specimen designed according to ISO 527-2 [37].

2.2. Methodology

This work was carried out in two parts. In the first part (Figure 4a), the tensile strength behavior of specimens combining different resins mixtures was analyzed. These specimens were manufactured using the ultrahigh-definition mode for XY axes, but with a layer height of 16 μm for productivity reasons. For this purpose, 3 specimens were manufactured with each of the possible VisiJet resin combinations (Table 1), equaling 42 specimens in total. After the MP process, these specimens were post-processed to remove the VisiJet S500 material with two 10-min thermal cycles in the furnace at 60 °C. During each of these cycles, the specimens were placed horizontally on a rack, analogous to the manufacturing orientation, so as to avoid possible warping and geometric distortions caused by heat. Between the two furnace cycles, the specimens were removed from the oven for a short period of time, during which the excess wax was manually removed using a piece of absorbent paper. Subsequently, after cooling to room temperature, they were tensile tested according to ISO527-1 recommendations [37], using a speed of 1 mm/min.

Once the good interface performance of the multi-material specimens analyzed in the first part of the work was verified, the influence of the post-processing used on the mechanical properties of these specimens was analyzed in the second part of the work (Figure 4b). In this part, the least and greatest strength specimens, CE-NT and CL-CR 200, respectively, were manufactured, as well as two additional specimens that had shown in the first part tests a strength equi-spaced in magnitude with respect to the previous ones. Three replicates of each of them were manufactured, making thirty-six specimens in total, as shown in Figure 5. These specimens were manufactured with the same machine configuration as the one used in the first part of the work. Then, the replicates were post-processed with each of the following procedures:

- F: two furnace cycles at 60 °C;
- OU: two ultrasonic cycles with mineral oil at 60 °C;
- IU: two ultrasound cycles with isopropyl alcohol at 60 °C.

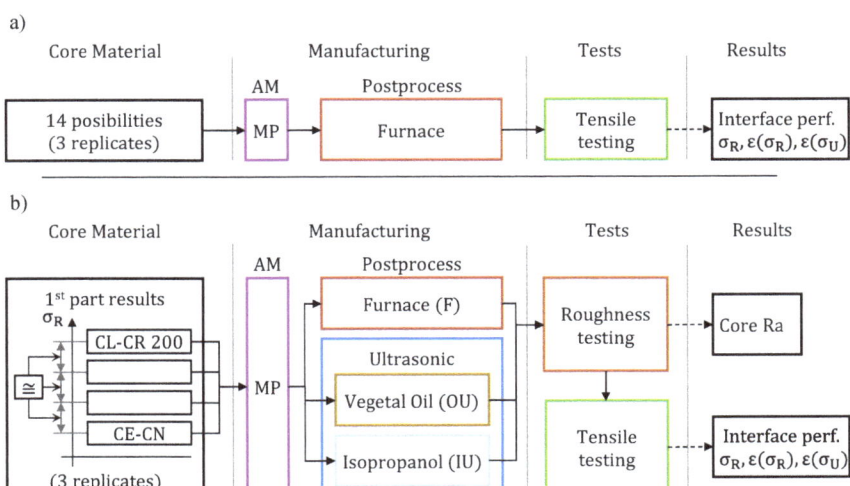

Figure 4. Diagram of procedure followed in the parts of the work: (**a**) first part and (**b**) second part.

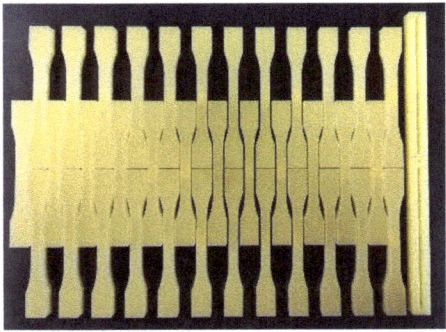

Figure 5. Specimens produced for the second part of the study before post-processing.

Each of the above cycles lasted 10 min, and the specimens were manually cleaned with absorbent paper between the two cycles, in each type of post-processing. During the ultrasonic cycles, a 40 kHz frequency was applied when each specimen was completely immersed in the treatment tank, which was filled with the appropriate fluid, i.e., mineral oil or isopropyl alcohol. After post-processing, the roughness of the core of the specimens was measured using a λ_c of 2.5 mm and λ_s of 8 µm, following the recommendations of the ISO 4288 standard [39], and subsequently a tensile test was performed under the same conditions as in the first part of the work.

To prevent environmental or machine-specific conditions from affecting the results, both the specimens tested in the first part and those tested in the second part were manufactured and tested within 24 h after manufacturing. In addition, the specimens were tested in random order [40]. All specimens were manufactured at the highest machine resolution and with the same material cartridges.

3. Results and Discussion

The 42 specimens analyzed in the first part of the work required 14 h of fabrication on the AM machine, consuming 46 g of VisiJet CE-NT, 196 g of VisiJet CR-CL, and 132 g of VisiJet S500. Post-processing took about 2 h. The results of the tensile test carried out on these specimens are shown in Figure 6. As can be seen, it follows a similar exponential behavior to that of the characteristics sheet provided by the manufacturer, but with higher resistances. For specimens of material from CE-NT to D70, the observed strength is several times that specified by the manufacturer, while for the D75 and CR-CL 200 specimens, 74.5% and 29.2% more strength were obtained, respectively.

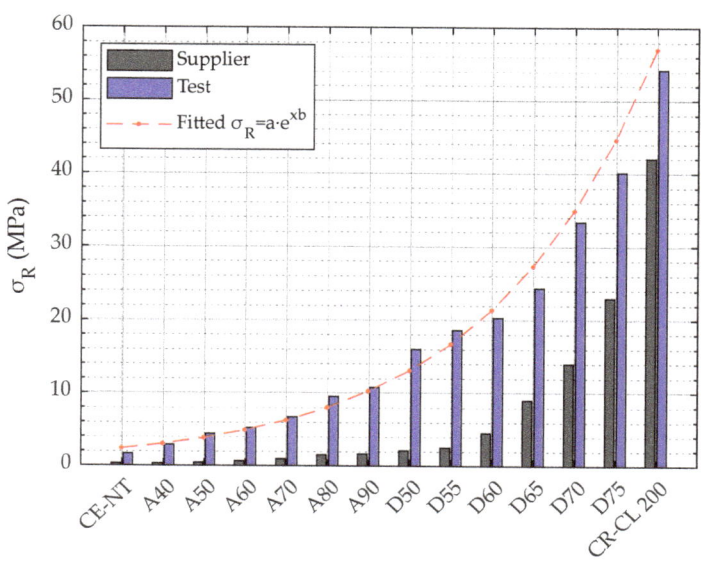

Figure 6. Tensile strength for different material mixtures.

In all cases, the interface between the core material or mixture and the head material, i.e., CR-CL 200, showed adequate behavior, with the rupture occurring in the central section of the core. This can be seen in the sequence of photographs captured during the testing of the A70 core material specimen (Figure 7).

Figure 6 shows the exponential behavior observed in the tensile strength of the specimens as a function of the specimen core composition. It can be approximated by the function represented in Equation (1) with an r-square of 98.02%, where P represents the ratio between CE-NT and CR-CL 200, so that 1 is equivalent to 100% CE-NT and 14 is equivalent to 100% CR-CL 200. This behavior, together with an improvement in the AM machine-control software, would allow more intermediate materials to be defined between the two extremes, so that they could be manufactured with a more suitable strength for each application.

$$\sigma_R = 1.8475 \cdot e^{0.2450 \cdot P} \text{ MPa, where } P \in [1, 14] \tag{1}$$

Figure 8 shows the elongation of specimens manufactured with different materials at tensile strength, $\varepsilon(\sigma_R)$, and at the moment of breaking, i.e., at ultimate strength $\varepsilon(\sigma_U)$. On the one hand, the similarity of both elongations can be highlighted, as well as the exponential decreasing behavior as a function of the CR-CL 200 material concentration. In the case of the specimen made entirely of this material, CR-CL 200 (in Figure 8), the elongation obtained in the test, i.e., 14%, is within the range indicated by the supplier [9].

Figure 7. Sequence of the tensile test on specimen A70.

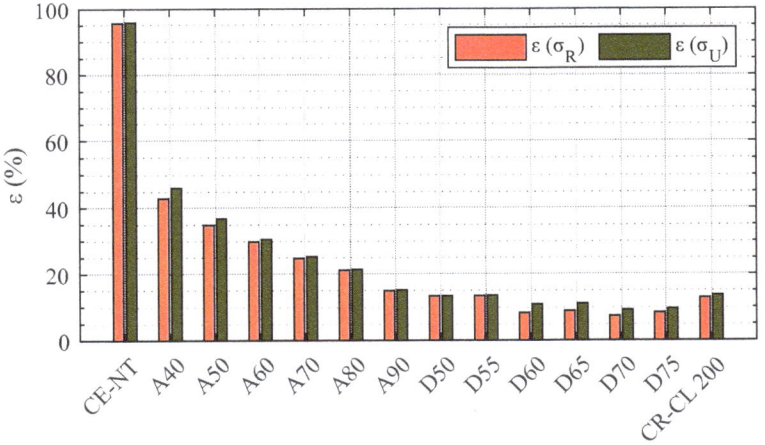

Figure 8. Elongation at tensile strength, $\varepsilon(\sigma_R)$, and at ultimate strength, $\varepsilon(\sigma_U)$, for the different materials and mixtures (reference length, l_1).

After verifying the good performance of the analyzed specimens, in the second part of this work, the roughness of the core area and the tensile behavior as a function of the type of post-processing were studied. This study was performed for those specimens with lower and higher strength and additionally two specimens with equally spaced tensile strength values: CE-NT, D60, D70, and CR-CL 200. From each of these specimens, 3 replicates were manufactured, in total 36 specimens, to be post-processed in three different ways (Figure 4b): F, IU, and OU.

The Ra roughness values obtained in the core area of the specimens are presented in Table 2 and represented in Figure 9. As can be seen, the specimen whose roughness is most altered depending on the post-processing used is the CE-NT specimen. In this specimen, a much higher roughness is obtained in the case of applying ultrasonic post-processing with isopropanol (IU), 8.87 µm on average in front of 2.03 µm from using mineral oil, and of 3.10 µm obtained with the furnace post-processing. This indicates a chemical interaction between isopropanol and the CE-NT material that alters its surface and, therefore, could also alter its integrity. In the rest of the specimens, no significant variation is observed when changing the post-processing, obtaining Ra values in the range of 2 µm.

Table 2. Ra measured in the core of the specimens: individual values of the replicates and mean values.

Post-Process	Material/Mixture	Ra (µm)			
		CE-NT	D60	D70	CR-CL 200
F	Replica 1	2.9	0.8	2.7	2.4
	Replica 2	3.4	1.5	2.9	1.5
	Replica 3	3.0	1.6	3.2	1.6
	Mean	3.10	1.30	2.93	1.83
OU	Replica 1	1.9	1.6	2.2	1.5
	Replica 2	2.2	1.7	1.9	1.4
	Replica 3	2.0	1.6	2.1	1.3
	Mean	2.03	1.63	2.07	1.40
IU	Replica 1	8.9	1.6	1.7	1.5
	Replica 2	8.9	1.7	2.5	1.3
	Replica 3	8.8	3.0	1.6	1.3
	Mean	8.87	2.10	1.93	1.37

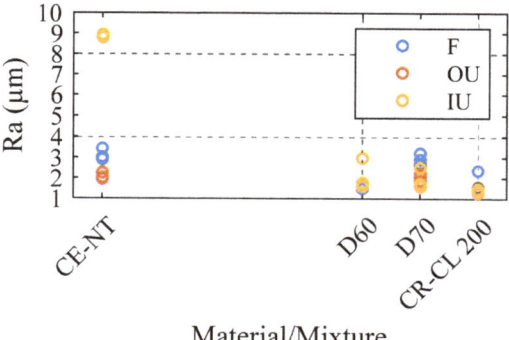

Figure 9. Ra values measured in the core of the specimens.

The results of tensile strength, strain at tensile strength, and strain at ultimate strength obtained in the tensile tests are shown in Figure 10a–c, respectively. Different statistical models were fitted to analyze these results in search of conclusions about the effect of the different post-processing on that tensile performance. The type of post-processing was used as the input parameter, and the following as output indicators were used: tensile strength, strain at tensile strength, and strain at ultimate strength. The p-values shown in Table 3 were obtained from these adjustments.

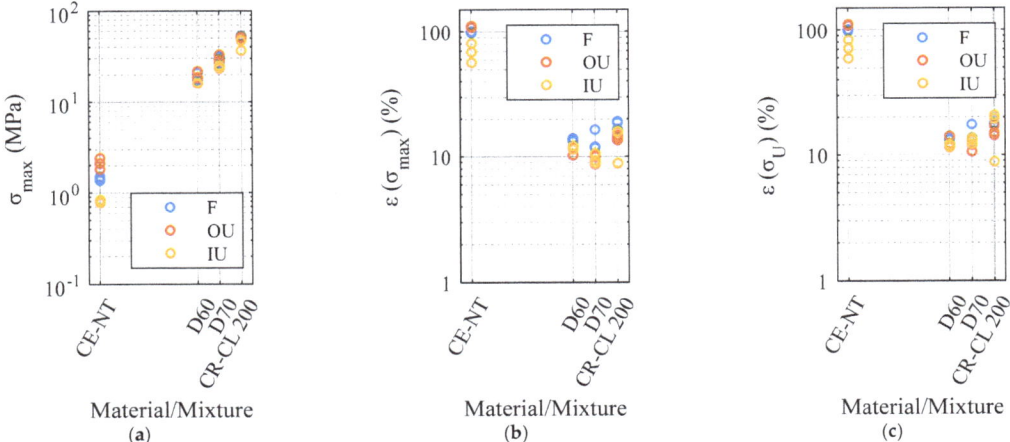

Figure 10. Tensile behavior as a function of post-processing type (F, OU, and IU): (**a**) tensile strength, (**b**) strain at tensile strength, and (**c**) strain at ultimate strength.

Table 3. P-values and correlations (p-value ≤ 0.05 criteria) between post-processing type and tensile performance (σ_R; $\varepsilon(\sigma_R)$; $\varepsilon(\sigma_U)$).

Material/Mixture	σ_R	$\varepsilon(\sigma_R)$	$\varepsilon(\sigma_U)$
CE-NT	0.000 (✓)	0.001 (✓)	0.001 (✓)
D60	0.048 (✓)	0.010 (✓)	0.054 (✗)
D70	0.031 (✓)	0.047 (✓)	0.132 (✗)
CR-CL 200	0.032 (✓)	0.110 (✗)	0.520 (✗)

As can be noticed, the type of post-processing influences the tensile strength reached by the material. Looking at Figure 10a, it can be seen that IU post-processing is the most detrimental. The most affected material is CE-NT with regard to the roughness study (Figure 9). For this material, it is interesting to use the OU post-processing to achieve the best properties, while for the rest of the materials, the F and OU post-processing, are practically equivalent (Figure 10a). In the case of the strain at tensile strength, the IU post-processing should again be avoided for CE-NT, with the effect of the other two processes being similar, while for the D60 and D70 materials, the F post-processing stands out as being beneficial in comparison with the other two (Figure 10b). Finally, regarding elongation at ultimate strength, a correlation was found only for the CE-NT material, with the IU post-processing again being detrimental (Figure 10c).

4. Conclusions

In the first part of this work, the tensile mechanical performance of specimens manufactured with Visijet CR-CL 200 material heads and core mixtures of this material in different proportions with Visijet CE-NT was analyzed. These specimens were manufactured using the material projection technique on a ProJet® MJP 5600 machine, whose manufacturer, 3DSystems, also supplies the material. After the design of a 1BA tensile specimen by following the recommendations of the ISO 527-2 standard, 42 of these specimens were manufactured and tensile tested (3 iterations for each of the 14 possible combinations). In these tests, it was verified that all the specimens break transversely through the core and with a resistance greater than that specified by the supplier, which demonstrated the good performance of the different materials' interfacial zones. These results demonstrate the feasibility of using combinations of these materials within the same component, with the associated advantages in terms of simplicity and reduction in the number of components.

In the second part of the work, the influence of the type of post-processing on the surface quality and tensile behavior was analyzed for several of the previously tested combinations. In view of the results achieved in the first part, the study was restricted to four combinations of Visijet CR-CL 200 with Visijet CE-NT, so that they were approximately equally spaced in terms of tensile strength: CE-NT, D60, D70, and CR-CL 200. In this way, the influence of three post-processing procedures could be assessed within the whole strength spectrum: furnace (F), ultrasonic with vegetable oil (OU), and ultrasonic with isopropyl alcohol (IU). Three iterations of the four combinations were used for each post-processing, in a total of 36 specimens. The results achieved in this part demonstrated the negative effect of using ultrasonics with isopropyl alcohol (IU) for the lower-strength specimens. This post-processing worsens the surface quality, limits the tensile strength, and reduces the deformation achievable by the material. In the case of the higher-strength specimens with lower CE-NT content, post-processing is less important for the elongation, but it has some influence on the strength. This behavior results from the detrimental effect of isopropyl alcohol on the CE-NT material. Therefore, for parts with a higher content of this material, post-processing in a furnace (F) or with ultrasonic and vegetable oil (OU) is recommended. For parts with a lower proportion of CE-NT, the furnace post-processing (F) stands out as the best option.

As future work, a study of all possible combinations of CR-CL 200 and CE-NT materials is proposed to carry out a more detailed characterization and, thus, to be able to identify in greater detail the combinations for which post-processing with isopropyl alcohol is detrimental. In addition, an in-depth analysis of the porosity of the multi-material parts using microscopy is proposed, so that the microscopic effect of the different post-processing on the material of the specimens can be studied. Thus, it will be possible to correlate the microscopic physical effect of the post-processing with the mechanical performance of the material in each case.

Author Contributions: Conceptualization, P.Z. and J.B.; methodology, P.R.-G.; formal analysis, A.I.F.-A.; investigation, P.R.-V.; resources, J.B.; data curation, P.Z.; writing—original draft preparation, P.R.-G.; writing—review and editing, P.Z.; project administration, J.B.; funding acquisition, Joaquín Barreiro. All authors have read and agreed to the published version of the manuscript.

Funding: This research received no external funding.

Institutional Review Board Statement: Not applicable.

Data Availability Statement: There are no shared data from this work.

Conflicts of Interest: The authors declare no conflict of interest.

References

1. Javaid, M.; Haleem, A. Additive manufacturing applications in medical cases: A literature based review. *Alex. J. Med.* **2018**, *54*, 411–422. [CrossRef]
2. Najmon, J.C.; Raeisi, S.; Tovar, A. Review of additive manufacturing technologies and applications in the aerospace industry. In *Additive Manufacturing for the Aerospace Industry*; Elsevier: Amsterdam, The Netherlands, 2019; pp. 7–31.
3. Nugraha, A.D.; Ruli; Supriyanto, E.; Rasgianti; Prawara, B.; Martides, E.; Junianto, E.; Wibowo, A.; Sentanuhady, J.; Muflikhun, M.A. First-rate manufacturing process of primary air fan (PAF) coal power plant in Indonesia using laser powder bed fusion (LPBF) technology. *J. Mater. Res. Technol.* **2022**, *18*, 4075–4088. [CrossRef]
4. Nugraha, A.D.; Syahril, M.; Muflikhun, M.A. Excellent performance of hybrid model manufactured via additive manufacturing process reinforced with GFRP for sport climbing equipment. *Heliyon* **2023**, *9*, e14706. [CrossRef] [PubMed]
5. Muflikhun, M.A.; Sentanu, D.A. Characteristics and performance of carabiner remodeling using 3D printing with graded filler and different orientation methods. *Eng. Fail. Anal.* **2021**, *130*, 105795. [CrossRef]
6. Patmonoaji, A.; Mahardika, M.A.; Nasir, M.; She, Y.; Wang, W.; Muflikhun, M.A.; Suekane, T. Stereolithography 3D Printer for Micromodel Fabrications with Comprehensive Accuracy Evaluation by Using Microtomography. *Geosciences* **2022**, *12*, 183. [CrossRef]
7. *ISO/ASTM 52900:2015*; Additive Manufacturing General principles—Terminology. International Organization for Standardization: Geneva, Switzerland, 1996.

8. Tyagi, S.; Yadav, A.; Deshmukh, S. Review on mechanical characterization of 3D printed parts created using material jetting process. *Mater. Today Proc.* **2022**, *51*, 1012–1016. [CrossRef]
9. VisiJet®Base Materials for the ProJet MJP 5600. Available online: https://es.3dsystems.com/multi-jet-printing (accessed on 16 May 2022).
10. Stratasys PolyJet Technology. Available online: https://www.stratasys.com/es/polyjet-technology (accessed on 16 May 2022).
11. Akbari, S.; Sakhaei, A.H.; Kowsari, K.; Yang, B.; Serjouei, A.; Yuanfang, Z.; Ge, Q. Enhanced multimaterial 4D printing with active hinges. *Smart Mater. Struct.* **2018**, *27*, 065027. [CrossRef]
12. Ge, Q.; Gu, G.; Wang, D.; Zhang, B.; Zhang, N.; Yuan, C.; Hingorani, H.; Ding, N.; Zhang, Y.-F. Fast-response, stiffness-tunable soft actuator by hybrid multimaterial 3D printing. *Adv. Funct. Mater.* **2019**, *29*, 1806698.
13. Dikshit, V.; Nagalingam, P.A.; Yap, L.Y.; Sing, L.S.; Yeong, Y.W.; Wei, J. Investigation of quasi-static indentation response of inkjet printed sandwich structures under various indenter geometries. *Materials* **2017**, *10*, 290. [CrossRef]
14. MultiJet Printing, Case of Study. Available online: https://es.3dsystems.com/motorsports/performance-wind-tunnel-testing?ind=motorsports (accessed on 16 May 2022).
15. Davoudinejad, A.; Khosravani, M.R.; Pedersen, D.B.; Tosello, G. Influence of thermal ageing on the fracture and lifetime of additively manufactured mold inserts. *Eng. Fail. Anal.* **2020**, *115*, 104694. [CrossRef]
16. León-Cabezas, M.A.; Martínez-García, A.; Varela-Gandía, F.J. Innovative advances in additive manufactured moulds for short plastic injection series. *Procedia Manuf.* **2017**, *13*, 732–737.
17. Dämmer, G.; Gablenz, S.; Hildebrandt, A.; Major, Z. Design and shape optimization of PolyJet bellows actuators. In Proceedings of the 2018 IEEE International Conference on Soft Robotics (RoboSoft), Livorno, Italy, 24–28 April 2018.
18. Kakogawa, A.; Ma, S.A. Differential Elastic Joint for Multi-linked Pipeline Inspection Robots. In Proceedings of the 2018 IEEE/RSJ International Conference on Intelligent Robots and Systems (IROS), Madrid, Spain, 1–5 October 2018.
19. Pazhamannil, R.V.; Govindan, P. Current state and future scope of additive manufacturing technologies via vat photopolymerization. *Mater. Today Proc.* **2021**, *43*, 130–136. [CrossRef]
20. Bezek, L.B.; Chatham, C.A.; Dillard, D.A.; Williams, C.B. Mechanical properties of tissue-mimicking composites formed by material jetting additive manufacturing. *J. Mech. Behav. Biomed. Mater.* **2022**, *125*, 104938. [CrossRef]
21. Khalid, G.A.; Bakhtiarydavijani, H.; Whittington, W.R.; Prabhu, R.; Jones, M.D. Material response characterization of three poly jet printed materials used in a high fidelity human infant skull. *Mater. Today Proc.* **2020**, *4*, 408–413. [CrossRef]
22. Wang, Y.C.; Chen, T.; Yeh, Y.L. Advanced 3D printing technologies for the aircraft industry: A fuzzy systematic approach for assessing the critical factors. *Int. J. Adv. Manuf. Technol.* **2019**, *105*, 4059–4069. [CrossRef]
23. Gao, W.; Zhang, Y.; Ramanujan, D.; Ramani, K.; Chen, Y.; Williams, C.B.; Wange, C.C.L.; Shin, Y.C.; Zhanga, S.; Zavattieri, P.D. The status, challenges, and future of additive manufacturing in engineering. *Comput. Aided Des.* **2015**, *69*, 65–89. [CrossRef]
24. Yap, Y.L.; Wang, C.; Sing, S.L.; Dikshit, V.; Yeong, W.Y.; Wei, J. Material jetting additive manufacturing: An experimental study using designed metrological benchmarks. *Precis. Eng.* **2017**, *50*, 275–285. [CrossRef]
25. Pugalendhi, A.; Ranganathan, R.; Ganesan, S. Impact of process parameters on mechanical behaviour in multi-material jetting. *Mater. Today Proc.* **2021**, *46*, 9139–9144. [CrossRef]
26. Sagbas, B.; Gümüş, B.E.; Kahraman, Y.; Dowling, D.P. Impact of print bed build location on the dimensional accuracy and surface quality of parts printed by multi jet fusion. *J. Manuf. Process.* **2021**, *70*, 290–299. [CrossRef]
27. Khoshkhoo, A.; Carrano, A.L.; Blersch, D.M. Effect of surface slope and build orientation on surface finish and dimensional accuracy in material jetting processes. *Procedia Manuf.* **2018**, *26*, 720–730. [CrossRef]
28. Chen, T.; Dilag, J.; Bateman, S. Surface topology modification of organic substrates using material jetting technologies. *Mater. Des.* **2020**, *196*, 109116. [CrossRef]
29. Stanković, T.; Mueller, J.; Shea, K. The effect of anisotropy on the optimization of additively manufactured lattice structures. *Addit. Manuf.* **2017**, *17*, 67–76. [CrossRef]
30. Sugavaneswaran, M.; Arumaikkannu, G. Analytical and experimental investigation on elastic modulus of reinforced additive manufactured structure. *Mater. Des.* **2015**, *66*, 29–36. [CrossRef]
31. Abayazid, F.F.; Ghajari, M. Material characterisation of additively manufactured elastomers at different strain rates and build orientations. *Addit. Manuf.* **2020**, *33*, 101160. [CrossRef]
32. Liu, W.; Song, H.; Huang, C. Maximizing mechanical properties and minimizing support material of PolyJet fabricated 3D lattice structures. *Addit. Manuf.* **2020**, *35*, 101257. [CrossRef]
33. Meisel, N.A.; Williams, C.B. Design for Additive Manufacturing: An investigation of key manufacturing considerations in multi-material PolyJet 3D printing. In Proceedings of the 25th Annual International Solid Freeform Fabrication Symposium, Austin, TX, USA, 4–6 August 2014.
34. Eren, O.; Sezer, H.K.; Yalçın, N. Effect of lattice design on mechanical response of PolyJet additively manufactured cellular structures. *J. Manuf. Process.* **2022**, *75*, 1175–1188. [CrossRef]
35. He, Y.; Zhang, F.; Saleh, E.; Vaithilingam, J.; Aboulkhair, N.; Begines, B.; Tuck, C.J.; Hague, R.J.M.; Ashcroft, I.A.; Wildman, R.D. A tripropylene glycol diacrylate-based polymeric support ink for material jetting. *Addit. Manuf.* **2017**, *16*, 153–161. [CrossRef]
36. Boopathy, V.R.; Sriraman, A.; Arumaikkannu, G. Energy absorbing capability of additive manufactured multi-material honeycomb structure. *Rapid Prototyp. J.* **2019**, *25*, 623–629. [CrossRef]

37. *ISO 527:2012*; Plastics Determination of Tensile Properties: Test Conditions for Moulding and Extrusion Plastics. International Organization for Standardization: Geneva, Switzerland, 1996.
38. Nguyen, C.H.P.; Kim, Y.; Choi, Y. Design for additive manufacturing of functionally graded lattice structures: A design method with process induced anisotropy consideration. *Int. J. Precis. Eng. Manuf.–Green. Technol.* **2021**, *8*, 29–45. [CrossRef]
39. *ISO 4288:1996*; Geometrical Product Specifications (GPS)—Surface Texture: Profile Method—Rules and Procedures for the Assessment of Surface Texture. International Organization for Standardization: Geneva, Switzerland, 1996.
40. Bass, L.; Meisel, N.A.; Williams, C.B. Exploring variability of orientation and aging effects in material properties of multi-material jetting parts. *Rapid Prototyp. J.* **2016**, *22*, 826–834. [CrossRef]

Disclaimer/Publisher's Note: The statements, opinions and data contained in all publications are solely those of the individual author(s) and contributor(s) and not of MDPI and/or the editor(s). MDPI and/or the editor(s) disclaim responsibility for any injury to people or property resulting from any ideas, methods, instructions or products referred to in the content.

Article

Prediction of Mechanical Properties for Carbon fiber/PLA Composite Lattice Structures Using Mathematical and ANFIS Models

Mustafa Saleh, Saqib Anwar *, Abdulrahman M Al-Ahmari and Abdullah Yahia AlFaify

Industrial Engineering Department, College of Engineering, King Saud University,
P.O. Box 800, Riyadh 11421, Saudi Arabia; msaleh3@ksu.edu.sa (M.S.); alahmari@ksu.edu.sa (A.M.A.-A.); aalfaify@ksu.edu.sa (A.Y.A.)
* Correspondence: sanwar@ksu.edu.sa

Abstract: This study investigates the influence of design, relative density (RD), and carbon fiber (CF) incorporation parameters on mechanical characteristics, including compressive modulus (E), strength, and specific energy absorption (SEA) of triply periodic minimum surface (TPMS) lattice structures. The TPMS lattices were 3D-printed by fused filament fabrication (FFF) using polylactic acid (PLA) and carbon fiber-reinforced PLA(CFRPLA). The mechanical properties of the TPMS lattice structures were evaluated under uniaxial compression testing based on the design of experiments (DOE) approach, namely, full factorial design. Prediction modeling was conducted and compared using mathematical and intelligent modeling, namely, adaptive neuro-fuzzy inference systems (ANFIS). ANFIS modeling allowed the 3D printing imperfections (e.g., RD variations) to be taken into account by considering the actual RDs instead of the designed ones, as in the case of mathematical modeling. In this regard, this was the first time the ANFIS modeling utilized the actual RDs. The desirability approach was applied for multi-objective optimization. The mechanical properties were found to be significantly influenced by cell type, cell size, CF incorporation, and RD, as well as their combination. The findings demonstrated a variation in the E (0.144 GPa to 0.549 GPa), compressive strength (4.583 MPa to 15.768 MPa), and SEA (3.759 J/g to 15.591 J/g) due to the effect of the studied variables. The ANFIS models outperformed mathematical models in predicting all mechanical characteristics, including E, strength, and SEA. For instance, the maximum absolute percent deviation was 7.61% for ANFIS prediction, while it was 21.11% for mathematical prediction. The accuracy of mathematical predictions is highly influenced by the degree of RD deviation: a higher deviation in RD indicates a lower accuracy of predictions. The findings of this study provide a prior prediction of the mechanical behavior of PLA and CFRPLA TPMS structures, as well as a better understanding of their potential and limitations.

Keywords: carbon fiber-reinforced PLA; composites; biodegradable polymer; additive manufacturing; FDM; TPMS lattice structures; compression testing; specific energy absorption; artificial intelligence; ANFIS

Citation: Saleh, M.; Anwar, S.; Al-Ahmari, A.M.; AlFaify, A.Y. Prediction of Mechanical Properties for Carbon fiber/PLA Composite Lattice Structures Using Mathematical and ANFIS Models. *Polymers* **2023**, *15*, 1720. https://doi.org/10.3390/polym15071720

Academic Editor: Matthew Blacklock

Received: 14 February 2023
Revised: 17 March 2023
Accepted: 22 March 2023
Published: 30 March 2023

Copyright: © 2023 by the authors. Licensee MDPI, Basel, Switzerland. This article is an open access article distributed under the terms and conditions of the Creative Commons Attribution (CC BY) license (https:// creativecommons.org/licenses/by/ 4.0/).

1. Introduction

Lattice structures are attracting attention in a wide range of industries, including aerospace, automotive, medical, and heat management, where lightweight and multifunctionality are required. Properties like thermal insulation, acoustic absorption, mechanical vibration damping, high stiffness-to-volume fraction ratio, and energy absorption are required within these structures [1]. They enable the enhancement of the performance-to-weight ratio, the creation of highly controlled architectures, and the distribution of impact shock across an object.

In comparison to bulk materials, lattice structures have a high number of design variables, which increases their complexity and limits their ability to be fabricated using conventional manufacturing processes. However, the advent of additive manufacturing

(AM) provides opportunities to fabricate such complex structures, e.g., lattice structures, and opens doors for creating and exploring more designs.

The fundamental advantage of the lattice structures is that materials can be placed only where they are needed for a particular application [2]. Thus, the inherent characteristics of the material, relative density (RD), and geometry variables are the most important factors that influence various lattices' performances, all of which can be tuned to satisfy the needed qualities. In other words, material, RD, and geometry are variables that can be varied to achieve specific lattice properties. In regards to material, AM allows for controlling material composition, e.g., incorporating particles/fibers as reinforcement, leading to uncovering new enhanced multiphase (composite) materials. This enhances the functionality and performance of the lattices in many aspects, such as mechanical [3–6], medical [6,7], electrical [8], and multifunctional characteristics [9]. In this regard, the FDM technique offers promising potential for developing highly reliable and mechanically strong composite lattice structures [10].

The most common lattice structures' unit cells are strut-based and surface-based, e.g., triply periodic minimal surfaces (TPMS) [11,12]. Strut-based structures are characterized by their structural members' joint and frame architecture [13]. Examples of strut-based cell types are body-centered cubic (BCC), face-centered cubic (FCC) [2], octet-truss [14], octagonal, and Kelvin [15]. On the other hand, surface-based structures have a sheet-like architecture. Examples of TPMS cell topologies are Diamond, Gyroid, Primitive, Fisher-Koch, and IWP [16]. Compared with strut-based structures, TPMS structures have shown better mechanical performance, such as high strength, high load-bearing capacity, high energy absorption capacity, and structural stability [17]. Furthermore, TPMS structures have a higher surface-to-volume ratio, enabling them to be used in medical fields (e.g., scaffolds [18,19]) and heat management [20].

Researchers are becoming increasingly interested in the design and 3D printing of TPMS structures in an attempt to enhance their performance. Spear and Palazotto [13] investigated the impact of different parameters including cell topologies (Diamond, Primitive, and I-WP), size and number of cells, and wall thickness on the mechanical performance of TPMS structures using factorial design. The samples were printed by selective laser melting (SLM) using INC718 material. The results indicated the importance of considering the main and combined variables when designing a lattice structure, especially for energy absorbing purposes. Kladovasilakis et al. [21] studied the mechanical behavior of three TPMS structures, namely Gyroid, Diamond, and Primitive, printed with PLA by FDM. The different structures were investigated at different RDs (10% to 30%). They found that Diamond structures showed the highest strength, whereas Primitive structures demonstrated the highest capacity for energy absorption. At low RDs, the mechanical properties of the considered TPMS structures deteriorate. Abueidda et al. [22] studied the effect of TPMS cell topologies (Primitive, Schoen IWP, and Neovius), structure size, and RD on the stiffness and strength of SLS-made TPMS polyamide-12 under compression testing. Results showed that both Neovius and IWP structures demonstrated improved strength and stiffness compared with Primitive structures. Shi et al. [23] researched the compression properties and energy absorption of four TPMS cell topologies (Gyroid, Diamond, IW, and Primitive). The TPMS structures were additively manufactured at different RDs from Ti6Al4V using SLM. Compression strength and modulus, and energy absorption were influenced by cell topologies and RDs, according to the findings. Ali et al. [24] investigated the effects of annealing heat treatment and various cell topologies, including Diamond, Primitive, Gyroid, Split-P, Kelvin, Octet, and Sea-Urchin Plus (SUP), on surface morphology, mechanical characteristics, and energy absorption. The results showed that among the studied structures, the SUP structure demonstrated the highest strength, while the Diamond lattice demonstrated the best energy absorption capacity.

Regarding the incorporation of reinforcements as a strategy for enhancing lattices' performance, very limited work has been reported particularly in TPMS structures. Zarei Zarei et al. [6] explored the effect of adding Ti6Al4V to PLA on the mechanical and biological performance of FDM-printed scaffolds. The results demonstrated that the ultimate compressive strength and compressive modulus of composite scaffolds were enhanced by the incorporation of 3–6 wt% Ti6Al4V. Qin et al. [17] studied the influence of adding reinforcements ($CaCO_3$ and TCP) to the PLA matrix on the compressive modulus, compressive strength, energy absorption, structure stability of Diamond (TPMS type) and cubic structures, additively manufactured by FDM. Results showed that the strength and compressive modulus of lattice structures were enhanced by adding $CaCO_3$ and TCP. Furthermore, with regard to cell topology, the Diamond lattice demonstrated better load-bearing capacity, energy absorption, and structure stability. Kaur et al. [3] performed compression testing to study the mechanical performance of octahedral and octet lattice structures printed with PLA and carbon fiber-reinforced composite of PLA (CFRPLA) using FDM. CFRPLA's octahedral and octet structures demonstrated higher modulus and energy absorption than PLA structures. Based on their observation, this could be due to shear forces acting on the polymer melt during extrusion causing fibers to align along the printing direction, enhancing the structures' mechanical stability. Stan et al. [10] investigated the mechanical behavior of FDM 3D-printed lattice structures under axial and transverse compression tests. Structures were 3D-printed from carbon fiber (CF) and glass fiber (GF) reinforced polyamide-12 (PA12) composites. Considering axial and transverse specific load and stiffness, the performance of the CF/PA12 structures was better than that of the PA12 or GF/PA12 structures. In [25], experiments and numerical simulations were used to study the bending properties of various lattice structures made of wood fiber/PLA. Hexagonal, squared, triangular, circular-cored hexagonal, and circular-core squared structures were considered. The results showed that the circular-cored hexagonal structure yielded the highest flexural strength and stiffness. However, this study did not compare the findings of wood fiber/PLA with neat PLA to show the influence of wood fiber incorporation.

It is critical to understand the single and combined effects of various variables, such as material characteristics, relative density, and design parameters on the performance of the FDM TPMS structures. Studying the single and combined effects of various variables needs to be conducted based on a design of experiments (DOE) scheme. This allows statistical exploration of the single and combined effects of factors and further generalization of the lattice's performance using model-based analysis. Developing models based on factorial design could be a beneficial method for designing customized lattice structures [26]. To the best of our knowledge, almost all reported studies of TPMS lattice structures have not employed a DOE to investigate the influence of design, materials, and relative density variables, which have the most impact on the lattice structure's properties. Only a few studies have used a DOE approach to investigate the performance of strut-based lattice structures [26–31] and focused only on single-phase materials. In addition, no study has reported using artificial intelligence modeling to predict the mechanical performance of 3D-printed lattice structures made from singular-phase or composite materials.

It is worth mentioning that most of the reported studies ignored the influence of RD when statistically investigating the influence of the design parameters on mechanical characteristics. For instance, the reports that studied the effect of design parameters, such as strut/wall thickness, cell type, and cell size directly induced RD variations in their results. In other words, it can be said that the results were affected by RD in an uncontrolled manner. This is because wall thickness, cell type, and cell size control the RD, and any combination of them will lead to a particular RD. For example, [26] performed a statistical analysis of the effect of cell size, cell type, and strut diameter on different mechanical characteristics. Similarly, the authors of [13] conducted a statistical study on the influence of different TPMS topologies, cell sizes, cell numbers, and surface thicknesses. In such cases, the statistical findings could be misleading since the influence may be attributed to the resulting RD rather than the investigated factors.

This study attempts to examine the influence of material composition, geometry (cell type and size), and RD variables on the FDM-printed TPMS lattice structures using the DOE approach. In this regard, three TPMS cell types, namely Diamond, Gyroid, and

Primitive, with various cell sizes (8 mm and 12 mm), CF incorporation (0% and 15%), and RDs (30% and 44%) were considered. The analysis was conducted based on a full factorial design. The actual relative densities of the printed samples were measured to ensure no significant difference between the designed and printed RDs. Uniaxial compression testing was used to evaluate the mechanical properties of 3D-printed samples, including compression modulus, strength, and SEA. Moreover, ANFIS modeling was also employed for predicting the performance of the TPMS structures, and the results were compared with the mathematical models developed based on the DOE approach. By using the ANFIS modeling predictions, we could accurately predict mechanical characteristics considering the inherent imperfections in 3D printing, such as an RD variation, which cannot be avoided. Due to the freedom in the ANFIS modeling approach, the actual RDs were used instead of the intended (designed) RDs, leading to improved prediction accuracy. Finally, the best parameter settings for maximizing the TPMS lattices' performance were determined using multi-objective optimization through the desirability function.

The following section discusses the materials and methods used in this study. The results are presented in Section 3. Section 4 provides the discussions. Finally, Section 5 presents the conclusions.

2. Materials and Methods

2.1. Materials

In this study, the TPMS structures were FDM-fabricated from two materials: (1) pure Polylactic acid (PLA) and (2) carbon fiber-reinforced PLA (CFRPLA) in which 15% high-modulus short carbon fibers CF were incorporated into the PLA matrix. The matrix of both materials is Natureworks 4043D PLA biopolymer grade. The materials (filaments) were provided by 3DXTech, USA, with a diameter of 1.75 mm, and used as received. For the pure PLA filament, the mechanical properties are 1.24 g/cm^3 (density, ρ), 56 MPa (Tensile strength at break), and 2865 MPa (Tensile Modulus) [32], while for the CFRPLA filament: 1.29 g/cm^3 (density, ρ), 48 MPa (Tensile strength at break), 4950 MPa (Tensile Modulus) [33].

2.2. Design and Relative Density

In the current research, three different TPMS cell topologies, including Gyroid (G), Diamond (D), and Primitive (P) were considered. Within the context of this study, the terms "cell topology" and "cell type" are used interchangeably. Figure 1a illustrates the G, D, and P cell topologies. The unit cell size (l) of all cell topologies was designed in a cubic unit cell with 8 mm and 12 mm edge lengths. The reason behind these particular lengths is to fit the whole lattice dimensions.

Relative density (RD) is one of the most factors that influence a lattice's performance. RD, also termed as volume fraction ($V_{lattice}/V_{overall}$), is defined as the lattice structure volume ($V_{lattice}$) divided by the overall structure volume ($V_{overall}$) [34]. This study employed 30% and 44% RDs, taking into account the minimum feasible wall thickness to meet FDM resolution, nozzle diameter, and cell sizes. The selected range of the RDs also considered that the actual densities of the printed samples showed a close agreement with the designed RDs. For each TPMS cell topology, the needed RD is determined by the wall thickness and length of the cell. Hence, the intended RD was controlled by adjusting the ratio of the cell wall thickness parameter (t) to the cell size (t/l) [16]. It should be mentioned that every cell topology has its function in relation to the t/l ratio because of the differences in surface areas. The t values were determined based on the CAD predictions. Table 1 presents the values of the t parameter for each cell topology at the designed RD and cell lengths.

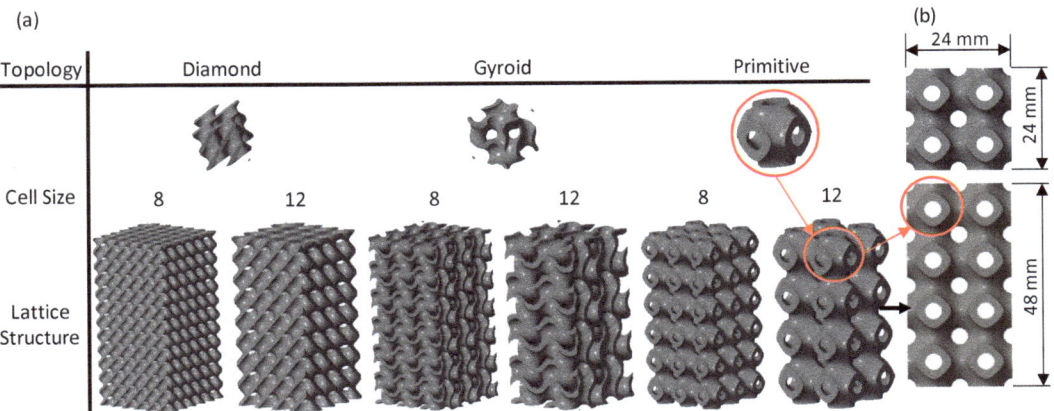

Figure 1. TPMS Lattice structures at 30% RD: (**a**) cell topologies, cell sizes, and the associated lattice structures and (**b**) lattice structure dimensions.

Table 1. Values of wall thickness parameter (*t*) for each TPMS topology at designed RDs and lengths (*l*).

Cell Topology	$l = 8$ mm		$l = 12$ mm	
	RD: 30	RD: 44	RD: 30	RD: 44
D	0.5374	0.7856	0.8065	1.1786
G	0.6826	0.995	1.024	1.493
P	0.772	1.131	1.1585	1.6965

The lattice structures were designed and printed in sizes of $24 \times 24 \times 48$ mm^3, see Figure 1b. The whole lattice dimensions were selected based on the aspect ratio (length to width ratio) of the ASTM D695-15 standard for compressive testing of rigid polymeric materials (), so that the length should be twice the sample width [17]. The number of cells to fit the selected dimensions depends on the cell size, whether 8 mm ($3 \times 3 \times 6$ cells) or 12 mm ($2 \times 2 \times 4$ cells). CREO 8.0 software was used to design the G, D, and P structures in the STL format. High-accuracy STL files were obtained for all investigated designs, with the number of generated triangles varying from 464,860 to 6,653,676 depending on the cell topology, cell size, and RD.

2.3. Experimental Design

The influence of the material composition (CF incorporation), cell topology, cell size, and relative density on the mechanical properties, including compression modulus (E) and strength, and specific energy absorption (SEA), was evaluated by the DOE approach. Table 2 illustrates the four parameters and their respective levels. A full factorial design was used with 24 runs. Each run was repeated 3 times resulting in a total of 72 experiments. Analysis of variance (ANOVA) was performed with a 95% confidence interval to study the significant influence of the variables and their interactions on E, compressive strength, and SEA. *p*-Values less than 0.05 imply that model terms (main factors and interactions) are statistically significant. Mathematical relationships between the investigated parameters and each of the output responses were developed. The developed mathematical models were further considered for optimization of the considered parameters. The best settings of the considered parameters were optimized through the desirability approach to simultaneously (multi-objective optimization) achieve the maximum E, compressive strength, and SEA. Design-Expert 13 software was used to systematically analyze the influence of the investigated parameters, develop mathematical (prediction) models, and conduct multi-objective optimization.

Table 2. Materials, design, and RD parameters and their levels.

Parameter	Levels		
	1	2	3
CF incorporation (CF), wt.%	0	15	-
Relative density (RD), %	30	44	-
Cell size, mm	8	12	-
Cell topology (Cell type)	D	G	P

2.4. FDM Printing of TPMS Lattices

An open-source FDM machine, Prusa FDM printer (Original Prusa i3 MK3S+, Czech Republic), equipped with a 0.4 mm nozzle, was used to additively manufacture the lattice structures. The STL files for the TPMS lattice structures were imported into slicer software (PrusaSlicer 2.4.2) in order to slice and generate "GCODE" files. Table 3 presents the printing parameters employed to 3D print the samples. All samples were printed without support structures. Examples of printed TPMS structures of different cell topologies are displayed in Figure 2.

Table 3. FDM printing parameters.

Printing Temperature	Bed Temperature	Printing Speed	Layer Thickness	Infill
220 °C	65 °C	45 mm/s	0.2 mm	100%

Figure 2. Examples of printed TPMS structures of PLA at 44% RD of different cell topologies.

2.5. Metrological Characterization

The printed samples' actual dimensions (e.g., length, width, and thickness) were determined by a Lab Profile Projector (VOM-2515), as shown in Figure 3a. Three measurements were made for each dimension, and the averaged values were used. Density was obtained by utilizing Archimedes' method, a common method used for determining the density of both porous and solid structures. First, the sample' density was determined by Archimedes' method by weighing the sample in air and distilled water. The volume of the lattice ($V_{lattice}$) was then determined from the obtained density. Then, the ratio between ($V_{lattice}/V_{overall}$), which represent the RD, was calculated. The overall volume ($V_{overall}$) was calculated from the actual sample' dimensions. Measurements were conducted on Shimadzu Analytical Balance (AUW220D, China) with a readability of 0.01 mg and a universal specific gravity kit (SGK-C, Mineralab, UK) as in [5]; see Figure 3b.

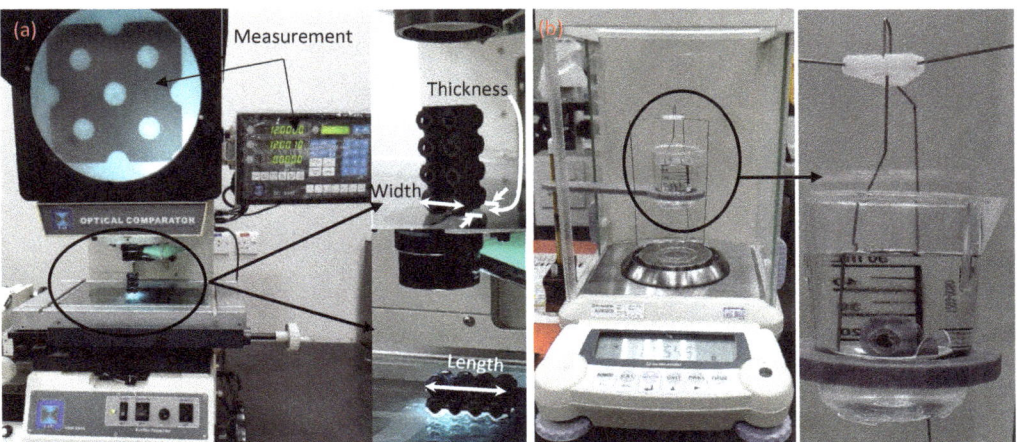

Figure 3. Measuring setups: (**a**) dimension measurements and (**b**) density measurements.

2.6. Mechanical Properties

The influence of the considered parameters on the mechanical characteristics of the TPMS structures was conducted under a uniaxial compression test. The compression tests were carried out following the ASTM D695-15 standard. Compression tests were performed along the build direction at a constant crosshead speed of 1.6 mm/min up to 60% strain on a Zwick Z100 testing machine equipped with a 100 KN load cell. During the tests, forces and displacements were recorded using testXpert II software.

Mechanical characteristics, including compressive modulus (E), compressive strength, and SEA, were gathered from the force-displacement curves. First, engineering compressive stress was used and calculated by dividing the recorded forces (F) by the measured original cross-sectional area. The strain was computed by dividing the recorded displacements (δ) during the tests by the original length of the sample. Then, compression modulus and strength were determined. Compression modulus was calculated by the slope of the tangent line at the linear portion of the stress-strain curves by testXpert II software. Compression strength (σ_{peak}) was considered as the peak strength, the maximum stress-value of the first peak (i.e., first local maximum) in the stress-strain curve [17,35].

Specific energy absorption (SEA) is a useful indicator for measuring a structure's capability to absorb energy per unit weight. The SEA of a lattice structure is represented by the area under the force-displacement curve divided by the structure's weight. The SEA was calculated as in Equation (1) [14]. The area under the force-displacement curve was calculated up to 55% strain (theoretical strain densification strain) using MATLAB R2022a.

$$SEA = \frac{\int_0^\delta F d\delta}{m} \quad (1)$$

where F is the force, δ is the displacement, and m is the structure's weight.

2.7. Adaptive Neuro-Fuzzy Inference System (ANFIS) Model

ANFIS is a hybrid neuro-fuzzy method for modeling complex systems. It integrates the best learning abilities of the artificial neural network (ANN) and inference capabilities of the fuzzy inference system (FIS) [36,37]. ANFIS accomplishes sample-based learning using the train data set to develop an efficient ANFIS structure for solving the associated problem. The developed ANFIS structure is being evaluated for its validity through a test data set. ANFIS uses a five-layer, feed-forward propagation structure [38]. Figure 4 shows an illustration of the ANFIS structure with two inputs, three membership functions (MFs), and one output. The layer explanation is described as follows [38–40]:

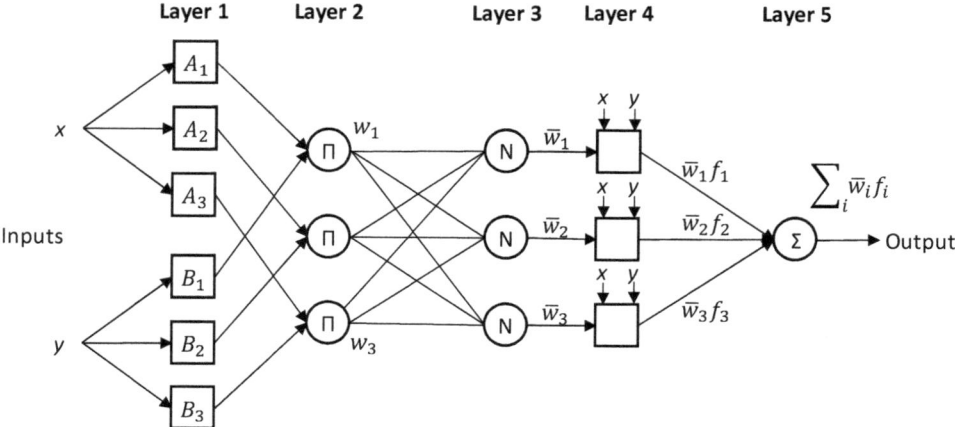

Figure 4. ANFIS structure.

Layer 1: input membership functions (MFs). In this layer, also called the fuzzification layer, the fuzzy membership value of each input, i.e., $\mu A_i(x)$ and $\mu B_i(y)$ is calculated by a proper membership function, e.g., trapezoidal, Gaussian, and triangular.

Layer 2: fuzzy rules. The firing strength (w_i), the weight for each rule's output is calculated in this layer. Each node's output is the product of all its input signals, which can be calculated using Equation (2).

$$w_i = \mu A_i(x) \times \mu B_i(y), i = 1, 2, 3 \qquad (2)$$

Layer 3: normalization. This layer represents the normalization of the firing strength ($\overline{w_i}$), as computed by Equation (3).

$$\overline{w_i} = \frac{w_i}{\sum_i w_i}, i = 1, 2, 3 \qquad (3)$$

Layer 4: defuzzification. The output of each node in this layer is calculated based on the function given in Equation (4).

$$\overline{w_i} \cdot f_i = \overline{w_i}(p_i x + q_i y + r_i), i = 1, 2, 3 \qquad (4)$$

where p_i, q_i, and r_i called a consequent parameter set.

Layer 5: output layer. This layer has only one node to calculate the system output, as in Equation (5).

$$Output = \sum_i \overline{w_i} \cdot f_i \qquad (5)$$

The ANFIS modeling was used for predicting the performance of the TPMS structures in terms of E, σpeak, and SEA. Using the ANFIS modeling predictions, we could accurately predict mechanical characteristics considering imperfections in 3D printing, such as an RD variation. This way, the actual RDs were used instead of the intended RDs, leading to improved prediction accuracy. Figure 5 depicts the methodology adopted in this study.

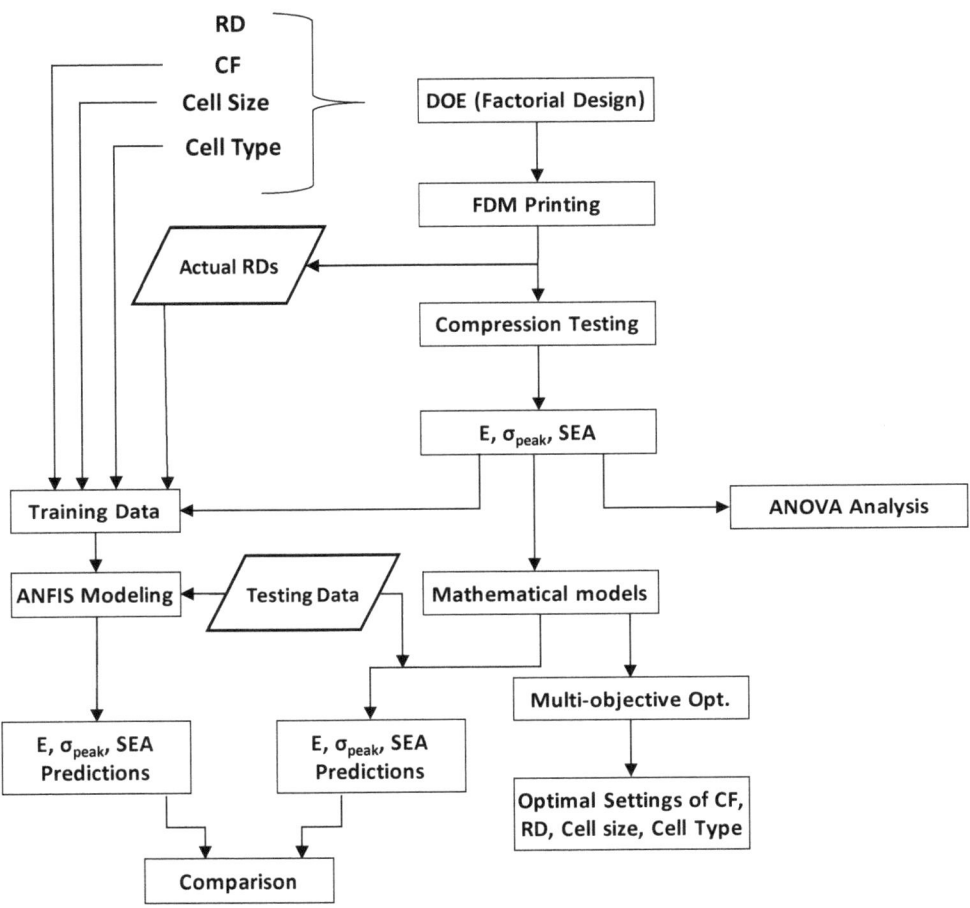

Figure 5. Methodology adopted in this study.

3. Results

3.1. Mechanical Characterization

Table 4 shows the average values of the actual RDs, E, σ_{peak}, and SEA of the 24 runs according to the full factorial DOE. The variabilty in the three repeated measured results is indicated by the standard deviation (SD). The results, including E, σ_{peak}, and SEA, were derived from the force-displacement curves. Typical stress-strain curves under the uniaxial compression test of the 24 runs listed in Table 4 are depicted in Figure 6. As can be seen in Figure 6, the mechanical behavior of the lattice structures is susceptible to changes in the CF incorporation, design, and RD factors. This influence is shown by how much the stress-strain curves vary in terms of either the stress range or the shape of the curves. For instance, the influence of the cell type in terms of the stress range and the deformation behavior is shown between Diamond-based structures (Figure 6a), Gyroid-based structures (Figure 6c), and Primitive-based structures (Figure 6e). For example, Diamond-based structures show higher load-bearing capacity and more uniformity in deformation in comparison with Primitive-based structures.

Table 4. Full factorial experiments along with the results (E, σ_peak, SEA, and actual RDs).

Run#	Variables				Responses			Actual RD (% ± SD)
	A: CF (%)	B: RD (%)	C: Cell Size (mm)	D: Cell Type	E (GPa ± SD)	σ_{peak} (MPa ± SD)	SEA (J/g ± SD)	
1	0	30	8	D	0.315 ± 0.004	10.057 ± 0.204	12.317 ± 0.532	32.36 ± 0.26
2	0	30	12	D	0.325 ± 0.010	9.225 ± 0.054	12.214 ± 0.298	28.50 ± 0.09
3	0	44	8	D	0.402 ± 0.005	13.891 ± 0.027	13.948 ± 0.062	42.65 ± 0.31
4	0	44	12	D	0.495 ± 0.017	15.352 ± 0.173	14.415 ± 0.287	42.58 ± 0.68
5	15	30	8	D	0.375 ± 0.007	9.929 ± 0.149	12.491 ± 0.24	31.18 ± 0.06
6	15	30	12	D	0.349 ± 0.005	8.581 ± 0.166	11.573 ± 0.418	28.11 ± 0.09
7	15	44	8	D	0.470 ± 0.021	13.316 ± 0.027	14.354 ± 0.321	42.35 ± 0.17
8	15	44	12	D	0.549 ± 0.028	15.768 ± 0.199	15.591 ± 0.292	43.33 ± 0.36
9	0	30	8	G	0.182 ± 0.011	6.612 ± 0.129	9.106 ± 0.360	28.91 ± 0.34
10	0	30	12	G	0.240 ± 0.002	7.575 ± 0.034	9.311 ± 0.083	30.48 ± 0.07
11	0	44	8	G	0.389 ± 0.006	13.099 ± 0.082	12.678 ± 0.058	44.99 ± 0.15
12	0	44	12	G	0.364 ± 0.016	11.411 ± 0.196	11.673 ± 0.118	41.68 ± 0.52
13	15	30	8	G	0.230 ± 0.006	6.342 ± 0.111	9.056 ± 0.102	28.51 ± 0.28
14	15	30	12	G	0.262 ± 0.002	6.679 ± 0.045	8.496 ± 0.124	30.18 ± 0.07
15	15	44	8	G	0.421 ± 0.008	12.367 ± 0.206	12.044 ± 0.074	44.15 ± 0.66
16	15	44	12	G	0.398 ± 0.032	11.214 ± 0.108	11.702 ± 0.108	42.55 ± 0.49
17	0	30	8	P	0.144 ± 0.005	5.164 ± 0.074	6.016 ± 0.216	28.50 ± 0.27
18	0	30	12	P	0.171 ± 0.001	5.580 ± 0.090	6.327 ± 0.051	28.59 ± 0.20
19	0	44	8	P	0.315 ± 0.010	11.778 ± 0.096	10.150 ± 0.061	43.48 ± 0.07
20	0	44	12	P	0.370 ± 0.006	11.541 ± 0.127	9.636 ± 0.156	42.99 ± 0.58
21	15	30	8	P	0.170 ± 0.004	4.583 ± 0.051	4.985 ± 0.089	28.11 ± 0.13
22	15	30	12	P	0.204 ± 0.005	4.957 ± 0.035	3.759 ± 0.164	28.55 ± 0.07
23	15	44	8	P	0.364 ± 0.008	10.643 ± 0.255	8.499 ± 0.237	42.51 ± 0.33
24	15	44	12	P	0.398 ± 0.006	11.262 ± 0.071	6.375 ± 0.188	43.37 ± 0.19

Deformation mechanisms are graphically illustrated in Figure 7. A bending–torsion coupled failure is evident for the Diamond-based structures; see Figure 7a. Figure 7b shows that the Gyroid-based structure exhibits bending and buckling mechanisms of failure. Figure 7c illustrates the deformation of the Primitive-based structure showing a layer-by-layer deformation mechanism as also reflected by the stress-strain curves; see Figure 6e,f.

3.2. ANOVA Analysis

The effect of the considered variables on the TPMS structure performances was studied statistically using ANOVA. Reduced ANOVA tables were utilized so that nonsignificant terms were eliminated using backward/forward methods to enhance the model accuracy without sacrificing the model fit. It is worth mentioning that the normality assumption was satisfied (see Figure 8). The R^2 of E, σ_{peak}, and SEA are 0.995, 0.999, and 0.97, respectively, which indicates an excellent representation of the variability of the data by the model terms. Table 5 shows the reduced ANOVA table of the compressive modulus, and p-values less than 0.05 indicate that model terms are significant. From the ANOVA table (Table 5), all considered factors, including CF incorporation, RD, cell type, and cell size, significantly influence the E. Two-source interactions, including the cell type and RD, and cell type and cell size, show a significant influence on the E. Furthermore, three-source interactions, namely RD, cell size, and cell type, have a significant influence on the E. The most significant effects are caused by changing the RD variable (58.95%), followed by cell type (31.32), and then the CF incorporation (3.51%). Figures 9 and 10 also provide a visual representation of the results, showing the influence direction of the main factors (Figure 9) and their interactions (Figure 10) on the compressive modulus. The Diamond cell type demonstrates the highest compressive modulus, followed by the Gyroid cell type, and finally, the Primitive cell type, which demonstrated the lowest compressive modulus performance. The other parameters—CF, RD, and cell size—have a proportional influence, meaning that a change from a low level to a high level of any of them can increase in the E. Figure 10b,c show how the interaction of the RD, cell size, and cell type affects the E. For instance, the E of the Diamond-based structures improved at high RD (44%) (Figure 10b) but decreased at low RD (30%) (Figure 10c) as the cell size increased.

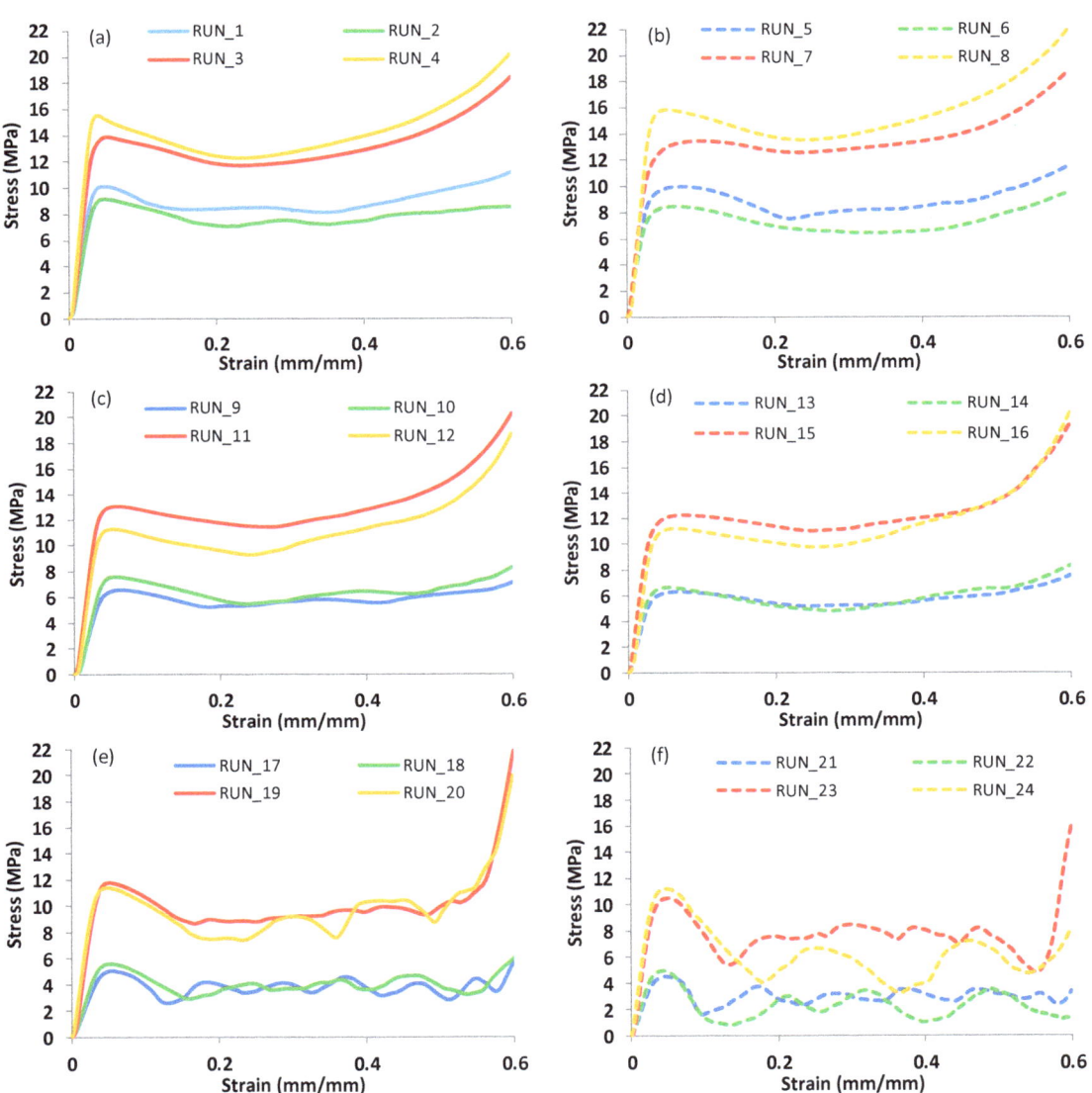

Figure 6. Typical stress-strain curves of the experimental runs listed in Table 4, classified based on the cell type and CF %: (**a**,**b**) Diamond-based samples at 0% CF and 15% CF, respectively; (**c**,**d**) Gyroid-based samples at 0% CF and 15% CF, respectively; (**e**,**f**) Primitive-based samples at 0% CF and 15% CF, respectively.

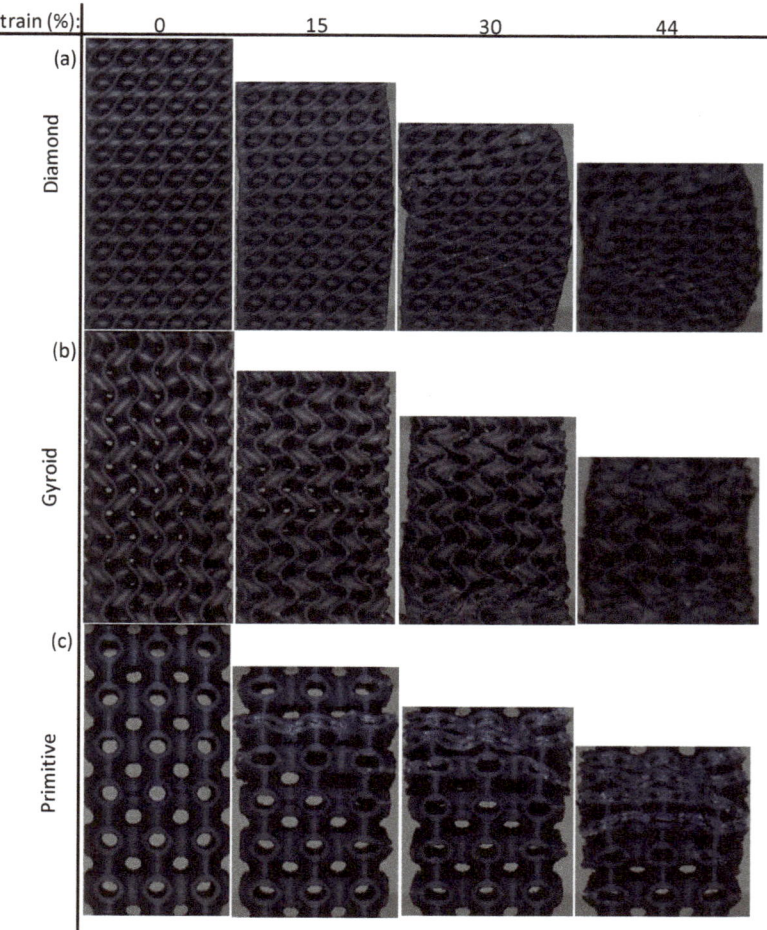

Figure 7. Failure mechanisms of different structures at different strain % of samples with 30% RD, 8 mm cell size, and 0% CF: (**a**) Diamond, (**b**) Gyroid, and (**c**) Primitive.

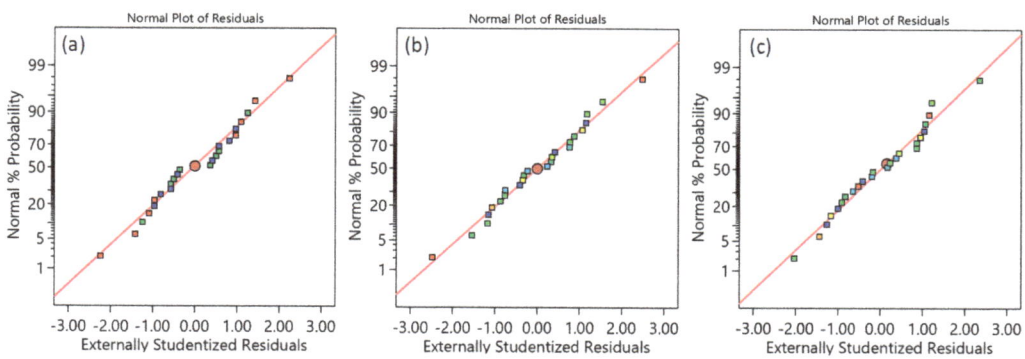

Figure 8. Normality plot of (**a**) E, (**b**) σ_{peak}, and (**c**) SEA.

Table 5. Reduced ANOVA table of E.

Source	Sum of Squares	Contribution %	df	Mean Square	F-Value	p-Value	
Model	0.2723	99.52	12	0.0227	190.15	<0.0001	significant
A-CF	0.0096	3.51	1	0.0096	80.43	<0.0001	
B-RD	0.1613	58.95	1	0.1613	1351.86	<0.0001	
C-Cell_Size	0.005	1.83	1	0.005	41.73	<0.0001	
D-Cell_Type	0.0857	31.32	2	0.0429	359.23	<0.0001	
BC	0.0003	0.11	1	0.0003	2.17	0.1686	
BD	0.0027	0.99	2	0.0014	11.32	0.0021	
CD	0.001	0.37	2	0.0005	4.37	0.0402	
BCD	0.0067	2.45	2	0.0033	27.88	<0.0001	
Residual	0.0013	0.48	11	0.0001			
Cor Total	0.2736	100	23				

R^2: 0.995, Adjusted R^2: 0.99, and Predicted R^2: 0.977.

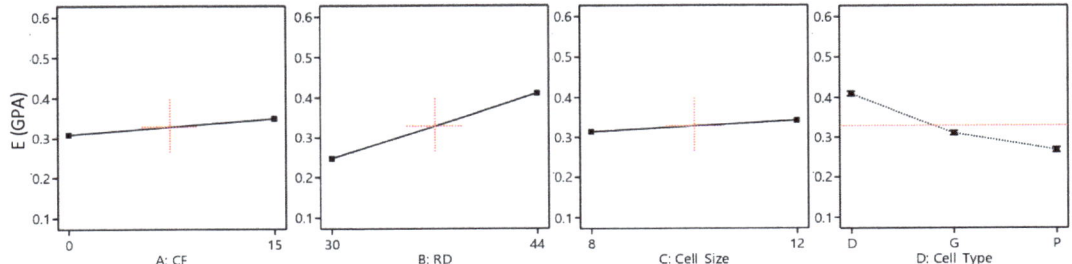

Figure 9. Effect of the main factors on the E.

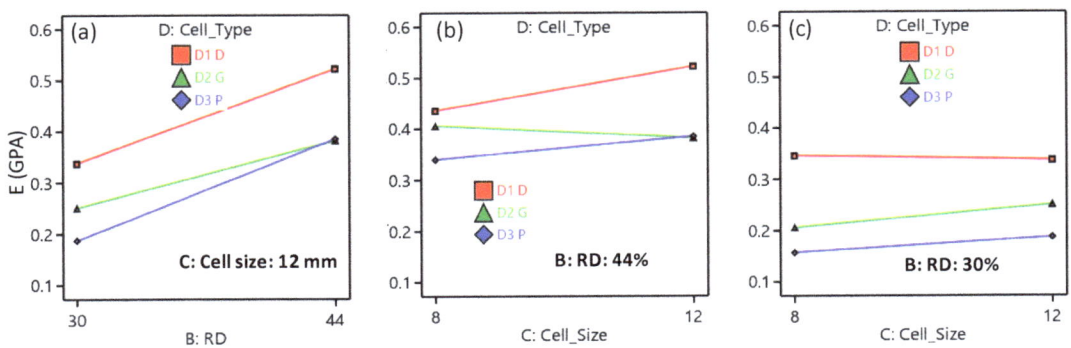

Figure 10. Interaction plots: (**a**) BD interaction, (**b**) CD interaction, and (**b**,**c**) BCD interaction.

Similarly, the reduced ANOVA table for the σ_{peak}, peak strength is presented in Table 6. According to the σ_{peak} ANOVA table, the factors that significantly influence σ_{peak} are CF incorporation, RD, and cell type. The two-source interaction terms, including RD and cell type, and cell size and cell type, have a significant influence on the σ_{peak}. Furthermore, RD and cell size interact significantly with the other two parameters, CF and cell type. The most significant effects are caused by changing the RD variable (71.68%), followed by cell type (23.87%). Figure 11 demonstrates the directional influence of the main components, whereas Figure 12 shows the two- and three-source interaction terms. Figure 11 shows that the TPMS structures with Diamond-cell type exhibit the highest σ_{peak}, whereas the Primitive-based cell type structures show the lowest. Compressive strength is proportionally influenced by RD; increasing RD results in an increase in σ_{peak} (Figure 11). Regarding CF, Figure 11 demonstrates that σ_{peak} decreases as CF increases. The interaction between cell type and size is illustrated in Figure 12b, which demonstrates that as cell size increases, the σ_{peak} of the Diamond and Primitive structures improve while the σ_{peak} of the Gyroid structures decreases. Figure 12d shows how the interaction of the RD, cell size, and cell type affects the σ_{peak}. For instance, the σ_{peak} of the Diamond-based structures improved at high RD (44%) but decreased at low RD (30%) as the cell size increased.

Table 6. Reduced ANOVA Table of σ_{peak}.

Source	Sum of Squares	Contribution %	df	Mean Square	F-Value	p-Value	
Model	255.66	99.88	15	17.04	448.56	<0.0001	significant
A-CF	1.33	0.52	1	1.33	34.97	0.0004	
B-RD	183.47	71.68	1	183.47	4828.63	<0.0001	
C-Cell_Size	0.0775	0.03	1	0.0775	2.04	0.191	
D-Cell_Type	61.1	23.87	2	30.55	804.05	<0.0001	
AB	0.0171	0.01	1	0.0171	0.4495	0.5215	
AC	0.0597	0.02	1	0.0597	1.57	0.2454	
BC	0.0994	0.04	1	0.0994	2.62	0.1444	
BD	1.5	0.59	2	0.75	19.74	0.0008	
CD	0.7668	0.3	2	0.3834	10.09	0.0065	
ABC	0.5298	0.21	1	0.5298	13.94	0.0058	
BCD	6.7	2.62	2	3.35	88.2	<0.0001	
Residual	0.304	0.12	8	0.038			
Cor Total	255.96	100	23				

R^2: 0.999, Adjusted R^2: 0.997, and Predicted R^2: 0.989.

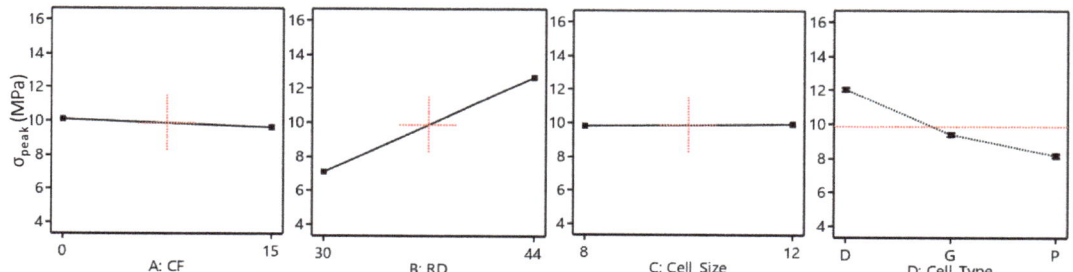

Figure 11. Influence of the main factors on the σ_{peak}.

The reduced ANOVA table for SEA is presented in Table 7. The results indicate that the cell type, CF incorporation, and RD have a significant influence on the SEA. In addition, the two-source interaction between the cell type and CF has a significant influence on the SEA. The most important effects are contributed by the cell type (70.51%), followed by RD (22.44%), and then the CF and cell type interaction (2.66%). The influence of the main factors is also graphically presented in Figure 13. Similar to the compressive modulus and strength, Diamond-based cell types exhibit the best SEA. SEA is proportionally influenced by RD, i.e., an increase in RD results in an upsurge in SEA. Figure 13 demonstrates that SEA decreases as CF increases. This influence, however, varies depending on the cell type, as illustrated in Figure 14. For instance, Diamond-based TPMS structures show an enhancement in SEA with an increase in CF (Figure 14a,b), whereas Primitive-based TPMS structures show a decrease in SEA (Figure 14a,d). Furthermore, Table 7 demonstrates the significant influence of the interaction between the CF and cell type (shown in Figure 14a) on SEA, providing further evidence that CF's influence on SEA is dependent on the cell topology.

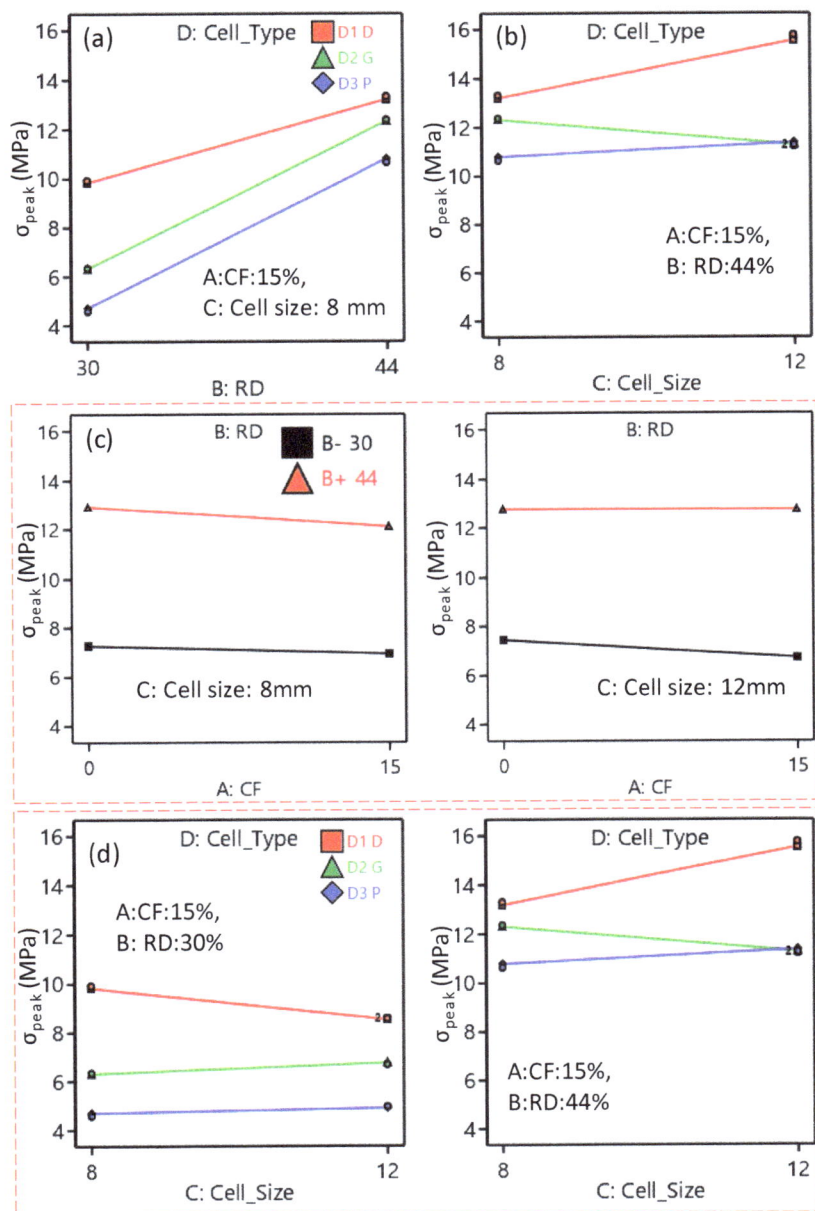

Figure 12. Interaction influence on the σ_{peak}: (**a**) BD interaction, (**b**) CD interaction, (**c**) ABC interaction, and (**d**) BCD interaction.

Table 7. Reduced ANOVA Table of SEA.

Source	Sum of Squares	Contribution %	df	Mean Square	F-Value	p-Value	
Model	225.92	97.02	6	37.65	92.36	<0.0001	significant
A-CF	3.28	1.41	1	3.28	8.03	0.0114	
B-RD	52.26	22.44	1	52.26	128.19	<0.0001	
D-Cell_Type	164.18	70.51	2	82.09	201.37	<0.0001	
AD	6.2	2.66	2	3.1	7.61	0.0044	
Residual	6.93	2.98	17	0.4077			
Cor Total	232.85	100	23				

R^2: 0.97, Adjusted R^2: 0.96, and Predicted R^2: 0.941.

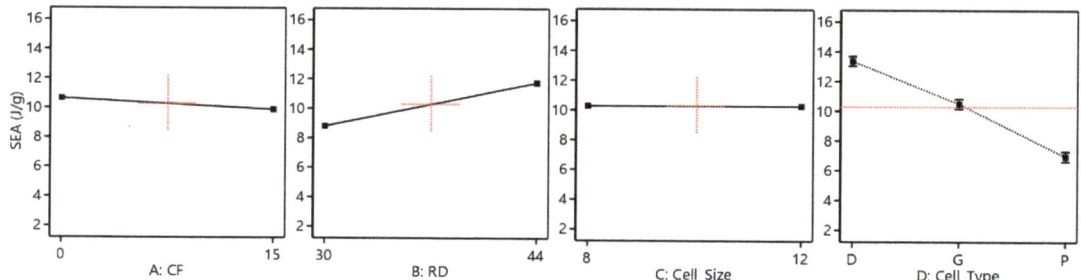

Figure 13. Influence of the main factors on the SEA.

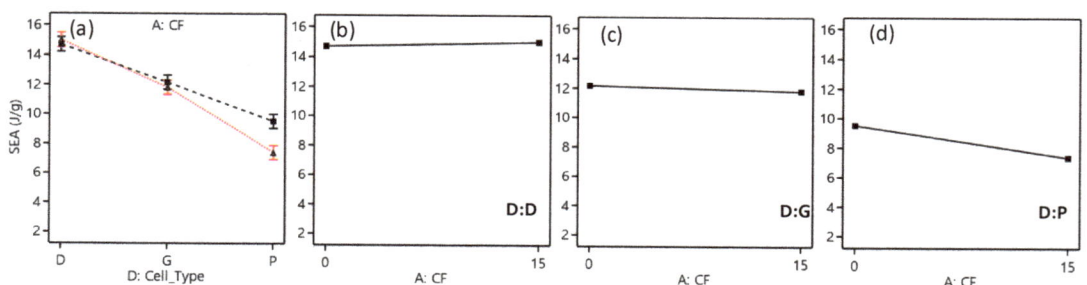

Figure 14. Varying impact of the CF % on the SEA based on cell type at a cell size of 12 mm and RD of 44%: (**a**) CF and cell type interaction, (**b**) Diamond cell type, (**c**) Gyroid Primitive cell type, and (**d**) Primitive cell type.

3.3. Mathematical and ANFIS Prediction Models

Mathematical relationships between the responses, including E, σ_{peak}, and SEA, and the investigated variables were developed based on the reduced models in the ANOVA analysis, e.g., Tables 5–7. The developed mathematical models are presented in Table 8.

Similarly, the ANFIS models were developed to predict the performance of the TPMS structures in terms of E, σ_{peak}, and SEA. An advantage of this type of modeling is that it eliminates the requirement to strictly adhere to predetermined settings for any variable in the DOE, where any change in that setting will most likely impact the results. For instance, the ANOVA analysis results (Section 3.2) proved that RD contributed the most to the mechanical properties, including E, σ_{peak}, and SEA. Thus, a variation in the actual RD from the designed RD will certainly lead to a variation in the mechanical properties. In this regard, actual RD should be used instead of using the designed one, particularly when there is a high variation, which cannot be avoided. Using the actual RDs makes the analysis and modeling more accurate. Variations between the actual and designed RDs are attributed to a number of reasons depending on the AM process. For instance, deviation in wall thickness and the presence of voids and cracks are reasons to deviate RD in the FDM process [5,41]. Excess porosity is an example of the source RD deviation in the SLS process, as elaborated by [42]. In laser-powder bed fusion (L-PBF), unmelted powder particles that have no pathway to get out of the structure, closed pores due to powder clog [43], deviation in wall thickness, voids and cracks, and surface roughness [16] are reasons for RD deviation. Therefore, a modeling approach such as ANFIS will be more accurate as it can consider the measured RDs instead of the designed ones.

Table 8. Mathematical prediction models of E, σ_{peak}, and SEA.

Response	Topology	Mathematical Model
E	G	$-0.625663 + 0.00266637A + 0.0240777B + 0.0481099C - 0.00123144BC$
	D	$0.551474 + 0.00266637A - 0.00698681B - 0.0525257C + 0.00168178BC$
	P	$-0.253648 + 0.00266637 + 0.011005B - 0.000032C + 0.000253839BC$
σ_{peak}	G	$-20.3991 + 0.44013A + 0.844853B + 1.63904C - 0.013642AB - 0.0490301AC - 0.0475763BC + 0.00141501ABC$
	D	$14.1892 + 0.44013A - 0.0749332B - 1.53656C - 0.013642AB - 0.0490301AC + 0.0437816BC + 0.00141501ABC$
	P	$-13.6707 + 0.44013A + 0.584075B + 0.575688C - 0.013642AB - 0.0490301AC - 0.0142519BC + 0.00141501ABC$
SEA	G	$2.89235 - 0.0244981A + 0.2108B$
	D	$5.42397 + 0.0185767A + 0.2108B$
	P	$0.232705 - 0.141851A + 0.2108B$

In this study, the ANFIS models were developed during the training phase based on the full factorial results (as training data) presented in Table 4. The ANFIS technique allows for the utilization of input data even if it does not comply with a specific DOE. It should be noted that the actual RDs were used instead of the designed RDs when developing ANFIS models in both the training and testing phases. Thus, more realistic models could be developed, resulting in improved prediction accuracy. The accuracy of the developed models was evaluated with the testing data set presented in Table 9. For each TPMS performance (e.g., E, σ_{peak}, and SEA), the accuracy of the developed ANFIS models was evaluated using the root mean square error (RMSE), which was calculated using Equation (6).

$$RMSE = \sqrt{\frac{1}{n}\sum_{i=1}^{n}(Exp_i - Pred_i)^2} \qquad (6)$$

where n is the number of testing data while Exp_i and $Pred_i$ are the experimental and predicted results of the ith testing experiment, respectively.

Table 9. Testing data set.

	Testing Data Set				
Run#	CF (%)	RD (%)	Cell Size (mm)	Cell Type	Act. RD
Test-1	0	37	8	P	37.07
Test-2	0	37	8	G	35.20
Test-3	0	37	8	D	35.62
Test-4	15	37	8	P	37.14
Test-5	15	37	8	G	34.91
Test-6	15	37	8	D	35.36

Furthermore, for each TPMS performance, RMSE was used during the training phase for tuning and selecting the fuzzy inference parameters to minimize the RMSE. The fuzzy inference parameters and their settings that resulted in the lowest RMSE are presented in Table 10. An illustration of the ANFIS structure of the σ_{peak} response is provided in Figure 15.

Table 10. Fuzzy inference parameters settings.

Response	Training Opt. Method	MF Type	Output Function	No. MFs	Epochs
E	Hybrid	trimf	Constant	2 2 3 4	210
σ_{peak}	Hybrid	pimf	Linear	2 2 2 3	30
SEA	Hybrid	trimf	Linear	2 2 2 2	30

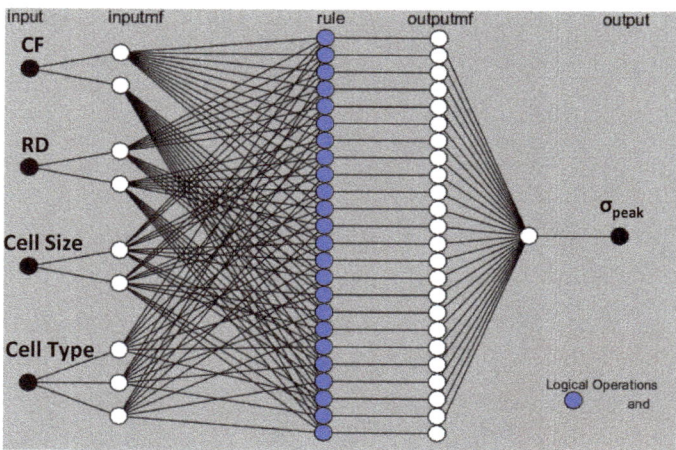

Figure 15. ANFIS structure network for modeling the σ_{peak} with "2 2 2 3" number of MFs.

In Figure 16, the predicted values for E, σ_{peak}, and SEA based on the mathematical and ANFIS models are shown alongside the experimental results. Figure 16a,c,e depict the experimental and predicted findings for the training data, demonstrating that experimental and predicted results obtained from both methods are comparable. Figure 17a–c show evidence of the close agreement between predicted results obtained by both models and the experimental results in terms of R^2 for E, σ_{peak}, and SEA. Similarly, Figure 16b,d,f depict the experimental and predicted findings for the testing data, showing that the predicted results obtained from ANFIS methods are much close to the experimental results. Furthermore, an enhancement is clearly shown in ANFIS predicting results for E, σ_{peak}, and SEA compared with mathematical results. For instance, the R^2 of ANFIS prediction for σ_{peak} is 0.977, which is significantly higher than that of mathematical prediction, which is 0.763.

Table 11 compares the RMSE performance of the mathematical and ANFIS models with respect to testing experiments, revealing that ANFIS models have outperformed mathematical models in all responses. Furthermore, a comparison between mathematical and ANFIS results for each test experiment in terms of absolute percent deviation ($Dev. = \left|\frac{Exp_i - Pred_i}{Exp_i}\right| * 100$) is presented in Table 12. From Table 12, the maximum deviation in the ANFIS' prediction results was 7.61% (the 3rd test experiment, σ_{peak}), while a deviation of 21.11% (the 5th test experiment, σ_{peak}) was found for the mathematical model. It should be noted that the maximum deviation in the mathematical prediction results (21.11%) occurred in the 5th testing experiment, which showed a high RD deviation (actual and designed RDs are 37% and 34.91%, respectively). Furthermore, mathematical models also displayed some deviations exceeding 10%, contrary to ANFIS deviation values.

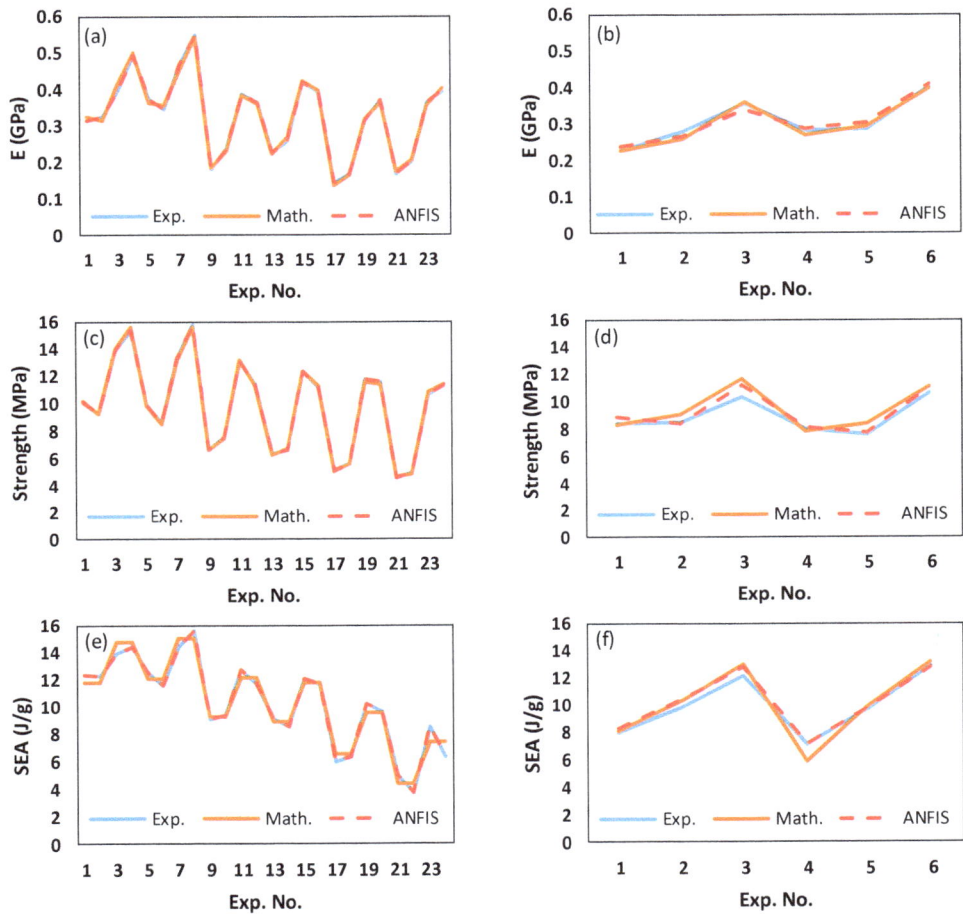

Figure 16. Comparison between experimental, mathematical, and ANFIS results for the training and testing data sets: (**a**) training data of E, (**b**) testing data of E, (**c**) training data of σ_{peak}, (**d**) testing data of σ_{peak}, (**e**) testing data of SEA, and (**f**) training data of SEA.

Table 11. RMSE of the mathematical and ANFIS prediction models based on the testing data set.

Response	Math. RMSE	ANFIS' RMSE
E	0.017	0.012
σ_{peak}	1.181	0.418
SEA	0.851	0.416

The results shown in Figures 16 and 17 and Tables 11 and 12 highlight the significance of employing AI models such as the ANFIS model for predicting the performances of TPMS lattice structures while accounting for the issues associated with 3D printing, such as RD deviation. Table 12 shows that whenever the actual RD is close to the designed RD, both models provide accurate predictions. However, for a high variation in RD, the ANFIS model is preferable. This indicates the ability of both modeling approaches to predict various mechanical properties of TPMS structures while considering different variables, including material composition, geometry, and RD.

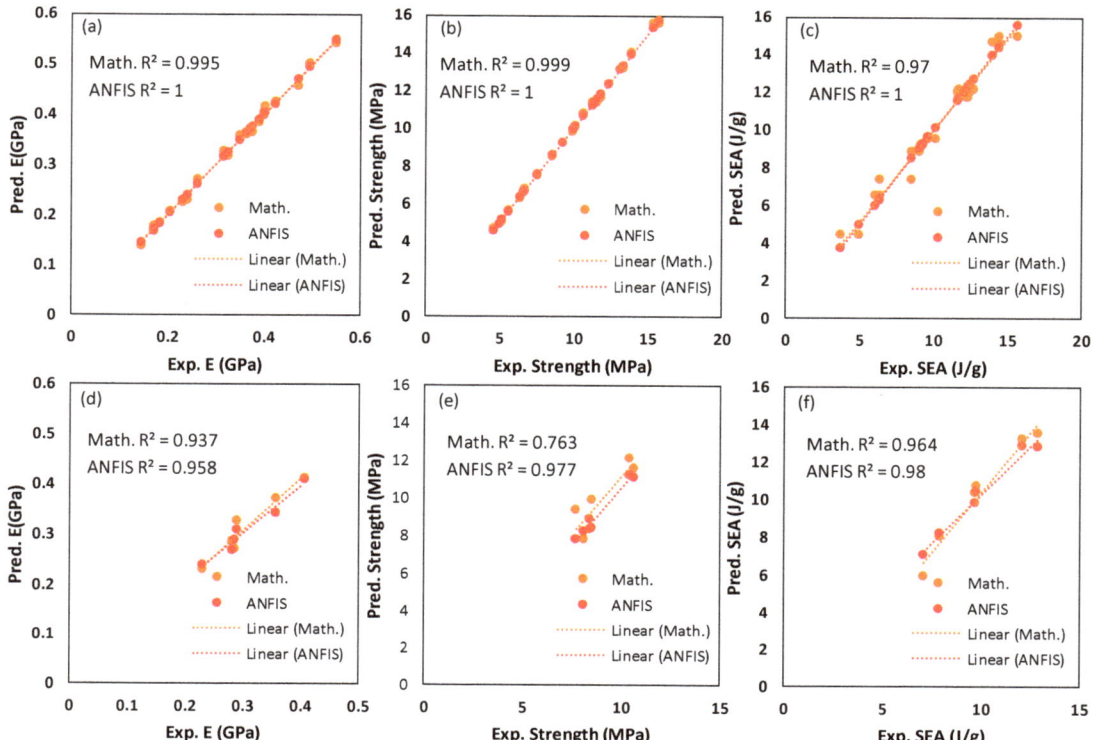

Figure 17. Experimental results vs. mathematical and ANFIS predicted results for the training and testing data sets: (**a**) training data of E, (**b**) training data of σ_{peak}, (**c**) training data of SEA, (**d**) testing data of E, (**e**) testing data of σ_{peak}, and (**f**) testing data of SEA.

Table 12. Experimental and predicted results of the testing data sets.

No. (RD Dev.%)	E (GPa)			σ_{peak} (MPa)			SEA (J/g)			Dev. (E)		Dev. (Str.)		Dev. (SEA)	
	Exp.	Math.	ANFIS	Exp.	Math.	ANFIS	Exp.	Math.	ANFIS	Math.	ANFIS	Math.	ANFIS	Math.	ANFIS
1(0.2)	0.23	0.23	0.24	8.37	8.33	8.84	7.92	8.03	8.19	0.13	3.93	0.56	5.62	1.37	3.39
2(4.9)	0.28	0.29	0.27	8.48	9.89	8.37	9.77	10.69	10.39	2.09	4.91	16.67	1.21	9.46	6.41
3(3.7)	0.36	0.37	0.34	10.40	12.08	11.19	12.10	13.22	12.83	3.91	4.30	16.23	7.61	9.33	6.08
4(0.4)	0.28	0.27	0.29	8.08	7.76	8.18	7.10	5.90	7.06	5.50	1.08	4.04	1.26	16.81	0.50
5(5.7)	0.29	0.33	0.31	7.70	9.32	7.75	9.67	10.32	9.82	12.52	6.14	21.11	0.69	6.76	1.51
6(4.4)	0.40	0.41	0.41	10.63	11.51	11.05	12.91	13.50	12.81	1.38	0.99	8.30	3.96	4.60	0.78

3.4. Multi-Objective Optimization

Desirability analysis was used to select the best settings of the CF incorporation, relative density, cell type, and cell size RD that led to maximizing the σ_{peak}, E, and SEA. Table 13 shows the optimal combination values of the considered variables for multi-objective optimization. Diamond topology, 12 mm cell size, 15% CF, and 44% RD should be used to achieve an overall desirability of 97.8%. Nevertheless, if, for whatever reason, either Gyroid or Primitive cell topologies are selected, the optimal combinations as well as the overall desirability values for both designs are illustrated in Table 14. Tables 13 and 14 present the significance of carefully selecting the cell topology, cell size, and CF incorporation, as different combinations of these factors result in varying performances.

Table 13. Optimal parameter settings obtained by multi-objective desirability analysis.

Cell Topology	Cell Size (mm)	CF (%)	RD (%)	E (GPa)	σ_{peak} (MPa)	SEA (J/g)	Desirability %
D	12	15	44	0.542	15.55	14.978	97.0

Table 14. Optimal parameter settings considering Gyroid and Primitive cell topologies.

Cell Topology	Cell Size (mm)	CF (%)	RD (%)	E (GPa)	σ_{peak} (MPa)	SEA (J/g)	Desirability %
G	8.178 (~8)	15	44	0.424	12.28	11.80	68.6
G	8.06 (~8)	0	44	0.385	13.112	12.168	68.6
P	12	0	44	0.364	11.412	9.508	54.4

Table 15 shows the validation experiments related to the multi-objective optimization findings reported in Tables 13 and 14, demonstrating a good agreement between the multi-objective optimization and experimental results.

Table 15. Validation experiments for multi-objective optimization.

Cell Topology	Cell Size (mm)	CF (%)	RD (%)	E (GPa)	σ_{peak} (MPa)	SEA (J/g)
D	12	15	44	0.549	15.768	15.591
G	8	15	44	0.421	12.367	12.044
G	12	0	44	0.364	11.411	11.673
P	12	0	44	0.370	11.541	9.636

4. Discussion

Stress-strain curves depicted in Figure 6 show evident variations in the mechanical response during the compression testing. Variations are clearly detected in terms of stress, compressive modulus, and deformation patterns. Primitive structures showed a wave pattern of deformation (Figure 6e,f (e.g., run #17–24)), while Diamond structures (Figure 6a,b) and Gyroid structures (Figure 6c,d) seemed to deform uniformly, giving them the advantage of accumulating the load-bearing capacity (e.g., run #1–8 for Diamond and run #9–16 for Gyroid). Compared with D (Figure 6a) and G (Figure 6c), samples with 0% CF, which exhibit a sharp reduction in stress following elastic deformation, D (Figure 6b) and G (Figure 6d) samples with 15% CF demonstrate more plastic deformation. However, Figure 6f shows that incorporating CF into the Primitive structures, particularly with samples of 12 mm cell sizes (run#22 and run#22 24), makes the deformation more wavy than in samples with 0% CF (Figure 6e).

From the results presented in Table 4, Diamond-based TPMS lattice structures showed the best mechanical properties, including compressive modulus, σ_{peak}, and SEA. For instance, the maximum compressive modulus (0.549 GPa), σ_{peak} (15.768 MPa), and SEA (15.591 J/g) were observed with Diamond, 15% CF, 44% RD, and 12 mm cell size (run #8). On the other hand, the worst mechanical responses were obtained by Primitive-based cell-type TPMS structures. The minimum compressive modulus (0.144 GPa) was obtained with Primitive, 0% CF, 30% RD, and 8 mm cell size (run #17), while the lowest σ_{peak} (4.583 MPa) was observed with Primitive, 15% CF, 30% RD, and 8 mm cell size (run #21). Similarly, the minimum SEA (3.759 J/g) was observed with Primitive, 15% CF, 30% RD, and 12 mm cell size (run #22). The walls' orientation and better material distribution within the geometry of the Diamond structures improve wall contact and reduce the empty spaces, making it less susceptible to fracture initiation [5,24]. For illustration, Figure 18 shows cross-sections of lattice structures with 12 mm cell size and 44% RD. It is evident from the cross sections that empty spaces between walls in the Diamond structure (Figure 18a) are less than those in Gyroid and Primitive structures; Figure 18b,c, respectively.

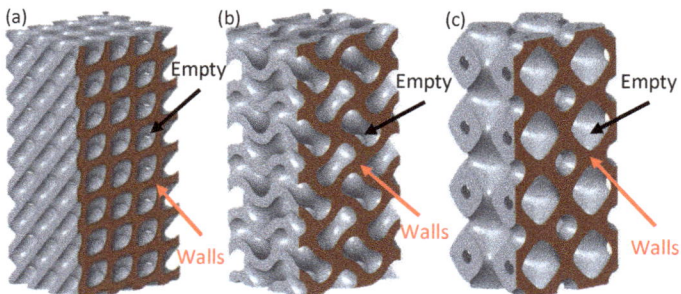

Figure 18. Cross-sectional of Lattice structures showing material distribution and empty spaces at 12 mm cell size and 44% RD: (**a**) Diamond, (**b**) Gyroid, and (**c**) Primitive.

The influence of all considered parameters, including cell topology and size, CF incorporation, and RD, as well as their different combinations on the compressive response of the TPMS structures, was statistically significant. Results showed that RD had the greatest impact on mechanical response among the four studied variables, followed by cell type and CF incorporation. Cell size had the least impact. The authors of [44] stated that RD is the key factor in mechanical performance, including elastic modulus and strength of a given lattice structure. This is in line with the statistical findings, which show that RD has a high influence on both the E and σ_{peak} (58.95% and 71.68%, respectively). The statistical analysis confirms that cell type and RD influence the compressive modulus, σ_{peak}, and SEA of the TPMS lattice structures. This finding is in line with previous studies that investigated the effect of TPMS cell topologies and RDs on E, σ_{peak}, and SEA characteristics, such that these properties were enhanced as RD increased [16]. Diamond structures showed the best performance, while Primitive structures were the worst [24]. In this regard, when statistically investigating the influence of design parameters, such as strut/wall thickness and cell type and size, RD has to be controlled; otherwise, the results could be misleading.

CF incorporation was found to be a statistically significant influence on the mechanical properties and SEA. These findings are consistent with [3]: CF-reinforced PLA lattices showed enhanced compressive modulus and energy absorption. According to results in [3], the shear forces acting on the polymer melt during extrusion cause fibers to align along the printing direction, enhancing the structures' mechanical stability. Furthermore, results reported in [4] stated that the tensile modulus and energy absorption at the break of chiral structures were significantly enhanced (by two times) when incorporating CF into PLA. The interaction of CF incorporation with the cell topology significantly influenced the SEA. This finding agrees with [5] in that the CF incorporation evidently enhanced the energy absorption in the octahedral lattices, while a slight influence was found on the octet lattices. This study confirms these previous findings by demonstrating that CF increases the SEA of Diamond-based structures while (Figure 14b) decreasing the SEA of Primitive-based structures (Figure 14d). This influence of CF and cell type interaction on the SEA is also depicted in Figure 19 using ANFIS 3D surface plot.

Cell size had the smallest effect, and the statistical analysis confirmed its influence only on the compressive modulus. However, its interactions with cell type and RD significantly influenced the compressive modulus and strength, as seen in Tables 5 and 6. Moreover, the influence of the combination of cell size, CF incorporation, and RD on the σ_{peak} was significant. Even though the combination of the three factors had no statistical influence on the SEA contrary to [28] (Truncated octahedron lattices), the negative influence on SEA was consistent with [28]. Findings in [41] stated that certain combinations of wall thickness and cell size (which control the structure RD) of Gyroid structures outperformed others, suggesting the need for predicting the optimal combination of wall thickness and cell size. This observation is noteworthy since it implies that different combinations of design, RD, and material composition (e.g., CF incorporation) factors could be used to attain the desired performance.

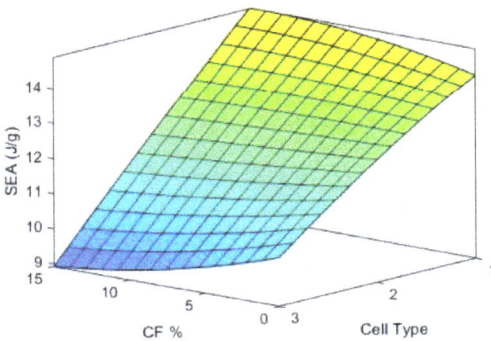

Figure 19. ANFIS 3D surface plot of SEA variation with the CF and cell type interaction at a cell size of 12 mm and RD of 44%; 1, 2, and 3 denote Diamond, Gyroid, and Primitive, respectively.

For the training data set, factorial design experiments presented in Table 4, both prediction models, mathematical and ANFIS models, performed well in predicting E, σ_{peak}, and SEA, as illustrated in Figures 16 and 17. However, regarding the testing data set presented in Table 9, ANFIS models clearly outperformed the mathematical models in terms of RMSE (Table 11) and the absolute percent deviation (Table 12). Furthermore, Tables 12 and 16 demonstrate that mathematical modeling predictions deviated more from experimental results compared with ANFIS predictions, notably for experiments having high RD deviation. This implies a connection between mathematical prediction performance and RD deviation. From the ANOVA analysis, Section 3.1, the RD influenced E, σ_{peak}, and SEA with a contribution of 58.95%, 71.68%, and 22.44%, respectively.

Table 16. Relationship between RD absolute percent deviation, RD contribution, and mathematical prediction absolute percent deviation.

Test No.	RD Dev. (%)	RD Contribution %		
		58.95	71.68	22.44
		E Math. Dev. (%)	σ_{peak} Math. Dev. (%)	SEA Math. Dev. (%)
1	0.2	0.13	0.56	1.37
2	4.9	2.09	16.67	9.46
3	3.7	3.91	16.23	9.33
4	0.4	5.50	4.04	16.81
5	5.7	12.52	21.11	6.76
6	4.4	1.38	8.30	4.60

A correlation was observed between the contribution percentage of the RD influence on the TPMS structure performance (e.g., E, σ_{peak}, and SEA) reported in ANOVA analysis (Tables 5–7) and the performance of the mathematical predictions. In other words, a high contribution percentage of RD influence on a TPMS structure's performance (e.g., E, σ_{peak}, or SEA) indicates a high error in mathematical modeling predictions for experiments having a high RD deviation. For instance, the higher mathematical modeling prediction deviations for σ_{peak} (16.67%, 16.23%, 21.11%, and 8.30%) were observed for the testing experiments (2, 3, 5, and 6, respectively), with relatively high RD deviations; see Table 16. The fourth testing experiment (i.e., regarding the mathematical prediction deviations for the E and SEA) was an exceptional case where the aforementioned phenomenon was not valid. Furthermore, Figure 20 depicts the relationship between the RD deviation and the mathematical prediction deviation for E, σ_{peak}, and SEA. A high correlation between the

RD deviation and mathematical prediction deviation in the case of σ_{peak} ($R^2 = 0.802$) can be observed, while a low correlation is in the case of SEA. Table 17 presents the results of the Spearman Rho correlation test, indicating a significant correlation (0.943) between the RD deviation and mathematical prediction performance in the case of σ_{peak}, which is highly influenced by the RD variable (71.68%).

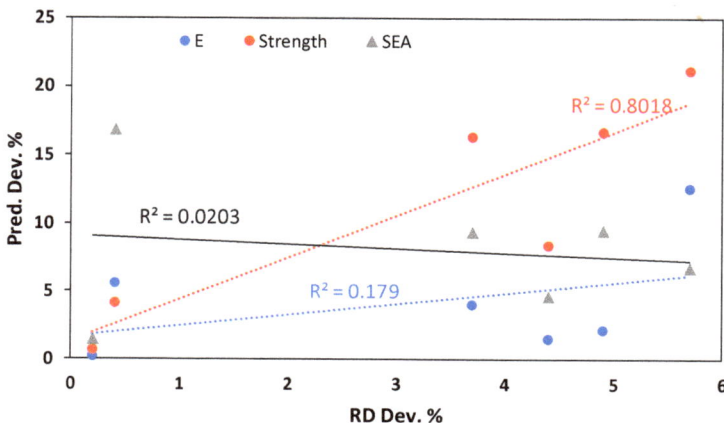

Figure 20. Relationship between RD absolute percent deviation, RD contribution, and mathematical prediction absolute percent deviation.

Table 17. Correlation between RD absolute percent deviation and mathematical prediction absolute percent deviation of E, σ_{peak}, and SEA.

		Math. Prediction Dev.		
		E	σ_{peak}	SEA
RD Dev.	Correlation	0.486	0.943	0.143
	p-value	0.329	0.005	0.787

5. Conclusions

In this study, the influence of different TPMS-based cell topologies with varying unit cell sizes, relative densities, and CF incorporation on the mechanical and specific energy absorption was statistically investigated. The FDM 3D-printed lattices were tested using a uniaxial compression testing, and their E, σ_{peak}, and SEA were evaluated. Prediction models were developed using ANFIS and mathematical modeling. By using the ANFIS models, we were able to predict mechanical characteristics considering imperfections in 3D printing, such as an RD variation. This was achieved by using the actual RDs instead of the designed RDs. Multi-objective optimization was conducted using the desirability approach with the objective of maximizing E, σ_{peak}, and SEA. The following inferences can be made from this study's findings:

- The findings demonstrated a change in the E (0.144 GPa to 0.549 GPa), σ_{peak} (4.583 MPa to 15.768 MPa), and SEA (3.759 J/g to 15.591 J/g) due to the impact of the considered variables.
- RD had a significant influence on both E and σ_{peak}, with a contribution of 58.95% and 71.68%, respectively. The cell type had the highest impact on the SEA, contributing to 70.51% of the total influence. In general, RD had the highest influence on mechanical responses among the four studied variables, followed by cell type, CF incorporation, and finally, cell size having the least impact.

- The findings revealed the importance of statistically evaluating the influence of the design, RD, and material composition (e.g., CF incorporation) parameters and their combination to attain the desired TPMS lattice structure performance.
- For the training data set (factorial design experiments), both mathematical and ANFIS models predicted E, σ_{peak}, and SEA well. However, when it comes to the testing data set (validation experiments), ANFIS models clearly outperformed mathematical models in terms of RMSE and absolute percent deviation in predicting all mechanical characteristics. For instance, the maximum absolute percent deviation was 7.61% for ANFIS prediction, while it was 21.11% for mathematical prediction.
- The accuracy of mathematical predictions is highly influenced by the degree of RD deviation; a higher deviation in RD results in lower accuracy of predictions. Furthermore, a correlation between the mathematical models' prediction accuracy and the RD deviation was found when the RD influence contribution on a TPMS performance was high. For instance, when RD accounted for 71.68% of the variation in σ_{peak}, there was a significant correlation (94.3%) between the accuracy of mathematical predictions and RD deviation. Therefore, for a high variation in RD, ANFIS models are preferable.
- Whenever the actual RD is close to the designed RD, both models provide accurate prediction models. This also indicates the ability of both models to predict different mechanical properties of TPMS structures, taking into account different variables, including material composition, geometry, and the RD.
- This study provides a better understanding of PLA and CFRPLA TPMS structures, as well as an ability to better predict their mechanical behavior. For instance, based on the multi-objective optimization using the desirability approach, the Diamond topology, 12 mm cell size, 15% CF, and 44% RD combination are the best settings that achieved an overall desirability of 97.8% for E, σ_{peak}, and SEA.

Author Contributions: M.S.: conceptualization, methodology, software, formal analysis, data curation, investigation, visualization, and writing—original draft, and writing—review and editing; S.A.: conceptualization, formal analysis, investigation, supervision, and writing—review and editing, validation; A.M.A.-A.: investigation, supervision, and writing—review and editing; A.Y.A.: investigation, resources, and funding acquisition. All authors have read and agreed to the published version of the manuscript.

Funding: The authors extend their appreciation to the Deputyship for Research & Innovation, "Ministry of Education" in Saudi Arabia for funding this research work through the project number (IFKSUDR_P114).

Data Availability Statement: The data presented in this study are available on request from the corresponding author.

Acknowledgments: The authors extend their appreciation to the Deputyship for Research & Innovation, Ministry of Education, Saudi Arabia for funding this research work through the project number (IFKSUDR_P114).

Conflicts of Interest: The authors declare no conflict of interest.

References

1. Červinek, O.; Werner, B.; Koutný, D.; Vaverka, O.; Pantělejev, L.; Paloušek, D. Computational approaches of quasi-static compression loading of SS316L lattice structures made by selective laser melting. *Materials* **2021**, *14*, 2462. [CrossRef]
2. Nazir, A.; Abate, K.M.; Kumar, A.; Jeng, J.-Y. A state-of-the-art review on types, design, optimization, and additive manufacturing of cellular structures. *Int. J. Adv. Manuf. Technol.* **2019**, *104*, 3489–3510. [CrossRef]
3. Kaur, M.; Yun, T.G.; Han, S.M.; Thomas, E.L.; Kim, W.S. 3D printed stretching-dominated micro-trusses. *Mater. Des.* **2017**, *134*, 272–280. [CrossRef]
4. Hu, C.; Dong, J.; Luo, J.; Qin, Q.-H.; Sun, G. 3D printing of chiral carbon fiber reinforced polylactic acid composites with negative Poisson's ratios. *Compos. Part B Eng.* **2020**, *201*, 108400. [CrossRef]
5. Saleh, M.; Anwar, S.; Al-Ahmari, A.M.; Alfaify, A. Compression performance and failure analysis of 3D-printed carbon fiber/PLA composite TPMS lattice structures. *Polymers* **2022**, *14*, 4595. [CrossRef]

6. Zarei, M.; Dargah, M.S.; Azar, M.H.; Alizadeh, R.; Mahdavi, F.S.; Sayedain, S.S.; Kaviani, A.; Asadollahi, M.; Azami, M.; Beheshtizadeh, N. Enhanced bone tissue regeneration using a 3D-printed poly(lactic acid)/Ti6Al4V composite scaffold with plasma treatment modification. *Sci. Rep.* **2023**, *13*, 3139. [CrossRef]
7. Belaid, H.; Nagarajan, S.; Teyssier, C.; Barou, C.; Barés, J.; Balme, S.; Garay, H.; Huon, V.; Cornu, D.; Cavaillès, V.; et al. Development of new biocompatible 3D printed graphene oxide-based scaffolds. *Mater. Sci. Eng. C* **2020**, *110*, 110595. [CrossRef]
8. Dong, K.; Panahi-Sarmad, M.; Cui, Z.; Huang, X.; Xiao, X. Electro-induced shape memory effect of 4D printed auxetic composite using PLA/TPU/CNT filament embedded synergistically with continuous carbon fiber: A theoretical & experimental analysis. *Compos. Part B Eng.* **2021**, *220*, 108994. [CrossRef]
9. Alam, F.; Varadarajan, K.M.; Kumar, S. 3D printed polylactic acid nanocomposite scaffolds for tissue engineering applications. *Polym. Test.* **2020**, *81*, 106203. [CrossRef]
10. Stan, F.; Sandu, I.-L.; Fetecau, C. 3D Printing and mechanical behavior of Anisogrid composite lattice cylindrical structures; American society of mechanical engineers digital collection, September 30 2022. In Proceedings of the ASME 2022 17th International Manufacturing Science and Engineering Conference, West Lafayette, IN, USA, 27 June–1 July 2022; Volume 85819, p. V002T05A044. [CrossRef]
11. Plocher, J.; Panesar, A. Effect of density and unit cell size grading on the stiffness and energy absorption of short fibre-reinforced functionally graded lattice structures. *Addit. Manuf.* **2020**, *33*, 101171. [CrossRef]
12. Riva, L.; Ginestra, P.S.; Ceretti, E. Mechanical characterization and properties of laser-based powder bed–fused lattice structures: A review. *Int. J. Adv. Manuf. Technol.* **2021**, *113*, 649–671. [CrossRef]
13. Spear, D.; Palazotto, A. Investigation and Statistical Modeling of the mechanical properties of additively manufactured lattices. *Materials* **2021**, *14*, 3962. [CrossRef] [PubMed]
14. Ling, C.; Cernicchi, A.; Gilchrist, M.D.; Cardiff, P. Mechanical behaviour of additively-manufactured polymeric octet-truss lattice structures under quasi-static and dynamic compressive loading. *Mater. Des.* **2019**, *162*, 106–118. [CrossRef]
15. Habib, F.N.; Iovenitti, P.; Masood, S.H.; Nikzad, M. Fabrication of polymeric lattice structures for optimum energy absorption using multi jet fusion technology. *Mater. Des.* **2018**, *155*, 86–98. [CrossRef]
16. AlMahri, S.; Santiago, R.; Lee, D.-W.; Ramos, H.; Alabdouli, H.; Alteneiji, M.; Guan, Z.; Cantwell, W.; Alves, M. Evaluation of the dynamic response of triply periodic minimal surfaces subjected to high strain-rate compression. *Addit. Manuf.* **2021**, *46*, 102220. [CrossRef]
17. Qin, D.; Sang, L.; Zhang, Z.; Lai, S.; Zhao, Y. Compression performance and deformation behavior of 3D-printed PLA-based lattice structures. *Polymers* **2022**, *14*, 1062. [CrossRef]
18. Zhang, X.Y.; Yan, X.C.; Fang, G.; Liu, M. Biomechanical Influence of Structural Variation Strategies on Functionally Graded Scaffolds Constructed with Triply Periodic Minimal Surface. *Addit. Manuf.* **2020**, *32*, 101015. [CrossRef]
19. Yoo, D.-J. Advanced porous scaffold design using multi-void triply periodic minimal surface models with high surface area to volume ratios. *Int. J. Precis. Eng. Manuf.* **2014**, *15*, 1657–1666. [CrossRef]
20. Qureshi, Z.A.; Al-Omari, S.A.B.; Elnajjar, E.; Al-Ketan, O.; Abu Al-Rub, R. Architected lattices embedded with phase change materials for thermal management of high-power electronics: A numerical study. *Appl. Therm. Eng.* **2023**, *219*, 119420. [CrossRef]
21. Kladovasilakis, N.; Tsongas, K.; Tzetzis, D. Mechanical and FEA-Assisted characterization of fused filament fabricated triply periodic minimal surface structures. *J. Compos. Sci.* **2021**, *5*, 58. [CrossRef]
22. Abueidda, D.W.; Bakir, M.; Abu Al-Rub, R.K.; Bergström, J.S.; Sobh, N.A.; Jasiuk, I. Mechanical properties of 3D printed polymeric cellular materials with triply periodic minimal surface architectures. *Mater. Des.* **2017**, *122*, 255–267. [CrossRef]
23. Shi, X.; Liao, W.; Li, P.; Zhang, C.; Liu, T.; Wang, C.; Wu, J. Comparison of compression performance and energy absorption of lattice structures fabricated by selective laser melting. *Adv. Eng. Mater.* **2020**, *22*, 2000453. [CrossRef]
24. Ali, M.; Sari, R.K.; Sajjad, U.; Sultan, M.; Ali, H.M. Effect of annealing on microstructures and mechanical properties of PA-12 lattice structures proceeded by multi jet fusion technology. *Addit. Manuf.* **2021**, *47*, 102285. [CrossRef]
25. Zarna, C.; Chinga-Carrasco, G.; Echtermeyer, A.T. Bending properties and numerical modelling of cellular panels manufactured from wood fibre/PLA biocomposite by 3D printing. *Compos. Part A Appl. Sci. Manuf.* **2023**, *165*, 107368. [CrossRef]
26. Silva, R.G.; Estay, C.S.; Pavez, G.M.; Viñuela, J.Z.; Torres, M.J. Influence of geometric and manufacturing parameters on the compressive behavior of 3D printed polymer lattice structures. *Materials* **2021**, *14*, 1462. [CrossRef]
27. Athanker, P.; Singh, A.K. Elastic and elasto-plastic analysis of Ti6Al4V micro-lattice structures under compressive loads. *Math. Mech. Solids* **2021**, *26*, 591–615. [CrossRef]
28. Rossiter, J.D.; Johnson, A.A.; Bingham, G.A. Assessing the design and compressive performance of material extruded lattice structures. *3D Print. Addit. Manuf.* **2020**, *7*, 19–27. [CrossRef]
29. El-Sayed, M.A.; Essa, K.; Ghazy, M.; Hassanin, H. Design optimization of additively manufactured titanium lattice structures for biomedical implants. *Int. J. Adv. Manuf. Technol.* **2020**, *110*, 2257–2268. [CrossRef]
30. De Pasquale, G.; Luceri, F.; Riccio, M. Experimental evaluation of selective laser melting process for optimized lattice structures. *Proc. Inst. Mech. Eng. Part E J. Process. Mech. Eng.* **2019**, *233*, 763–775. [CrossRef]
31. Ben Ali, N.; Khlif, M.; Hammami, D.; Bradai, C. Optimization of structural parameters on hollow spherical cells manufactured by fused deposition modeling (FDM) using Taguchi method. *Cell. Polym.* **2021**, *41*, 3–20. [CrossRef]
32. PLA Filament. Available online: https://www.3dxtech.com/product/ecomax-pla/ (accessed on 8 November 2022).

33. CarbonXTM PLA+CF 3D Filament | Carbon Fiber Reinforced. Available online: https://www.3dxtech.com/product/carbonx-pla-cf/ (accessed on 8 November 2022).
34. Guo, H.; Takezawa, A.; Honda, M.; Kawamura, C.; Kitamura, M. Finite element simulation of the compressive response of additively manufactured lattice structures with large diameters. *Comput. Mater. Sci.* **2020**, *175*, 109610. [CrossRef]
35. Mishra, A.K.; Chavan, H.; Kumar, A. Effect of material variation on the uniaxial compression behavior of FDM manufactured polymeric TPMS lattice materials. *Mater. Today Proc.* **2021**, *46*, 7752–7759. [CrossRef]
36. Karaboga, D.; Kaya, E. Adaptive network based fuzzy inference system (ANFIS) training approaches: A comprehensive survey. *Artif. Intell. Rev.* **2019**, *52*, 2263–2293. [CrossRef]
37. Deshwal, S.; Kumar, A.; Chhabra, D. Exercising hybrid statistical tools GA-RSM, GA-ANN and GA-ANFIS to optimize FDM process parameters for tensile strength improvement. *CIRP J. Manuf. Sci. Technol.* **2020**, *31*, 189–199. [CrossRef]
38. Xia, C.; Pan, Z.; Polden, J.; Li, H.; Xu, Y.; Chen, S. Modelling and prediction of surface roughness in wire arc additive manufacturing using machine learning. *J. Intell. Manuf.* **2021**, *33*, 1467–1482. [CrossRef]
39. Buragohain, M.; Mahanta, C. A novel approach for ANFIS modelling based on full factorial design. *Appl. Soft Comput.* **2008**, *8*, 609–625. [CrossRef]
40. Nasr, M.M.; Anwar, S.; Al-Samhan, A.M.; Ghaleb, M.; Dabwan, A. Milling of graphene reinforced Ti6Al4V nanocomposites: An artificial intelligence based industry 4.0 approach. *Materials* **2020**, *13*, 5707. [CrossRef]
41. Maconachie, T.; Tino, R.; Lozanovski, B.; Watson, M.; Jones, A.; Pandelidi, C.; Alghamdi, A.; Almalki, A.; Downing, D.; Brandt, M.; et al. The compressive behaviour of ABS gyroid lattice structures manufactured by fused deposition modelling. *Int. J. Adv. Manuf. Technol.* **2020**, *107*, 4449–4467. [CrossRef]
42. Graziosi, S.; Ballo, F.M.; Libonati, F.; Senna, S. 3D Printing of bending-dominated soft lattices: Numerical and experimental assessment. *Rapid Prototyp. J.* **2022**, *28*, 51–64. [CrossRef]
43. Zhang, C.; Banerjee, A.; Hoe, A.; Tamraparni, A.; Felts, J.R.; Shamberger, P.J.; Elwany, A. Design for laser powder bed additive manufacturing of AlSi12 periodic mesoscale lattice structures. *Int. J. Adv. Manuf. Technol.* **2021**, *113*, 3599–3612. [CrossRef]
44. Zhao, J.; Liu, H.; Zhou, Y.; Chen, Y.; Gong, J. Effect of relative density on the compressive properties of Ti6Al4V diamond lattice structures with shells. *Mech. Adv. Mater. Struct.* **2021**, *29*, 3301–3315. [CrossRef]

Disclaimer/Publisher's Note: The statements, opinions and data contained in all publications are solely those of the individual author(s) and contributor(s) and not of MDPI and/or the editor(s). MDPI and/or the editor(s) disclaim responsibility for any injury to people or property resulting from any ideas, methods, instructions or products referred to in the content.

Article

Effect of Manufacture-Induced Interfaces on the Tensile Properties of 3D Printed Polyamide and Short Carbon Fibre-Reinforced Polyamide Composites

Yingwei Hou * and Ajit Panesar

Department of Aeronautics, Imperial College London, London SW7 2AZ, UK
* Correspondence: y.hou17@ic.ac.uk

Abstract: This study aims to elucidate the structure–property–process relationship of 3D printed polyamide and short carbon fibre-reinforced polyamide composites. The macroscopic properties (tensile modulus) of the 3D printed samples are quantitatively correlated to the printing process-induced intrinsic microstructure with multiple interfaces. The samples were printed with different layer thicknesses (0.1, 0.125 and 0.2 mm) to obtain the varied number of interface densities (number of interfaces per unit sample thickness). The result shows that the printed short carbon fibre-reinforced polyamide composites had inferior partially bonded interfaces compared to the printed polyamide, and consequently exhibited interface-dependent elastic performance. The tensile modulus of 3 mm thick composites decreased up to 18% as a function of interface density, whilst the other influencing aspects including porosity, crystallinity and fibre volume fraction (9%) were the same. Injection moulding was also employed to fabricate samples without induced interfaces, and their tensile properties were used as a benchmark. Predictions based on the shear-lag model were in close agreement (<5%) with the experimental data for the injection-moulded composites, whereas the tensile modulus of the printed composites was up to 38% lower than the predicted modulus due to the partial bonded interfaces.

Keywords: additive manufacturing; defects; short fibre-reinforced polymer; tensile properties

Citation: Hou, Y.; Panesar, A. Effect of Manufacture-Induced Interfaces on the Tensile Properties of 3D Printed Polyamide and Short Carbon Fibre-Reinforced Polyamide Composites. *Polymers* **2023**, *15*, 773. https://doi.org/10.3390/polym15030773

Academic Editor: Yang Li

Received: 26 December 2022
Revised: 28 January 2023
Accepted: 30 January 2023
Published: 2 February 2023

Copyright: © 2023 by the authors. Licensee MDPI, Basel, Switzerland. This article is an open access article distributed under the terms and conditions of the Creative Commons Attribution (CC BY) license (https://creativecommons.org/licenses/by/4.0/).

1. Introduction

Additive manufacturing (AM), also known as 3D printing, is a fabrication process which adds materials through a successive deposition method (layer by layer) until a final 3D object is fabricated. Compared to traditional subtractive manufacturing methods, AM technology has advantages such as negligible material wastage and manufacturing complex structures without using a mould or assemblies [1]. This leads to significant cost-saving, and therefore AM technology attracts increasing interest from the aerospace [2] and automotive sectors [3].

The material extrusion (ME) method is one of the most-used AM technologies due to its simplicity, and especially recent developments in materials and printers. Available material for ME is mainly thermoplastics such as polylactide (PLA) [4–7], acrylonitrile butadiene styrene (ABS) [8–11], polycarbonate [12–14] and polyamide [15–17]. However, the mechanical performance of the ME-printed thermoplastics shown in Table 1 may not be satisfactory for structural parts. Specifically, the highest tensile strength and tensile modulus of ME-printed PLA reported so far are 40–62 MPa and 3.4~4.7 GPa, respectively. The ME-printed ABS also has a relatively low maximum tensile strength (20~35 MPa) and tensile modulus (1.8~2.2 GPa). The tensile modulus of ME-printed polyamide is up to 0.9 GPa. The unsatisfactory mechanical performance may limit the wide application of ME-printed parts as structural parts in the industry.

Table 1. The mechanical properties of ME-printed thermoplastic polymer reported in the literature.

Authors	Materials	Test Type	Standard	Results *
Song et al. [18]	PLA	Tensile; Compressive	Not reported (NR)	σ_t: 55 MPa; E_t: 4.0 GPa; σ_c: 98 MPa; E_c: 4.7 GPa
Yao et al. [4]	PLA	Tensile; Flexural	ISO 527; ISO 14125	σ_t: 46 MPa; σ_f: 82 MPa
Ning et al. [19]	ABS	Tensile	ASTM D638	σ_t: 34 MPa; E_t: 1.9 GPa
Love et al. [20]	ABS	Tensile	ASTM D638	σ_t: 35 MPa; E_t: 2.2 GPa
Omuro et al. [21]	PLA	Tensile; Flexural	NR	σ_t: 40 MPa; E_t: 4.7 GPa; σ_f: 65 MPa; E_f: 2.5 GPa
Tian et al. [22]	PLA	Tensile; Flexural	GB/T 1447; GB/T 1449	σ_t: 62 MPa; E_t: 4.2 GPa; σ_f: 100 MPa; E_f: 4.0 GPa
Van Der Klift et al. [23]	Polyamide	Tensile	JIS K 7073	E_t: 0.9 GPa;
Tymrak et al. [8]	ABS; PLA	Tensile	ASTM D638	ABS: σ_t: 29 MPa; E_t: 1.8 GPa; PLA: σ_t: 57 MPa; E_t: 3.4 GPa
Cantrell et al. [24]	ABS; PC	Tensile	ASTM D638	ABS: σ_t: 30 MPa; E_t: 2.0 GPa; PC: σ_t: 54 MPa; E_t: 1.9 GPa
McLouth et al. [25]	ABS	Fracture Toughness	ASTM D5045	1.97 MPa m$^{1/2}$
D'Amico et al. [9]	ABS	Tensile; Flexural	ASTM D638; ASTM D790	σ_t: 20 MPa; σ_f: 21 MPa;
Rahmatabadi et al. [26]	PLA-polyurethane	Tensile; Compressive; Flexural	ASTM D638; ISO604:2002; ASTM D790	σ_t: 54 MPa; σ_c: 43 MPa σ_f: 124 MPa;
Rahmatabadi et al. [27]	Poly vinyl chloride	Tensile; Compressive; Flexural	ASTM D638; ISO604:2002; ASTM D790	σ_t: 77 MPa; E_t: 0.7 GPa σ_c: 57 MPa; E_c: 0.8 GPa σ_f: 201 MPa; E_f: 1.3 GPa
Moradi et al. [28]	polyamide	Tensile	ASTM D638	Elongation: 596%

* σ_t, σ_f, σ_c refer to tensile strength, flexural strength and compressive strength, respectively. E_t, E_f, E_c refer to tensile modulus, flexural modulus and compressive modulus, respectively.

Fibre reinforcement is an effective approach to alleviate the drawback of ME-printed thermoplastic applied in the industry. Recent developments in available materials for ME enabled the printing of carbon nanotube [29,30], graphite [31,32], short carbon fibre [19,33–35], short glass fibre [36,37], short basalt fibre [38,39] and continuous fibre [40–43] - reinforced polymer composites. The fibre reinforcements significantly improve the mechanical properties of ME-printed parts. The improvements reported in the literature are summarised in Table 2. For example, the tensile strengths of PLA and ABS with the inclusion of the short carbon fibre reinforcement are seen to improve by up to almost 220% and 240%, respectively. The tensile modulus of continuous carbon fibre-reinforced polyamide is more than an order of magnitude higher than that of the polyamide matrix.

Table 2. The mechanical improvement of fibre-reinforced polymer manufactured via ME.

Authors	FRPs	Fibre Fraction	Test Type	Standard	Mechanical Improvements *
Ning et al. [19]	Short carbon fibre/ABS	5 wt%	Tensile	ASTM D638	σ_t: 24%; E_t: 32%;
Love et al. [20]	Short carbon fibre/ABS	13 wt%	Tensile	ASTM D638	σ_t: 236%; E_t: 427%;
Mahajan and Cormier [44]	Short carbon fibre/epoxy	15 wt%	Tensile	ASTM D638	σ_t: 41%; E_t: 45%;
Omuro et al. [21]	Continuous carbon fibre/PLA	30 vol%	Tensile; Flexural	NR	σ_t: 1389%; E_t: 1356%; σ_f: 1012%; E_f: 242%
Ferreira et al. [33]	Short carbon fibre/PLA	15 wt%	Tensile; Shear	ASTM D638; ASTM D3518	σ_t: 220%; σ_s: 5%; E_s: 116%;
Tekinalp et al. [45]	Short carbon fibre/ABS	30 wt%	Tensile	ASTM D638	σ_t: 115%; E_t: 700%;
Tian et al. [22]	Continuous carbon fibre/PLA	9 vol%	Tensile; Flexural	GB/T 1447; GB/T 1449	σ_t: 313%; E_t: 390%; σ_f: 260%; E_f: 230%
Hinchcliffe et al. [46]	Continuous flax fibre/PLA	NR	Tensile	ASTM D638	σ_t: 116%; E_t: 62%; σ_f: 14%; E_f: 10%
Matsuzaki et al. [47]	Continuous carbon fibre/PLA	6 vol%	Tensile	JIS K 7162	σ_t: 363%; E_t: 400%
Shofner et al. [29]	Nanocarbon fibre/ABS	10 wt%	Tensile	ASTM D638	σ_t: 39%; E_t: 40%
Dutra et al. [48]	Continuous carbon fibre/polyamide	30 vol%	Tensile	ASTM D3039	E_t: 894%
Caminero et al. [49]	Continuous carbon fibre/polyamide; Continuous Kevlar fibre/polyamide; Continuous glass fibre/polyamide	50 wt%	Impact	ASTM D6110	Impact resistance: Continuous carbon fibre/polyamide: 181%; Continuous Kevlar fibre/polyamide/: 513%; Continuous glass fibre/polyamide: 1225%

Table 2. Cont.

Authors	FRPs	Fibre Fraction	Test Type	Standard	Mechanical Improvements *
Naranjo-Lozada et al. [50]	Short carbon fibre/polyamide; Continuous carbon fibre/polyamide	NR; 50 wt%	Tensile	ASTM D638	Short fibre: σ_t: 46%; E_t: 115% Continuous fibre: σ_t: 2826%; E_t: 3848%
Dickson et al. [51]	Continuous carbon fibre/polyamide; Continuous Kevlar fibre/polyamide; Continuous glass fibre/polyamide	8–11 vol%	Tensile; Flexural	ASTM D3039; ASTM D7264	Continuous carbon fibre/polyamide: σ_t: 254%; E_t: 1358%; σ_f: 260%; E_f: 1128% Continuous Kevlar fibre/polyamide: σ_t: 169%; E_t: 725%; σ_f: 200%; E_f: 527% Continuous glass fibre/polyamide: σ_t: 238%; E_t: 608%; σ_f: 369%; E_f: 297%

* σ_t, σ_f, σ_c, σ_s refer to the improvements of tensile strength, flexural strength, compressive strength and shear strength, compared to printed polymer matrix, respectively. E_t, E_f, E_c, E_s refer to the improvements of tensile modulus, flexural modulus, compressive modulus and shear modulus, compared to printed polymer matrix, respectively.

Although the mechanical performance of ME-printed fibre-reinforced polymers (FRPs) is improved, it is not comparable to that of FRPs manufactured via the traditional process, e.g., autoclaves [52]. Firstly, the fibre fraction is relatively low as it is limited by the ME process. The reported maximum fibre contents of short carbon fibre and continuous fibre are 30–40 wt% [45,53] and 50 wt% [21,48], respectively. Increasing fibre content increases the viscosity of the composite, typically leading to nozzle clog interrupting the printing process [45,54,55]. Furthermore, manufacturing defects such as voids may result in high porosity and partial bonded filaments. The porosity of ME-printed FRPs reported so far varies from 7% [56] to 22% [41], and the voids may cause premature failure. Ferreira et al. [33] found noticeable voids in printed short fibred carbon fibre-reinforced PLA and the tensile strength was 53.4 MPa, which was similar to that of pure PLA (54.7 MPa). Zhang et al. [57] found that the tensile strength of short carbon fibre-reinforced ABS was even lower than that of the matrix due to the high porosity. Furthermore, the partial bonded interfaces due to the imperfect coalescence of adjacent filaments may affect the strain distribution of a single layer. Increased strain was found by Christensen et al. [58] at the interfaces between adjacent filaments of transversely printed single-layer sodium alginate relative to the tensile load direction.

The porosity or the partial bonded interfaces may be sensitive to ME parameters such as nozzle temperature and printing speed [59–61]. The Markforged© series are provided with an optimised ME process for specific materials, while limiting users' access to ME parameters. Therefore, the printing process is stable, and the quality of fabricated samples is consistent. Due to this advantage, the Markforged company has an estimated USD 2.1 billion value, and their desktops printers have been adopted widely by the industry [62]. Despite the limited access to ME parameters, there are several options for layer thickness which determine the number of layers printed for a certain sample thickness. As interfaces take place between adjacent layers, the number of interfaces of printed samples can also be determined by layer thickness accordingly.

S. Sommacal et al. [63] analysed the microstructure of ME-printed short fibre-reinforced polyether ether ketone using micro-CT scanning. They found voids in the printed samples aligned in rows parallel to the printing direction. The porosity of printed parts ranged from 19% to 21%, which was independent of the printing parameters. The authors also found the internal microstructure, i.e., voids' distribution, and the number of interfaces between layers were determined by key printing parameters such as layer thickness and printing temperature. However, the relationship between the microstructure and the mechanical properties of ME-printed samples is not well understood. A quantitative investigation is important for designing and predicting the mechanical performance of ME-printed parts with partial bonded interfaces.

In this paper, the relationship between the ME process-induced microstructure and the properties of printed samples is elucidated by investigating the tensile properties of printed (short carbon fibre-reinforced) polyamide with a different number of interfaces. A Markforged desktop, i.e., Mark Two was used to print polyamide and short fibre-reinforced polyamide samples. The structure of the printed samples was obtained by cryofracture and was observed by scanning electron microscopy (SEM). Prior to the tensile test, influencing factors including porosity, crystallinity and fibre volume fraction were measured and compared. Injection moulding was used to fabricate samples without interfaces, and their tensile performance was used as a benchmark in investigating the effect of interfaces.

2. Methods

2.1. Materials

The materials were purchased from Markforged© and the polyamide (brand name: nylon) was PA6 indicated by the manufacturer datasheet [64]. Polyamide and short fibre-reinforced nylon (SFRN) filaments (brand name: Onyx) of 1.75 mm diameter from one spool were used to print all specimens via a Mark Two (Markforged, Somerville, MA, USA). The fibre volume fraction of the SFRN filament was about 9% determined by the densities of the matrix and composites measured by a He psycnometry (Accupyc II 1340, Micromeritics Ltd., Hexton, UK). The filaments were conditioned in a vacuum oven at 60 °C for at least 24 h prior to printing as the polyamide is sensitive to moisture [65].

2.2. Preparation of Dog-Bone Polyamide and SFRN Samples

2.2.1. ME-Printed Dog-Bone Samples

Tensile test specimens with dog-bone geometry were printed in accordance with the ASTM D638-14 type V [66]. The dimension of the gauge section was 9.53 × 3.18 mm^2. Consecutive layers were printed in alternating +45 and −45 (+135) degrees relative to the X-axis due to the default printing directions as shown in Figure 1. Specifically, the first layer was printed in +45 degrees, while the second layer was printed in −45 degrees. The specimens were printed in a rectangular infill pattern with 100% infill density, which may provide the highest tensile properties [50]. The printing temperature was pre-set at 275 °C and the default printing speed was estimated to be 17 mm/s. The setting values were kept constant during the printing process for all the specimens.

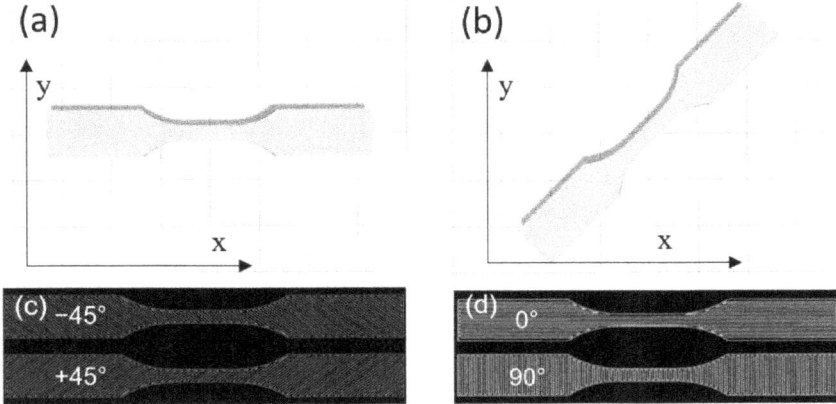

Figure 1. Schematic of sample layouts on the printing platform: (**a**) [+45, −45] and (**b**) [0, 90]. The raster's arrangement within two consecutive layers: (**c**) [+45, −45] and (**d**) [0, 90].

The above-mentioned printing directions were consistent and set by the printer-control software, i.e., Eiger. However, two raster patterns, i.e., [+45, −45] and [0, 90] relative to the tensile load direction, could be printed by rotating the specimen layout on the printing platform by 0° and 45°, respectively. Figure 1 shows the schematic of the specimen layout in Eiger and the raster patterns within the specimens. Specimens printed with the [+45, −45] and [0, 90] patterns had interfaces at +45/−45 and 0/90 degrees relative to the tensile load direction, respectively.

Layer thickness values were also set by Eiger, and all available options (0.1, 0.125 and 0.2 mm) were chosen to obtain different interface densities, i.e., the number of interfaces per unit sample thickness. The thickness values of samples were 2, 3 and 4 mm to obtain the integer number of total printed layers which equalled the sample thickness divided by layer thickness, and meanwhile the sample thickness was under 4 mm for the type V specimen for the tensile test in accordance with ASTM D638-14. The sample thickness was changed to investigate the effect of the interface density on the tensile properties of printed SFRN with the different total number of layers. The polyamides printed with 0.1, 0.125 and 0.2 mm layer thicknesses were named polyamide_0.1, polyamide_0.125 and polyamide_0.2, respectively. Similarly, the SFRN samples were named SFRN_0.1, SFRN_0.125 and SFRN_0.2.

2.2.2. Injection-Moulded Dog-Bone Samples

Materials from the same spool for the ME-printed samples were also used for injection moulding (HaakeMinijet II, ThermoFisher Scientific, Hampshire, UK). Pellets cut from the spools were melted in a barrel at 260 °C for 120 s prior to injection and then injected into a mould at 80 °C under the pressure of 600 bar for 10 s. A post-pressure of 100 bar was maintained for 60 s after the injection process. The injection-moulded samples were 3 mm thick and had the same geometry and dimensions as the ME-printed samples.

2.3. Porosity Measurement of Filaments and Fabricated Samples

The polyamide and the SFRN filaments were cylindrical, and they were cut into segments of about 1 m length. The cross-section diameter (d) of the segments (five specimens) was measured using a calliper. The bulk density (ρ_1) was calculated after weighing the mass (m) of the segments following Equation (1):

$$\rho_1 = \frac{m}{v} = \frac{m}{1 \times \pi \times \left(\frac{d}{2}\right)^2} \tag{1}$$

The bulk density of fabricated samples for the tensile test was measured by dividing the weighed sample mass by their envelope volume, which was evaluated by a computer-aided design software (SolidWorks® 2018). The density of pellets cut from the spools using a pelletizer machine (VariCut Pelletizer, Thermo Fisher Scientific Inc., Waltham, MA, USA) was measured by He pycnometry (Accupyc II 1340, Micromeritics Ltd., Hexton, UK). The pellets were assumed to have no voids, and their density referred to the true density (ρ_t) of the materials. The porosity (P) of the samples was measured due to the difference between the measured bulk densities (ρ_b) and the true density. The measurement was conducted as follows:

$$P\,[\%] = \left(1 - \frac{\rho_b}{\rho_t}\right) \times 100 \tag{2}$$

2.4. Crystallinity of Filaments and Fabricated Samples

The crystallinity and the melting temperature of the filaments and the fabricated samples for the tensile tests were measured by differential scanning calorimetry (DSC, Discovery DSC, TA Instruments, Newcastle, UK). Samples consisting of five specimens with about 3–5 mg were heated in a nitrogen atmosphere from 20 °C to 275 °C, and then cooled to 20 °C prior to a second heating to 275 °C. Both the heating and the cooling rates were 5 °C min^{-1}. The crystallinity (χ_c) was measured as below:

$$\chi_c = \frac{\Delta H_m}{(1-\alpha)\Delta H^\varnothing} \tag{3}$$

where ΔH_m refers to the melting enthalpy of the samples measured by calculating the area under the endothermic peak shown on the heating curves. $\Delta H^\varnothing$ denotes the melting enthalpy of pure polyamide, taken as 230 J/g [65]. The symbol α refers to the mass fraction of carbon fibre, and it was taken as zero for the polyamide samples.

2.5. Tensile Properties of Fabricated Polyamide and SFRN

Tensile tests on the fabricated dog-bone samples were conducted on an Instron universal machine (Model 5960, Instron, Norwood, MA, USA) equipped with a 10 kN load cell following the ASTM D638-14. A speckle pattern was applied to the gauge section of the specimens using an ink stamp. The strain of the specimens subjected to 10 mm min^{-1} displacement was measured by monitoring the patterns' movement using a non-contact video extensometer (iMetrum Video Gauge, Bristol, UK). Each sample consisted of five specimens, and the tensile test was repeated twice for each sample.

2.6. Shear-Lag Model

The tensile properties of the short fibre-reinforced composites can be described by the shear-lag theory developed by Cox and Krenchel [67,68]. The assumptions of the theory are as follows: (1) both fibre and matrix deform elastically, (2) the fibre/matrix interface is intact and (3) no load at the fibre ends. The shear-lag theory generally underestimates the stiffness of short fibre composites as it neglects stress at the fibre ends [69]. The predicted tensile modulus (E_c) of composites based on the shear-lag model is given by

$$E_c = \eta_0 \eta_L V_f E_f + \left(1 - V_f\right) E_m \tag{4}$$

where E_m and E_f refer to the tensile modulus of matrix and fibre, respectively. V_f refers to the fibre volume fraction of composites. η_L is a length correction factor due to the finite length of fibre written as

$$\eta_L = 1 - \frac{\tanh\left(\frac{\beta L}{2d}\right)}{\frac{\beta L}{2d}} \tag{5}$$

$$\beta = \sqrt{\frac{2E_m}{E_f(1+\nu_m)\ln\left(\frac{1}{V_f}\right)}} \tag{6}$$

where L and d refer to the length and the diameter of the fibre, respectively. ν_m is the Poisson's ratio of matrix. η_0 is a fibre-orientation factor referring to the fraction of fibre ($V_{f,n}$) aligning at angle θ_n relative to the tensile load written as

$$\eta_0 = \frac{\sum_n V_{f,n} \cos^4 \theta_n}{\sum_n V_{f,n}} \tag{7}$$

2.7. Structure Morphology of Fabricated Polyamide and SFRN

The fabricated samples were cryofractured by placing them in liquid nitrogen for 10 min, and then the cross-section structure was observed by scanning electron microscopy (SEM, Hitachi S-3700N, Tokyo, Japan). The samples were firstly mounted onto aluminum stubs with carbon tabs and then coated with Au (Agar Auto Sputter Coater, Essex, UK) with a coating current of 40 mA for 60 s. The short carbon fibre in SFRN filaments was observed by optical microscope (Axioscope, Zeiss, Germany), and its dimension was measured by ImageJ software (Rasband, W.S., ImageJ, USA).

3. Results and Discussion

3.1. Thermal Properties of Filaments and Fabricated Samples

The crystallinity of the filaments and the fabricated samples was measured based on the DSC curves shown in Figure 2. The DSC results for the ME-printed samples are shown by a representative curve for simplicity, as the DSC curves of the printed samples are similar. All the heating curves show an endothermic peak corresponding to the melting process of polyamide. The peak point is about 200 °C referring to the melting temperature of the samples, and the value is close to the results from the literature for polyamide 6 [65,70].

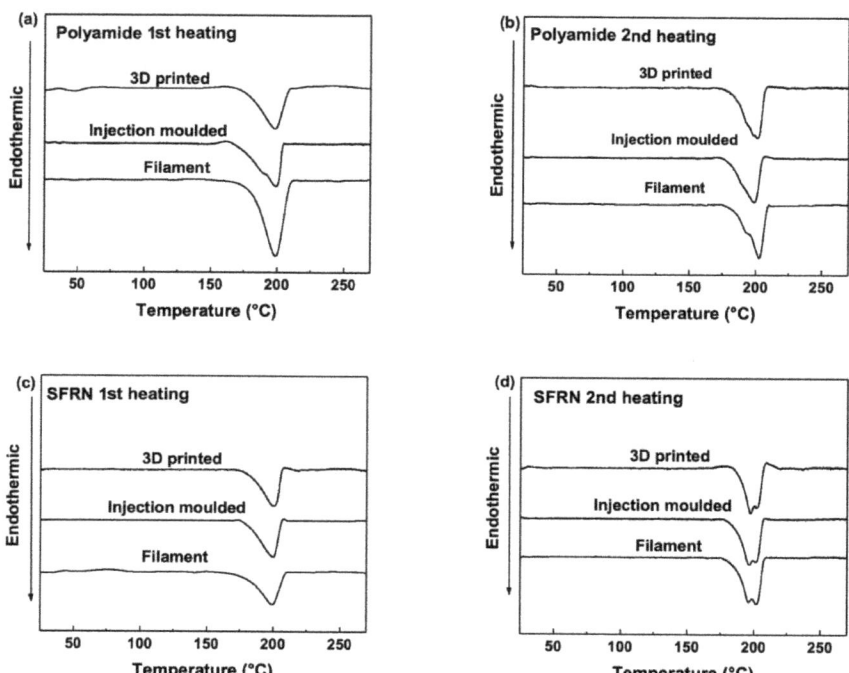

Figure 2. DSC curves of polyamide during (**a**) first heating and (**b**) second heating; SFRN composites during (**c**) first heating and (**d**) second heating.

Table 3 summarises the crystallinity and the melting temperature (T_m) of the filaments and the fabricated samples based on the heating curves. Both the ME-printed samples and the injection-moulded samples have similar crystallinity. Specifically, the crystallinity of the ME-printed samples based on the 1st heating curves is 16–19%, whereas that of the injection-moulded samples is 18%. Furthermore, the polyamide filaments have 35% crystallinity, which is higher than that of the printed polyamide. The ME process is a rapid manufacturing process and the printed samples may cool down to ambient temperature in seconds [71], resulting in insufficient time for crystallisation. After the thermal history induced by the manufacturing process was removed by the 1st heating, the printed samples had similar crystallinity as that of filaments on the 2nd heating curves.

Table 3. The crystallinity and melting temperature of filaments, fabricated polyamide and SFRN samples.

Raster Pattern	Layer Thickness (mm)		Polyamide		SFRN	
			T_m (°C)	Crystallinity (%)	T_m (°C)	Crystallinity (%)
[0, 90]	0.1	1st heating	200	18	199	17
		2nd heating	202	20	200	18
	0.125	1st heating	200	16	200	17
		2nd heating	201	17	200	17
	0.2	1st heating	199	16	199	17
		2nd heating	200	19	200	18
[+45, −45]	0.1	1st heating	199	19	197	17
		2nd heating	201	20	200	20
	0.125	1st heating	200	17	197	16
		2nd heating	201	18	200	18
	0.2	1st heating	201	18	197	18
		2nd heating	202	19	199	20
Filament		1st heating	200	35	199	21
		2nd heating	201	19	201	20
Injection-moulded		1st heating	199	18	199	18
		2nd heating	198	20	199	20

The first heating curves reveal the crystallinity determined by the manufacturing process. The similar crystallinity of the printed samples indicates the crystallisation is not influenced by the raster patterns and the layer thickness. All the ME-printed samples and the injection-moulded samples have similar crystallinity, indicating that the factor of crystallinity would not result in any variation in the tensile properties of the fabricated samples.

3.2. The Structure of the Fabricated Samples

Firstly, the porosity of the polyamide and the SFRN filaments is 2–3%, while the printed polyamide and the SFRN samples have the porosities of 5% and 10%, respectively. The higher porosity indicates the printing process may induce extra voids in the printed samples. Therefore, the structure of the printed samples was analysed to investigate the voids induced by the printing process.

Figure 3 shows the cryofracture surface of the printed 3 mm thick [0, 90] polyamide samples with 0.1, 0.125 and 0.2 mm layer thicknesses. Triangular voids with 29 μm to 47 μm edge length were found between adjacent filaments due to their partial coalescence. The reason for this incomplete fusion of filaments is the rapid cooling and solidification of the extruded materials. Therefore, the printed samples have partial bonded interfaces between filaments induced by the manufacturing process. The partial bonded interfaces may exist along the filament's longitudinal direction, thereby resulting in micro-size channels. The number of the interfaces is a function of the number of printed layers, which is determined by layer thickness. Specifically, 0.1 mm layer thickness has the greatest number of interfaces when compared to 0.125 or 0.2 mm layer thickness for the same sample thickness, leading to the greatest interface density.

Compared with the printed polyamide samples, the structure of the printed [0, 90] SFRN samples (see Figure 4) shows large-size voids (edge length: 75~93 μm) taking place between the filaments. Voids are not only triangular in shape but form quadrilateral shapes resulting from the joining of two triangular voids. This indicates the manufacturing process-induced interfaces of the printed SFRN samples are inferior to those of the printed polyamide. Consequently, the printed SFRN samples have higher porosity (10%) than that (5%) of the printed polyamide. The SFRN filaments in 0° direction are more distinguishable due to the inferior interfaces. The number of 0° filaments at the gauge section (width: 3.18 mm) is eight, which is independent of sample thickness and layer thickness. Therefore, interfaces within a layer are consistent, and the interface density is only determined by layer thickness. Furthermore, fibres with round cross-sections were observed in 0° SFRN filaments. This indicates fibres relatively align along filaments' longitudinal direction (printing direction).

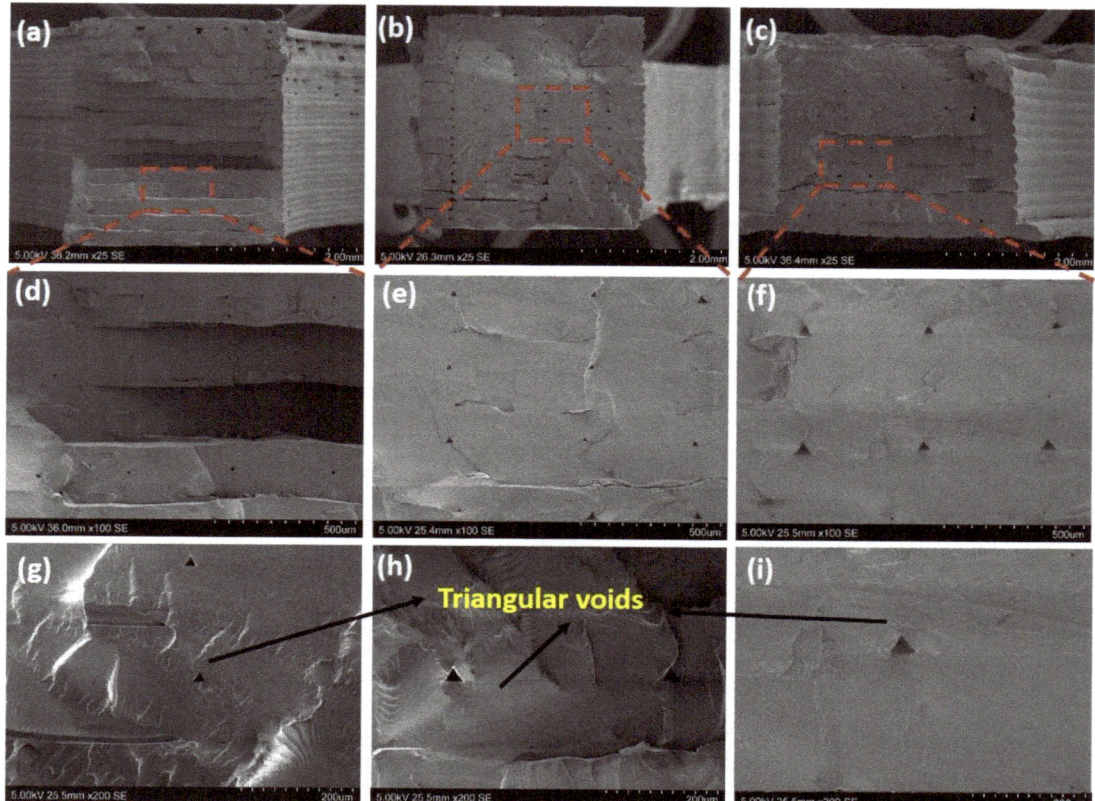

Figure 3. The structure of the printed [0, 90] polyamide samples at different magnifications: 25×: (**a**) polyamide_0.1, (**b**) polyamide_0.125, (**c**) polyamide_0.2; 100×: (**d**) polyamide_0.1, (**e**) polyamide_0.125, (**f**) polyamide_0.2; 200×: (**g**) polyamide_0.1, (**h**) polyamide_0.125, (**i**) polyamide_0.2.

It is worth mentioning that the porosity of the printed SFRN with the same sample thickness is independent of layer thickness (interface density). On the one hand, the lower layer thickness determines the greater number of layers printed within the sample, which may result in the greater number of heating cycles. The thermal conduction from the upper layers can improve the filaments' fusion of bottomed layers [71,72] and lead to lower porosity of the bottom part [63]. However, the greater number of partial bonded interfaces could take place when more layers are printed.

The structure of printed SFRN samples with 2 mm and 4 mm shown in Figure 5 also shows the inferior interfaces in triangular and quadrilateral geometry. The porosity of 2 mm and 4 mm thick samples is 12% and 8%, respectively, which was measured using the method described in Section 2.3. The lower porosity might have resulted from the slower cooling of the 4 mm thick sample, as the cooling rate depends on the sample thickness. The slower cooling rate may lead to the better fusion and coalescence of the printed filaments [71]. The influences of key printing parameters on material fusion and the resulting porosity of printed parts need more research efforts.

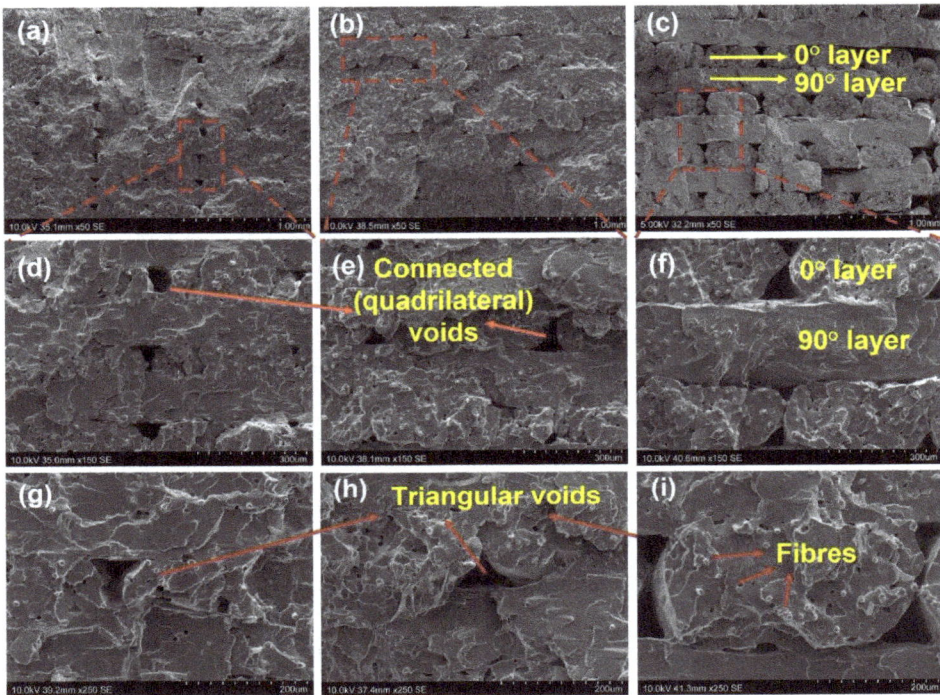

Figure 4. The structure of the printed [0, 90] composite samples with different layer thickness: 50×: (a) 0.1 mm, (b) 0.125 mm, (c) 0.2 mm; 150×: (d) 0.1 mm, (e) 0.125 mm, (f) 0.2 mm; 250×: (g) 0.1 mm, (h) 0.125 mm, (i) 0.2 mm.

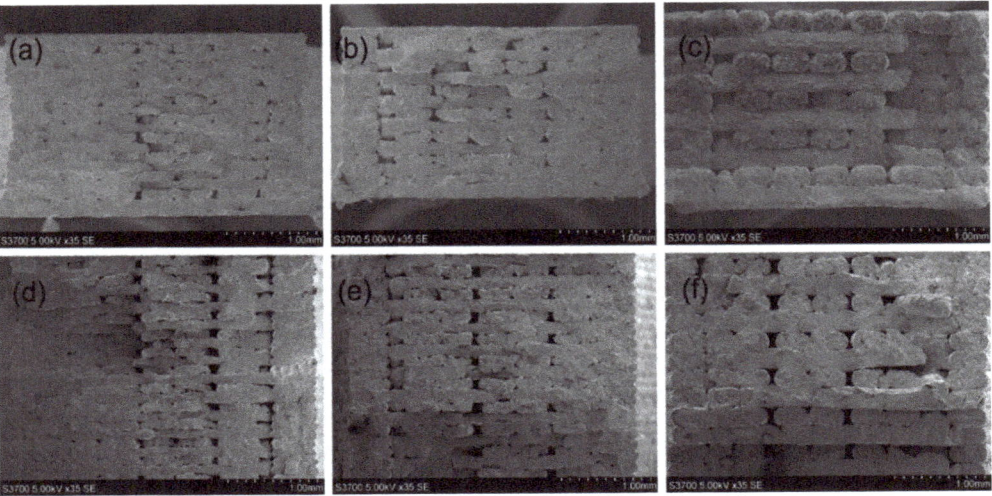

Figure 5. SEM images (35×) of the structure of the printed [0, 90] composite samples with different thickness: 2 mm: (a) SFRN_0.1, (b) SFRN_0.125, (c) SFRN_0.2; 4 mm: (d) SFRN_0.1, (e) SFRN_0.125, (f) SFRN_0.2.

The porosities of 2 mm and 4 mm thick polyamide samples are similar (about 5%). The printed polyamide has superior interfaces due to better filaments' fusion, and therefore the porosity is less dependent on the sample thickness or layer thickness. The structure of injection-moulded samples (see Figure 6) was also investigated for comparison with the printed samples. Voids as well as the partial bonded interfaces were not found in the samples and the resulting porosity is less than 1%. Therefore, the number of interfaces of injection-moulded samples would be assumed as zero.

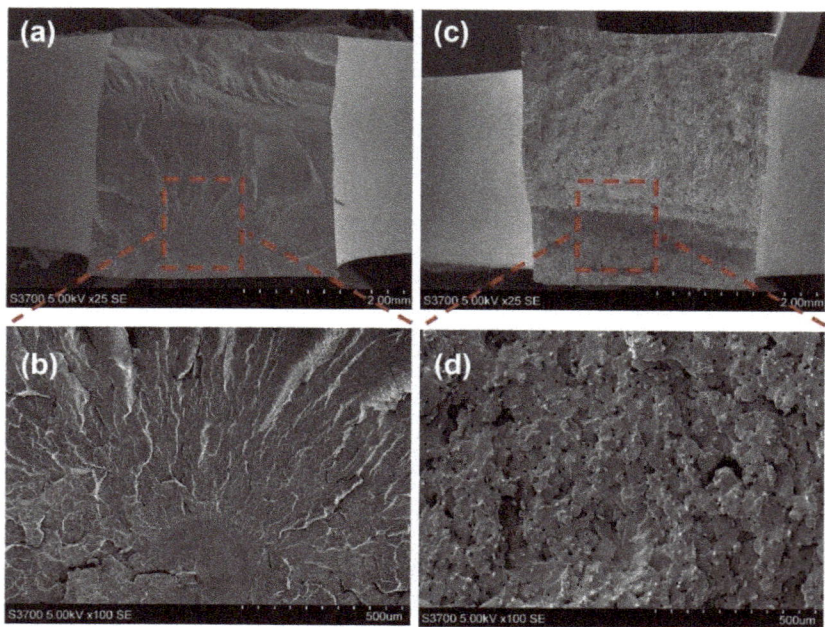

Figure 6. The structure of injection-moulded polyamide: (**a**) 25× and (**b**) 100×; injection-moulded SFRN: (**c**) 25× and (**d**) 100×.

3.3. The Tensile Properties of the Fabricated Samples

3.3.1. Comparison between the Injection-Moulded Samples and the Printed Samples

The stress–strain curves of the fabricated samples with 3 mm thickness are plotted in Figure 7. The standard deviation of each sample (six specimens) is presented by the upper and lower bounds of the stress–strain bands. Strain values obtained from the video extensometer and tensile properties (yield stress and tensile modulus) were determined from the stress–strain curves.

The yield stress is considered as the point on the stress–strain curve where the stress did not increase with increasing strain (zero slope). The yield stress of SFRN refers to yield strength at the yield point, and all SFRN specimens broke at the gauge section area. The tensile modulus is determined by the slope of linear regression lines in the strain range of 0.05% and 0.25%. The measured tensile properties of the fabricated samples are summarised in Table 4.

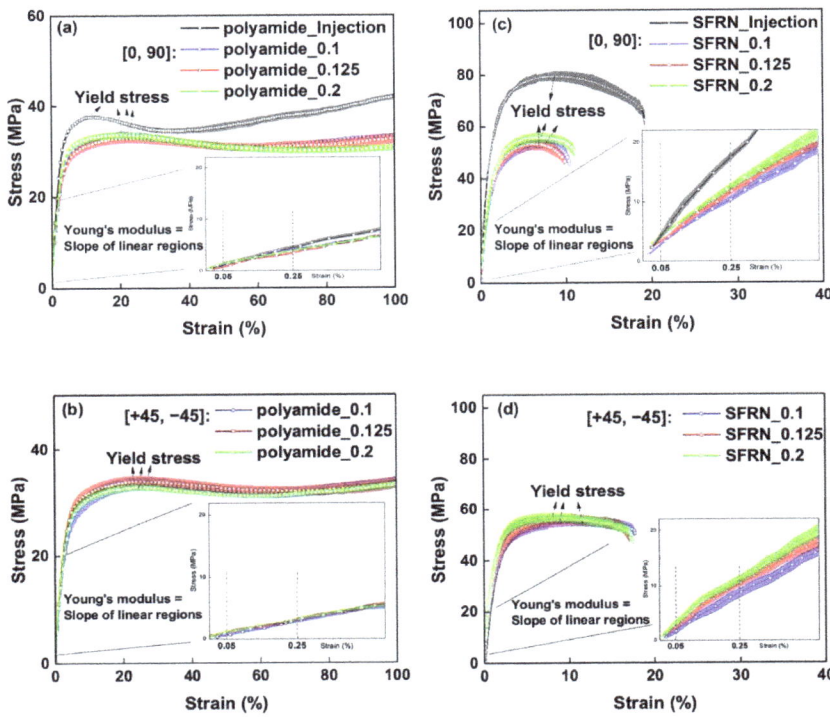

Figure 7. The strain–stress curves of the fabricated samples with 3 mm thick: (**a**) injection-moulded and [0, 90] polyamide; (**b**) [+45, −45] polyamide; (**c**) injection-moulded and [0, 90] SFRN; (**d**) [+45, −45] SFRN.

Table 4. The tensile properties of the printed and injection-moulded samples.

			Polyamide		SFRN	
Thick (mm)	Layer Thickness (mm)	Raster Pattern	Yield Stress (MPa)	Tensile Modulus (GPa)	Yield Stress (MPa)	Tensile Modulus (GPa)
2	0.1	[0, 90]	32.5 ± 0.6	1.19 ± 0.04	51.5 ± 3.6	3.25 ± 0.10
		[+45, −45]	30.3 ± 0.2	1.11 ± 0.05	53.7 ± 1.2	3.00 ± 0.08
	0.125	[0, 90]	32.3 ± 0.7	1.17 ± 0.02	53.8 ± 1.6	3.57 ± 0.19
		[+45, −45]	30.4 ± 0.5	1.16 ± 0.10	57.3 ± 2.1	3.16 ± 0.08
	0.2	[0, 90]	33.8 ± 0.4	1.21 ± 0.07	55.2 ± 1.3	4.08 ± 0.21
		[+45, −45]	31.2 ± 0.9	1.14 ± 0.07	57.5 ± 2.6	3.56 ± 0.10
3	0.1	[0, 90]	33.5 ± 1.1	1.12 ± 0.06	54.5 ± 2.6	3.88 ± 0.18
		[+45, −45]	34.3 ± 0.9	1.18 ± 0.11	56.8 ± 1.6	3.66 ± 0.17
	0.125	[0, 90]	33.2 ± 0.8	1.13 ± 0.09	54.7 ± 1.9	4.03 ± 0.22
		[+45, −45]	34.0 ± 0.9	1.12 ± 0.09	56.7 ± 2.1	3.82 ± 0.24
	0.2	[0, 90]	33.3 ± 1.5	1.17 ± 0.07	56.8 ± 2.4	4.57 ± 0.27
		[+45, −45]	34.1 ± 0.4	1.16 ± 0.11	58.5 ± 1.7	4.11 ± 0.21
	Injection-moulded	-	37.6 ± 0.4	1.43 ± 0.22	78.1 ± 3.5	6.67 ± 0.22
4	0.1	[0, 90]	34.3 ± 0.5	1.19 ± 0.05	55.7 ± 3.7	4.25 ± 0.19
		[+45, −45]	32.9 ± 0.4	1.19 ± 0.09	58.8 ± 1.9	4.13 ± 0.30
	0.125	[0, 90]	32.9 ± 0.7	1.20 ± 0.03	56.6 ± 1.7	4.48 ± 0.25
		[+45, −45]	32.5 ± 0.4	1.11 ± 0.07	60.1 ± 2.4	4.25 ± 0.24
	0.2	[0, 90]	32.9 ± 0.7	1.16 ± 0.08	59.5 ± 3.7	4.90 ± 0.23
		[+45, −45]	34.1 ± 0.4	1.21 ± 0.08	59.8 ± 4.2	4.57 ± 0.21

Firstly, the tensile properties of the injection-moulded samples are compared to the printed samples. The yield stress and the tensile modulus of the injection-moulded polyamide are 37.6 MPa and 1.43 GPa, respectively. The yield stress and the tensile modulus of the printed polyamide are 9–12% and 17–22% lower compared to the injection-moulded sample, respectively. The printed polyamide has higher porosity and the partial bonded interfaces, which may contribute to the lower tensile properties as interfaces could cause non-uniform strain distribution and lead to premature failure. Furthermore, the failure strain of the injection-moulded SFRN is about 20 ± 1%, whereas the printed [0, 90] and [+45, −45] SFRN samples fail at about 10 ± 1% and 17 ± 2%, respectively. The lower failure strain of printed samples indicates premature failure took place due to the induced interfaces. The result also indicates the interfaces in [0, 90] SFRN have a more significant impact on premature failure compared to the interface in [+45, −45] SFRN. The transverse interfaces relative to tensile load direction in the printed [0, 90] may act as cracks leading to stress concentration and premature failure.

The yield stress and the tensile modulus of the injection-moulded SFRN are 78.1 MPa and 6.67 MPa, respectively. Compared to the injection-moulded samples, the yield stress and the tensile modulus are 25–30% and 31–45% lower, respectively. The printed SFRN samples have a more significant reduction in tensile properties than the printed polyamide, due to the inferior partial bonded interfaces. The contact area of adjacent SFRN filaments is reduced by the larger-size interfaces as well as the connected interfaces, resulting in a more substantial decrease in load-transfer efficiency. Therefore, the influence of the partial bonded interfaces is more significant on the tensile properties of the printed SFRN.

Table 4 also summarises the tensile properties of the printed samples with 2 mm and 4 mm thickness. The yield stress and the tensile modulus of the printed polyamide are 30.3–34.1 MPa and 1.11–1.21 GPa, respectively. The tensile properties of the printed polyamide are close and not influenced by sample thickness. However, the tensile modulus of the printed SFRN increases with the sample thickness. For example, the tensile modulus of [0, 90] SFRN_0.2 with 4 mm thickness is 29.6% higher than that of [0, 90] SFRN_0.2 with 2 mm thickness. The reason might be the porosity of the printed SFRN decreasing with the sample thickness. The higher porosity of 2 mm thick SFRN samples may result in the lower tensile modulus.

3.3.2. Shear-Lag Theory Analysis

The shear-lag effect happens in discontinuous fibre composites in which the fibre and polymer matrix have an apparent mismatch in the modulus. According to shear-lag theory, the efficiency of short carbon fibre is limited due to its discontinuity, and axial stress is transferred to fibre by shear stress at the fibre/matrix interface. Consequently, the stiffness of composites degrades due to the inefficiency of short fibre.

The measured fibre volume fraction of fabricated SFRN is about 9%. The length and the diameter of short fibre are 140 ± 10 μm and 7 ± 0.5 μm based on the microscope images of fibres shown in Figure 8. The $_0$ values for printed SFRN and injection SFRN are taken as 0.56 and 0.80 based on the literature [69,73], respectively. The tensile modulus of carbon fibre is 230 GPa [74] and the tensile modulus of polyamide is 1.4 GPa due to the tensile result of the injection-moulded polyamide. The Poisson's ratio of polyamide is assumed to be 0.39 as suggested in the literature [75].

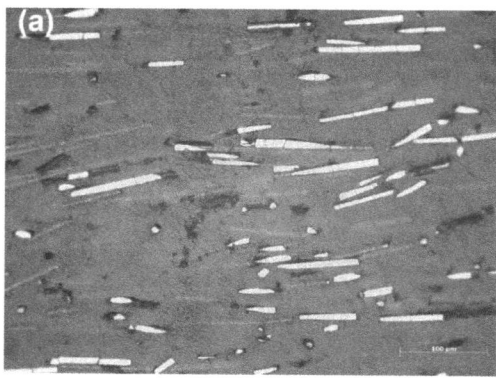 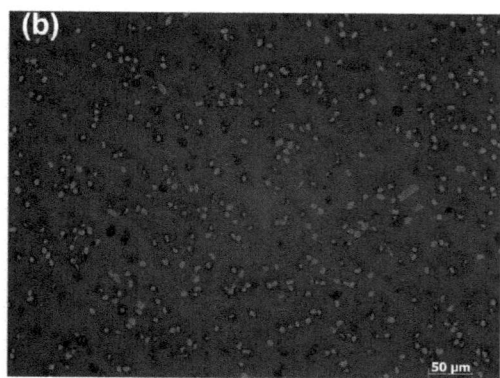

Figure 8. The microscope images of short carbon fibres in SFRN filaments: (**a**) longitudinal; (**b**) axial.

The predicted tensile modulus of the injection-moulded SFRN based on the shear-lag model is 6.36 GPa, which is in good agreement (<5% difference) with the experimental value (6.67 GPa). This result indicates the loss on the tensile modulus of the fabricated SFRN due to the limited fibre length and the reduced efficiency of fibre reinforcement. However, the predicted tensile modulus (4.84 GPa) is greater than the experimental measured modulus of the printed SFRN, except the 4 mm SFRN_0.2 sample. The tensile modulus of 2 mm [+45, −45] SFRN_0.1 is almost 38% lower than the predicted modulus. The reasons could be the printing-induced partial bonded interfaces in SFRN which are not considered in the theory. Furthermore, the shear-lag effect between the fibre/matrix may not be sufficient in explaining the variation on the tensile modulus of the printed SFRN with multiple partial bonded interfaces.

3.3.3. The Effect of Interfaces on the Tensile Modulus of the Printed Samples

As discussed above, inferior interfaces with a larger size were found between adjacent SFRN filaments compared to the polyamide, resulting in a more significant reduction in the tensile properties. The partial bonded interfaces may continuously distribute along the filament direction, and therefore the printed [0, 90] and [−45, +45] samples may consist of 90° and 45° interfaces relative to the tensile load, respectively. The load transfer in laminates subject to the tensile load is degraded due to the off-axis pre-existing defects where the load carried by the layers is transferred to the neighbouring layers. The degraded load transfer resulting in the reduction in stiffness is described by various approaches such as shear-lag models [76–78], McCartney's models [79] and crack-opening-displacement models [80]. The schematic of the transverse interfaces in [0, 90] samples is shown in Figure 9. The transverse interfaces may also influence the load transfer of the printed samples subject to tensile load, and then lead to the reduction in the tensile modulus.

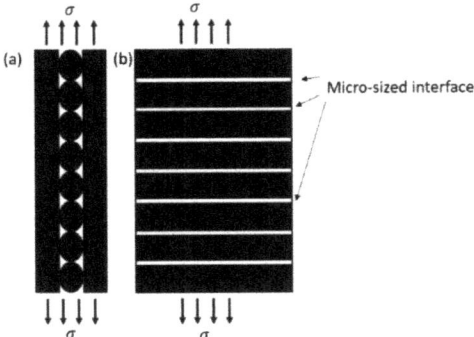

Figure 9. The schematic of transverse interfaces in [0, 90] samples: (**a**) edge view; (**b**) top view.

The average tensile modulus of the printed polyamide summarised in Table 4 is about 1.15 GPa, which is independent of the layer thickness and raster pattern. However, the tensile modulus of the printed SFRN samples increases with the layer thickness. For example, the tensile modulus of 3 mm thick [0, 90] SFRN_0.1 is 4.9% and 17.8% lower than that of SFRN_0.125 and SFRN_0.2, respectively. The tensile modulus of 2 mm and 4 mm thick SFRN samples also increases with the layer thickness.

Layer thickness determines the number of printed layers for certain sample thicknesses, i.e., more interfaces (higher interface density) may take place. The interfaces may result in the reduction in tensile modulus, and a higher interface density may lead to a more significant reduction. The number of interfaces as well as interface density are summarised in Table 5. The relationship between interface density and tensile modulus is quantitatively correlated in Figure 10. The correlation coefficients are summarised in Table 5.

Table 5. The interface density of the printed SFRN samples.

Sample Thick (T_S, mm)	Layer Thickness (T_L, mm)	Number of Interfaces (T_S/T_L-1)	Interface Density ($\frac{1}{T_L}-\frac{1}{T_S}$), /mm	Linear Correlation [0, 90]	Linear Correlation [+45, −45]
2	0.1	20	9.50	0.999	0.983
	0.125	16	7.50		
	0.2	10	4.50		
3	0.1	30	9.67	0.960	0.998
	0.125	24	7.67		
	0.2	15	4.67		
4	0.1	40	9.75	0.997	0.980
	0.125	32	7.75		
	0.2	20	4.75		

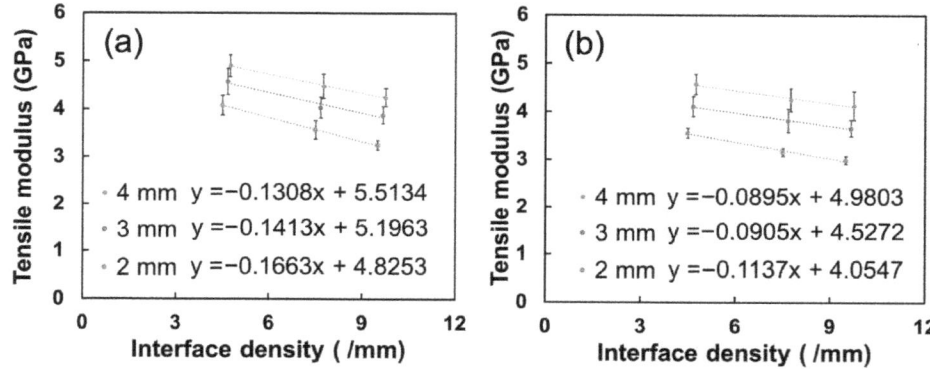

Figure 10. The tensile modulus of the printed SFRN samples as a function of interface density: (a): [0, 90] SFRN; (b): [+45, −45] SFRN.

The correlation suggests that the tensile modulus of the printed SFRN samples decreases linearly with the interface density as described below:

$$E = \alpha \cdot I_d + E_0 \tag{8}$$

where E and E_0 refer to the tensile modulus of the printed SFRN samples and the tensile modulus of the printed SFRN having no interfaces or porosity, respectively. I_d refers to interface density and α is a coefficient. Firstly, the coefficient α for [0, 90] and [+45, −45] SFRN is about −0.16~−0.13 and −0.09~−0.11, respectively. The negative coefficients reveal the interface density has a negative influence on the tensile modulus of the SFRN samples. The [0, 90] SFRN samples have a relatively higher coefficient α compared to [+45, −45]. This indicates the transverse partial bonded interfaces relative to load direction in [0, 90] samples could degrade the load transfer more significantly, resulting in the greater loss of stiffness. Secondly, the y-intercept refers to the tensile modulus (E_0) of printed samples without interfaces. The E_0 of [+45, −45] SFRN samples ranges from about 4.1 GPa to 5.0 GPa, which is relatively lower compared to [0, 90] SFRN (4.8 GPa to 5.5 GPa), which may have resulted from the anisotropic performance due to the filament orientations. Fires were found to relatively align along the filaments' orientation, and therefore the printed SFRN samples may exhibit an anisotropic performance. The tensile modulus of the fibre-reinforced polymer composite with [0, 90] layup sequence has been reported to be higher compared to [−45, +45] layup [74].

It is worthy to mention that the E_0 of the printed SFRN decreases with sample thickness due to the increasing porosity. The 2, 3 and 4 mm SFRNs have 7%, 10% and 12% void contents, respectively. This also explains that the E_0 of the printed SFRN is still lower than that of the injection-moulded SFRN (6.67 GPa) as the samples have less than 1% porosity. As the correlations are based on the limited range of interface density, more research work would be helpful to improve accuracy and applicability.

4. Conclusions

This study investigates the influence of printing-induced defects at the interface on the tensile properties of (short carbon fibre-reinforced) polyamide. A correlation between the interface density and the tensile properties is analysed and quantified. The main achievements are summarised below:

- Firstly, relying on the commercial Markforged printer with limited access to processing parameters except layer thickness, the printed samples exhibit consistent quality including porosity, crystallinity and fibre volume fraction.
- Secondly, printing process-induced interfaces are found in both printed polyamide and SFRN samples. The partial bonded interfaces are distributed at the interface between printed filaments. The interface density increases when layer thickness decreases from 0.2 mm to 0.1 mm. Compared to the printed polyamide, the printed SFRN samples have inferior interfaces with a larger size.
- Consequently, the tensile properties of the printed SFRN are more significantly lower than those of the injection-moulded SFRN. The printed polyamide exhibits a relatively lower yield stress (9–12%) and tensile modulus (17–22%) compared to the injection-moulded sample, whereas the yield stress and the tensile modulus of the printed SFRN are 25–30% and 31–45% lower, respectively.
- Furthermore, the tensile modulus of the printed SFRN decreases as a function of interface density, while the tensile modulus of the printed polyamide is independent of interface density. The tensile modulus of the 3 mm thick [0, 90] SFRN_0.1 is 4.9% and 17.8% lower compared to SFRN_0.125 and SFRN_0.2, respectively. A shear-lag model is found to predict the tensile modulus in good agreement with the experimentally measured modulus of the injection-moulded SFRN. However, the experimental modulus of the printed SFRN is lower than the predicted modulus due to the printing-induced interfaces in SFRN.
- Lastly, the quantitative correlation between the tensile modulus of the SFRN and the interface density is analysed. An empirical model is developed based on data fitting, and the model shows that the tensile modulus of the printed SFRN decreases with interface density following a linear function. This result suggests that the quantitative degradation of the stiffness due to interfaces should be considered when designing 3D printed parts for engineering applications. The microstructure can be improved to achieve a maximum and interface-independent mechanical performance.

Author Contributions: Conceptualisation, Y.H. and A.P.; methodology, Y.H.; validation, Y.H. and A.P.; formal analysis, Y.H.; investigation, Y.H.; resources, A.P.; data curation, Y.H.; writing—original draft preparation, Y.H.; writing—review and editing, Y.H. and A.P.; visualization, Y.H.; supervision, A.P.; project administration, Y.H. All authors have read and agreed to the published version of the manuscript.

Funding: This research received no external funding.

Institutional Review Board Statement: Not applicable.

Informed Consent Statement: Not applicable.

Data Availability Statement: The data presented in this study are available on request from the corresponding author.

Conflicts of Interest: The authors declare no conflict of interest.

References

1. Brenken, B.; Barocio, E.; Favaloro, A.; Kunc, V.; Pipes, R.B. Fused Filament Fabrication of Fiber-Reinforced Polymers: A Review. *Addit. Manuf.* **2018**, *21*, 1–16. [CrossRef]
2. Joshi, S.C.; Sheikh, A.A. 3D Printing in Aerospace and Its Long-Term Sustainability. *Virtual Phys. Prototyp.* **2015**, *10*, 175–185. [CrossRef]
3. Shahrubudin, N.; Lee, T.C.; Ramlan, R. An Overview on 3D Printing Technology: Technological, Materials, and Applications. *Procedia Manuf.* **2019**, *35*, 1286–1296. [CrossRef]
4. Yao, X.; Luan, C.; Zhang, D.; Lan, L.; Fu, J. Evaluation of Carbon Fiber-Embedded 3D Printed Structures for Strengthening and Structural-Health Monitoring. *Mater. Des.* **2017**, *114*, 424–432. [CrossRef]
5. Chacón, J.M.; Caminero, M.A.; García-Plaza, E.; Núñez, P.J. Additive Manufacturing of PLA Structures Using Fused Deposition Modelling: Effect of Process Parameters on Mechanical Properties and Their Optimal Selection. *Mater. Des.* **2017**, *124*, 143–157. [CrossRef]
6. Yu, N.; Sun, X.; Wang, Z.; Zhang, D.; Li, J. Effects of Auxiliary Heat on the Interlayer Bonds and Mechanical Performance of Polylactide Manufactured through Fused Deposition Modeling. *Polym. Test.* **2021**, *104*, 107390. [CrossRef]
7. Cicala, G.; Giordano, D.; Tosto, C.; Filippone, G.; Recca, A.; Blanco, I. Polylactide (PLA) Filaments a Biobased Solution for Additive Manufacturing: Correlating Rheology and Thermomechanical Properties with Printing Quality. *Materials* **2018**, *11*, 1191. [CrossRef]
8. Tymrak, B.M.; Kreiger, M.; Pearce, J.M. Mechanical Properties of Components Fabricated with Open-Source 3-D Printers under Realistic Environmental Conditions. *Mater. Des.* **2014**, *58*, 242–246. [CrossRef]
9. D'Amico, A.A.; Debaie, A.; Peterson, A.M. Effect of Layer Thickness on Irreversible Thermal Expansion and Interlayer Strength in Fused Deposition Modeling. *Rapid Prototyp. J.* **2017**, *23*, 943–953. [CrossRef]
10. Ziemian, S.; Okwara, M.; Ziemian, C.W. Tensile and Fatigue Behavior of Layered Acrylonitrile Butadiene Styrene. *Rapid Prototyp. J.* **2015**, *21*, 270–278. [CrossRef]
11. Lee, J.; Huang, A. Fatigue Analysis of FDM Materials. *Rapid Prototyp. J.* **2013**, *19*, 291–299. [CrossRef]
12. Zhou, Y.G.; Zou, J.R.; Wu, H.H.; Xu, B.P. Balance between Bonding and Deposition during Fused Deposition Modeling of Polycarbonate and Acrylonitrile-Butadiene-Styrene Composites. *Polym. Compos.* **2020**, *41*, 60–72. [CrossRef]
13. Salazar-Martín, A.G.; Pérez, M.A.; García-Granada, A.A.; Reyes, G.; Puigoriol-Forcada, J.M. A Study of Creep in Polycarbonate Fused Deposition Modelling Parts. *Mater. Des.* **2018**, *141*, 414–425. [CrossRef]
14. Puigoriol-Forcada, J.M.; Alsina, A.; Salazar-Martín, A.G.; Gomez-Gras, G.; Pérez, M.A. Flexural Fatigue Properties of Polycarbonate Fused-Deposition Modelling Specimens. *Mater. Des.* **2018**, *155*, 414–421. [CrossRef]
15. Toro, E.V.D.; Sobrino, J.C.; Mart, A.; Ayll, J. Investigation of a Short Carbon Fibre-Reinforced Polyamide and Comparison of Two Manufacturing Processes: Fused Deposition Modelling (FDM) and Polymer Injection Moulding (PIM). *Materials* **2020**, *13*, 672. [CrossRef]
16. Zhang, X.; Fan, W.; Liu, T. Fused Deposition Modeling 3D Printing of Polyamide-Based Composites and Its Applications. *Compos. Commun.* **2020**, *21*, 100413. [CrossRef]
17. Gao, X.; Zhang, D.; Wen, X.; Qi, S.; Su, Y.; Dong, X. Fused Deposition Modeling with Polyamide 1012. *Rapid Prototyp. J.* **2019**, *25*, 1145–1154. [CrossRef]
18. Song, Y.; Li, Y.; Song, W.; Yee, K.; Lee, K.Y.; Tagarielli, V.L. Measurements of the Mechanical Response of Unidirectional 3D-Printed PLA. *Mater. Des.* **2017**, *123*, 154–164. [CrossRef]
19. Ning, F.; Cong, W.; Qiu, J.; Wei, J.; Wang, S. Additive Manufacturing of Carbon Fiber Reinforced Thermoplastic Composites Using Fused Deposition Modeling. *Compos. Part B Eng.* **2015**, *80*, 369–378. [CrossRef]
20. Love, L.J.; Kunc, V.; Rios, O.; Duty, C.E.; Elliott, A.M.; Post, B.K.; Smith, R.J.; Blue, C.A. The Importance of Carbon Fiber to Polymer Additive Manufacturing. *J. Mater. Res.* **2014**, *29*, 1893–1898. [CrossRef]
21. Omuro, R.; Ueda, M.; Matsuzaki, R.; Todoroki, A.; Hirano, Y. Three-Dimensional Printing of Continuous Carbon Fiber Reinforced Thermoplastics By in-Nozzle Impregnation With Compaction Roller. In Proceedings of the 21st International Conference on Composite Materials, Xi'an, China, 20–25 August 2017.
22. Tian, X.; Liu, T.; Wang, Q.; Dilmurat, A.; Li, D.; Ziegmann, G. Recycling and Remanufacturing of 3D Printed Continuous Carbon Fiber Reinforced PLA Composites. *J. Clean. Prod.* **2017**, *142*, 1609–1618. [CrossRef]
23. Van Der Klift, F.; Koga, Y.; Todoroki, A.; Ueda, M.; Hirano, Y.; Matsuzaki, R. 3D Printing of Continuous Carbon Fibre Reinforced Thermo-Plastic (CFRTP) Tensile Test Specimens. *Open J. Compos. Mater.* **2016**, *6*, 18–27. [CrossRef]
24. Cantrell, J.T.; Rohde, S.; Damiani, D.; Gurnani, R.; DiSandro, L.; Anton, J.; Young, A.; Jerez, A.; Steinbach, D.; Kroese, C.; et al. Experimental Characterization of the Mechanical Properties of 3D-Printed ABS and Polycarbonate Parts. *Rapid Prototyp. J.* **2017**, *23*, 811–824. [CrossRef]
25. McLouth, T.D.; Severino, J.V.; Adams, P.M.; Patel, D.N.; Zaldivar, R.J. The Impact of Print Orientation and Raster Pattern on Fracture Toughness in Additively Manufactured ABS. *Addit. Manuf.* **2017**, *18*, 103–109. [CrossRef]
26. Rahmatabadi, D.; Ghasemi, I.; Baniassadi, M.; Abrinia, K.; Baghani, M. 3D Printing of PLA-TPU with Different Component Ratios: Fracture Toughness, Mechanical Properties, and Morphology. *J. Mater. Res. Technol.* **2022**, *21*, 3970–3981. [CrossRef]

27. Rahmatabadi, D.; Soltanmohammadi, K.; Aberoumand, M.; Soleyman, E.; Ghasemi, I.; Baniassadi, M.; Abrinia, K.; Bodaghi, M.; Baghani, M. Development of Pure Poly Vinyl Chloride (PVC) with Excellent 3D Printability and Macro- and Micro-Structural Properties. *Macromol. Mater. Eng.* **2022**, *307*, 2200568. [CrossRef]
28. Moradi, M.; Aminzadeh, A.; Rahmatabadi, D.; Rasouli, S.A. Statistical and Experimental Analysis of Process Parameters of 3D Nylon Printed Parts by Fused Deposition Modeling: Response Surface Modeling and Optimization. *J. Mater. Eng. Perform.* **2021**, *30*, 5441–5454. [CrossRef]
29. Shofner, M.L.; Lozano, K.; Rodrı, F.J. Nanofiber-Reinforced Polymers Prepared by Fused Deposition Modeling. *J. Appl. Polym. Sci.* **2003**, *89*, 3081–3090. [CrossRef]
30. Patanwala, H.S.; Hong, D.; Vora, S.R.; Bognet, B.; Ma, A.W.K. The Microstructure and Mechanical Properties of 3D Printed Carbon Nanotube-Polylactic Acid Composites. *Polym. Compos.* **2018**, *39*, E1060–E1071. [CrossRef]
31. Przekop, R.E.; Kujawa, M.; Pawlak, W.; Dobrosielska, M.; Sztorch, B.; Wieleba, W. Graphite Modified Polylactide (PLA) for 3D Printed (FDM/FFF) Sliding Elements. *Polymers* **2020**, *12*, 1250. [CrossRef]
32. Jia, Y.; He, H.; Geng, Y.; Huang, B.; Peng, X. High Through-Plane Thermal Conductivity of Polymer Based Product with Vertical Alignment of Graphite Flakes Achieved via 3D Printing. *Compos. Sci. Technol.* **2017**, *145*, 55–61. [CrossRef]
33. Ferreira, R.T.L.; Amatte, I.C.; Dutra, T.A.; Bürger, D. Experimental Characterization and Micrography of 3D Printed PLA and PLA Reinforced with Short Carbon Fibers. *Compos. Part B Eng.* **2017**, *124*, 88–100. [CrossRef]
34. Yang, Z.; Yang, Z.; Chen, H.; Yan, W. 3D Printing of Short Fiber Reinforced Composites via Material Extrusion: Fiber Breakage. *Addit. Manuf.* **2022**, *58*, 103067. [CrossRef]
35. Ma, S.; Yang, H.; Zhao, S.; He, P.; Zhang, Z.; Duan, X.; Yang, Z.; Jia, D.; Zhou, Y. 3D-Printing of Architectured Short Carbon Fiber-Geopolymer Composite. *Compos. Part B Eng.* **2021**, *226*, 109348. [CrossRef]
36. Zhong, W.; Li, F.; Zhang, Z.; Song, L.; Li, Z. Short Fiber Reinforced Composites for Fused Deposition Modeling. *Mater. Sci. Eng. A* **2001**, *301*, 125–130. [CrossRef]
37. Sodeifian, G.; Ghaseminejad, S.; Yousefi, A.A. Preparation of Polypropylene/Short Glass Fiber Composite as Fused Deposition Modeling (FDM) Filament. *Results Phys.* **2019**, *12*, 205–222. [CrossRef]
38. Ghabezi, P.; Flanagan, T.; Harrison, N. Short Basalt Fibre Reinforced Recycled Polypropylene Filaments for 3D Printing. *Mater. Lett.* **2022**, *326*, 132942. [CrossRef]
39. Sang, L.; Han, S.; Li, Z.; Yang, X.; Hou, W. Development of Short Basalt Fiber Reinforced Polylactide Composites and Their Feasible Evaluation for 3D Printing Applications. *Compos. Part B Eng.* **2019**, *164*, 629–639. [CrossRef]
40. Hao, W.; Liu, Y.; Zhou, H.; Chen, H.; Fang, D. Preparation and Characterization of 3D Printed Continuous Carbon Fiber Reinforced Thermosetting Composites. *Polym. Test.* **2018**, *65*, 29–34. [CrossRef]
41. Le Duigou, A.; Castro, M.; Bevan, R.; Martin, N. 3D Printing of Wood Fibre Biocomposites: From Mechanical to Actuation Functionality. *Mater. Des.* **2016**, *96*, 106–114. [CrossRef]
42. Galos, J.; Hu, Y.; Ravindran, A.R.; Ladani, R.B.; Mouritz, A.P. Electrical Properties of 3D Printed Continuous Carbon Fibre Composites Made Using the FDM Process. *Compos. Part A Appl. Sci. Manuf.* **2021**, *151*, 106661. [CrossRef]
43. Hu, Y.; Ladani, R.B.; Brandt, M.; Li, Y.; Mouritz, A.P. Carbon Fibre Damage during 3D Printing of Polymer Matrix Laminates Using the FDM Process. *Mater. Des.* **2021**, *205*, 109679. [CrossRef]
44. Mahajan, C.; Cormier, D. 3D Printing of Carbon Fiber Composites with Preferentially Aligned Fibers. In Proceedings of the 2015 Industrial and Systems Engineering Research Conference, Nashville, Tennessee, 30 May—2 June 2015.
45. Tekinalp, H.L.; Kunc, V.; Velez-Garcia, G.M.; Duty, C.E.; Love, L.J.; Naskar, A.K.; Blue, C.A.; Ozcan, S. Highly Oriented Carbon Fiber-Polymer Composites via Additive Manufacturing. *Compos. Sci. Technol.* **2014**, *105*, 144–150. [CrossRef]
46. Hinchcliffe, S.A.; Hess, K.M.; Srubar, W.V. Experimental and Theoretical Investigation of Prestressed Natural Fiber-Reinforced Polylactic Acid (PLA) Composite Materials. *Compos. Part B Eng.* **2016**, *95*, 346–354. [CrossRef]
47. Matsuzaki, R.; Ueda, M.; Namiki, M.; Jeong, T.K.; Asahara, H.; Horiguchi, K.; Nakamura, T.; Todoroki, A.; Hirano, Y. Three-Dimensional Printing of Continuous-Fiber Composites by in-Nozzle Impregnation. *Sci. Rep.* **2016**, *6*, 23058. [CrossRef]
48. Dutra, T.A.; Ferreira, R.T.L.; Resende, H.B.; Guimarães, A. Mechanical Characterization and Asymptotic Homogenization of 3D-Printed Continuous Carbon Fiber-Reinforced Thermoplastic. *J. Braz. Soc. Mech. Sci. Eng.* **2019**, *41*, 133. [CrossRef]
49. Caminero, M.A.; Chacón, J.M.; García-Moreno, I.; Rodríguez, G.P. Impact Damage Resistance of 3D Printed Continuous Fibre Reinforced Thermoplastic Composites Using Fused Deposition Modelling. *Compos. Part B Eng.* **2018**, *148*, 93–103. [CrossRef]
50. Naranjo-Lozada, J.; Ahuett-Garza, H.; Orta-Castañón, P.; Verbeeten, W.M.H.; Sáiz-González, D. Tensile Properties and Failure Behavior of Chopped and Continuous Carbon Fiber Composites Produced by Additive Manufacturing. *Addit. Manuf.* **2019**, *26*, 227–241. [CrossRef]
51. Dickson, A.N.; Barry, J.N.; McDonnell, K.A.; Dowling, D.P. Fabrication of Continuous Carbon, Glass and Kevlar Fibre Reinforced Polymer Composites Using Additive Manufacturing. *Addit. Manuf.* **2017**, *16*, 146–152. [CrossRef]
52. Zhuo, P.; Li, S.; Ashcroft, I.; Jones, A.; Pu, J.; Science, C.; Division, T. 3D Printing of Continuous Fibre Reinforced Thermoplastic Composite. In Proceedings of the 21st International Conference on Composite Materials, Xi'an, China, 20–25 August 2017.
53. Griffini, G.; Invernizzi, M.; Levi, M.; Natale, G.; Postiglione, G.; Turri, S. 3D-Printable CFR Polymer Composites with Dual-Cure Sequential IPNs. *Polymer* **2016**, *91*, 174–179. [CrossRef]
54. Kariz, M.; Sernek, M.; Obućina, M.; Kuzman, M.K. Effect of Wood Content in FDM Filament on Properties of 3D Printed Parts. *Mater. Today Commun.* **2018**, *14*, 135–140. [CrossRef]

55. Compton, B.G.; Lewis, J.A. 3D-Printing of Lightweight Cellular Composites. *Adv. Mater.* **2014**, *26*, 5930–5935. [CrossRef]
56. Blok, L.G.; Longana, M.L.; Yu, H.; Woods, B.K.S. An Investigation into 3D Printing of Fibre Reinforced Thermoplastic Composites. *Addit. Manuf.* **2018**, *22*, 176–186. [CrossRef]
57. Zhang, W.; Cotton, C.; Sun, J.; Heider, D.; Gu, B.; Sun, B.; Chou, T.W. Interfacial Bonding Strength of Short Carbon Fiber/Acrylonitrile-Butadiene-Styrene Composites Fabricated by Fused Deposition Modeling. *Compos. Part B Eng.* **2018**, *137*, 51–59. [CrossRef]
58. Christensen, K.; Davis, B.; Jin, Y.; Huang, Y. Effects of Printing-Induced Interfaces on Localized Strain within 3D Printed Hydrogel Structures. *Mater. Sci. Eng. C* **2018**, *89*, 65–74. [CrossRef]
59. Bellehumeur, C.; Li, L.; Sun, Q.; Gu, P. Modeling of Bond Formation between Polymer Filaments in the Fused Deposition Modeling Process. *J. Manuf. Process.* **2004**, *6*, 170–178. [CrossRef]
60. Gurrala, P.K.; Regalla, S.P. Part Strength Evolution with Bonding between Filaments in Fused Deposition Modelling: This Paper Studies How Coalescence of Filaments Contributes to the Strength of Final FDM Part. *Virtual Phys. Prototyp.* **2014**, *9*, 141–149. [CrossRef]
61. Li, L.; Sun, Q.; Bellehumeur, C.; Gu, P. Investigation of Bond Formation in FDM Process Investigation of Bond Formation in FDM Process. *Trans. N. Am. Manuf. Res. Inst. SME* **2019**, *31*, 400–407.
62. Everett, H. Markforged Merges with One to Go Public as a $2.1 Billion 3D Printing Business. Available online: https://3dprintingindustry.com/news/markforged-merges-with-one-to-go-public-as-a-2-1-billion-3d-printing-business-185077 (accessed on 24 February 2021).
63. Sommacal, S.; Matschinski, A.; Drechsler, K.; Compston, P. Characterisation of Void and Fiber Distribution in 3D Printed Carbon-Fiber/PEEK Using X-Ray Computed Tomography. *Compos. Part A Appl. Sci. Manuf.* **2021**, *149*, 106487. [CrossRef]
64. MarkForged FFF Nylon Filament, Safety Data Sheet. Available online: https://markforged.com/materials/plastics/nylon (accessed on 20 January 2022).
65. Millot, C.; Fillot, L.A.; Lame, O.; Sotta, P.; Seguela, R. Assessment of Polyamide-6 Crystallinity by DSC: Temperature Dependence of the Melting Enthalpy. *J. Therm. Anal. Calorim.* **2015**, *122*, 307–314. [CrossRef]
66. *ASTM International D638-14 Standard Test Method for Tensile Properties of Plastics*; ASTM International: West Conshohocken, PA, USA, 2015.
67. Cox, H.L. The Elasticity and Strength of Paper and Other Fibrous Materials. *Br. J. Appl. Phys.* **1952**, *3*, 72. [CrossRef]
68. Krenchel, H. *Fibre Reinforcement; Theoretical and Practical Investigations of the Elasticity and Strength of Fibre-Reinforced Materials*; AGRIS: Rome, Italy, 1964.
69. O'Regan, D.F.; Akay, M.; Meenan, B. A Comparison of Young's Modulus Predictions in Fibre-Reinforced-Polyamide Injection Mouldings. *Compos. Sci. Technol.* **1999**, *59*, 419–427. [CrossRef]
70. Parodi, E.; Govaert, L.E.; Peters, G.W.M. Glass Transition Temperature versus Structure of Polyamide 6: A Flash-DSC Study. *Thermochim. Acta* **2017**, *657*, 110–122. [CrossRef]
71. Sun, Q.; Rizvi, G.M.; Bellehumeur, C.T.; Gu, P. Effect of Processing Conditions on the Bonding Quality of FDM Polymer Filaments. *Rapid Prototyp. J.* **2008**, *14*, 72–80. [CrossRef]
72. Sood, A.K.; Ohdar, R.K.; Mahapatra, S.S. Parametric Appraisal of Mechanical Property of Fused Deposition Modelling Processed Parts. *Mater. Des.* **2010**, *31*, 287–295. [CrossRef]
73. Sauer, M.J. *Evaluation of the Mechanical Properties of 3D Printed Carbon Fiber Composites*; South Dakota State University: Brookings, SD, USA, 2018; ISBN 035597438X.
74. Performance Composites Ltd. *Mechanical Properties of Carbon Fibre Composite Materials, Fibre/Epoxy Resin (120 °C Cure)*; Performance Composites Ltd.: Compton, CA, USA, 2009.
75. Greaves, G.N.; Greer, A.L.; Lakes, R.S.; Rouxel, T. Poisson's Ratio and Modern Materials. *Nat. Mater.* **2011**, *10*, 823–837. [CrossRef]
76. Nairn, J.A.; Mendels, D.A. On the Use of Planar Shear-Lag Methods for Stress-Transfer Analysis of Multilayered Composites. *Mech. Mater.* **2001**, *33*, 335–362. [CrossRef]
77. Carraro, P.A.; Quaresimin, M. A Stiffness Degradation Model for Cracked Multidirectional Laminates with Cracks in Multiple Layers. *Int. J. Solids Struct.* **2015**, *58*, 34–51. [CrossRef]
78. Kashtalyan, M.; Soutis, C. Modelling Off-Axis Ply Matrix Cracking in Continuous Fibre-Reinforced Polymer Matrix Composite Laminates. *J. Mater. Sci.* **2006**, *41*, 6789–6799. [CrossRef]
79. McCartney, L.N. Theory of Stress Transfer in a 0°–90°–0° Cross-Ply Laminate Containing a Parallel Array of Transverse Cracks. *J. Mech. Phys. Solids* **1992**, *40*, 27–68. [CrossRef]
80. Gudmundson, P.; Weilin, Z. An Analytic Model for Thermoelastic Properties of Composite Laminates Containing Transverse Matrix Cracks. *Int. J. Solids Struct.* **1993**, *30*, 3211–3231. [CrossRef]

Disclaimer/Publisher's Note: The statements, opinions and data contained in all publications are solely those of the individual author(s) and contributor(s) and not of MDPI and/or the editor(s). MDPI and/or the editor(s) disclaim responsibility for any injury to people or property resulting from any ideas, methods, instructions or products referred to in the content.

Article

A Comparative Analysis of Chemical, Plasma and In Situ Modification of Graphene Nanopleteletes for Improved Performance of Fused Filament Fabricated Thermoplastic Polyurethane Composites Parts

Xiaojie Zhang [1], Jianhua Xiao [2], Jinkuk Kim [3] and Lan Cao [4,*]

1. School of Materials Science Engineering, Nanchang Hangkong University, Nanchang 330000, China
2. School of Chemistry and Chemical Engineering, Shanghai University of Engineering Science, Shanghai 201620, China
3. Department of Materials Engineering and Convergence Technology, Engineering Research Institute, Gyeongsang National University, Jinju 52828, Republic of Korea
4. School of Polymer Science and Engineering, Qingdao University of Science and Technology, Qingdao 266042, China
* Correspondence: lancao@qust.edu.cn

Citation: Zhang, X.; Xiao, J.; Kim, J.; Cao, L. A Comparative Analysis of Chemical, Plasma and In Situ Modification of Graphene Nanopleteletes for Improved Performance of Fused Filament Fabricated Thermoplastic Polyurethane Composites Parts. *Polymers* 2022, 14, 5182. https://doi.org/10.3390/polym14235182

Academic Editor: Matthew Blacklock

Received: 18 October 2022
Accepted: 23 November 2022
Published: 28 November 2022

Publisher's Note: MDPI stays neutral with regard to jurisdictional claims in published maps and institutional affiliations.

Copyright: © 2022 by the authors. Licensee MDPI, Basel, Switzerland. This article is an open access article distributed under the terms and conditions of the Creative Commons Attribution (CC BY) license (https://creativecommons.org/licenses/by/4.0/).

Abstract: The limited number of materials and mechanical weakness of fused deposition modeling (FDM) parts are deficiencies of FDM technology. The preparation of polymer composites parts with suitable filler is a promising method to improve the properties of the 3D printed parts. However, the agglomerate of filler makes its difficult disperse in the matrix. In this work, graphene nanoplatelets (GnPs) were surface modified with chemical, low-temperature plasma and in situ methods, in order to apply them as fillers for thermoplastic polyurethane (TPU). Following its modification, the surface chemical composition of GnPs was analyzed. Three wt% of surface-modified GnPs were incorporated into TPU to produce FDM filaments using a melting compounding process. Their effects on rheology properties and electrical conductivity on TPU/GnPs composites, as well as the dimensional accuracy and mechanical properties of FDM parts, are compared. The images of sample facture surfaces were examined by scanning electron microscope (SEM) to determine the dispersion of GnPs. Results indicate that chemical treatment of GnPs with zwitterionic surfactant is a good candidate to significantly enhance TPU filaments, when considering the FDM parts demonstrated the highest mechanical properties and lowest dimensional accuracy.

Keywords: FDM; graphene nanoplatelets; low-temperature plasma; surface modification

1. Introduction

Three-dimensional (3D) printing is revolutionizing different important industrial areas, as it is distinguished from traditional processing techniques by its ability to manufacture components from prototypes to complex geometries with great design flexibility, high recyclability and less material waste [1,2]. Among the different 3D printing techniques, fused deposition modeling (FDM) has been widely used because of its easy handling, relative inexpensiveness and low chemical toxicity [3].

Although there is abundant availability of FDM machines, in many cases, FDM processes lack in providing consistency and reliability in terms of part properties, accuracy and finish, which limits the widespread application of FDM. These deficiencies depend mainly on the materials used, process parameters and post-processing techniques. Several studies and reviews of the literature have confirmed the effectiveness of optimizing the geometry, operation-specific parameters and annealing treatment for improving the manufactured parts [4–7]. With regard to the materials, till now, only a few commercial thermoplastic filaments have been available for FDM techniques, such polylactic acid (PLA), polyamide (Nylon), acrylonitrile butadiene styrene (ABS), thermoplastic polyurethane (TPU), polyetheretherketone (PEEK), etc. [8]. High-performance materials are highly demanded due to the versatility of FDM applications, such as engineering, automotive composites and aviation fields. The incorporation of fillers such as glass fiber, carbon fiber, ceramic or carbon-based nano-size are common approaches to attain this goal [9].

In recent years, graphene and its derivatives have received considerable attention due to their excellent properties. However, low-cost, high-quality and eco-friendly processes for manufacturing graphene are still challenging [10]. Compared with graphene oxide and other graphene-related materials, graphene nanoplatelets (GnPs) can be easily produced in large scale with physical methods and are commercially available with different particle sizes at a relatively low cost and have been identified as a substitute for graphene. GnPs exhibit exciting properties such as light weight and high electrical conductivity and mechanical strength. Polymer/GnPs composites exhibit more efficient improvement in terms of strength and thermal stability as compared with reduced graphene oxide (rGO), graphene oxide (GO), and multiwalled carbon nanotubes (MWCNT) [11,12]. Consequently, polymer/GnPs nanocomposites are popular and widely researched for many applications including 3D printing materials, low-cost composites films, wearable devices, medical hydrogels and many more [13]. Masarra and Batistella et al. [14] fabricated an electrically conductive circuit using PLA/PLC/GnPs by FDM. Misra and Ostadhossein et al. [15] constructed a multidrug-eluting stent using direct 3D printing from polycaprolactone/GnPs biodegradable composites. Jing and Chen et al. [16] prepared high thermal conductive polyethylene/GnPs nanocomposites for heat diffusion application of some advanced electronic devices. Li et al. [17] FDM printed a wearable pressure sensor based on TPU/GnPs nanocomposites. However, the platelet structure of GnPs exhibits some disadvantages. During the preparation of nanocomposites, GnPs tend to agglomerate, because of a considerable extent of π–π stacking interactions and their weak interactions with the polymer chains [18]. Agglomerates decrease the surface area of GnPs and, correspondingly, deteriorate reinforcement performance [19–21]. Attempts have been made to develop the dispersion of GnPs in nanocomposites, including utilizing chemical and physical surfactants modification, covalent functionalization and low-temperature plasma modification [18,22–24].

Based on the above, this work aims to investigate a series of approaches to treat GnPs, including chemical, low-temperature plasma and in situ modification, providing a comparative overview of the effect of surface modification on the quality of FDM-printed TPU/GnPs, in terms of dimensional accuracy and mechanical performance.

2. Materials and Methods

2.1. Materials

Thermoplastic polyurethane (TPU), LM-95A, was purchased from The Lubrizol Corporation (Wickliffe, OH, USA). Graphene nanoplatelets (GnPs), HGP-10, were supplied by Qingdao Yanhai Carbon Materials Co., Ltd. (Qingdao, China), with the average diameter around 8 μm and 1~15 nm thickness of graphene stacked into GnPs particle. 12-aminododecanoic acid (ADA), 98% purity, was bought from Shanghai Macklin Biochemical Co., Ltd. (Shanghai, China). Dimethylformamide (DMF), 97% purity, was purchased from Xilong Scientific Co., Ltd. (Shantou, China).

2.2. GnPs Surface Modification

2.2.1. Low-Temperature Plasma Modification (PGnPs)

Plasma treatment was conducted in a plasma cleaner, model JS-P200, Hefei Jieshuo Vacuum Technology (Hefei, China). GnPs were exposed to air plasma for 50 min under powder condition of 100 W. During the treatment, the gas flow and gas low rate were kept as 40 Pa and 5 sccm.

2.2.2. Chemical Modification of GnPs (CGnPs)

The GnPs and the zwitterionic surfactant ADA (GnPs:ADA = 10:1, weight ratio) were dispersed in DMF followed by stirring overnight at 85 °C, then the dispersion was sonicated for 3 h using an ultrasonic cleaner (SB-3200DTD, Ningbo Scientz Biotechnology, Ningbo, China) at frequency of 37 kHz. After that, the GnPs was separated from the suspension and washed by the DMF. Finally, the GnPs was dried for 48 h under vacuum at 90 °C [13].

2.2.3. In Situ Modification (IGnPs)

The GnPs were premixed with the ADA prior to the extrusion process. The ratio of ADA to GnPs was 1:10 (wt/wt). Then, melt mixing of TPU and the GnPs/ADA hybrids (3 wt%) composites were processed using a signal-screw extruder (Wuhan Yiyang Plastic Machinery, Wuhan, China). The screw speed was 25 rpm. Extruder barrel temperatures were set as follows: 195 °C, 190 °C and 185 °C.

2.3. Filament Fabrication

To produce the desired filament size (1.75 mm) for use in FDM 3D printer, a twin-screw extruder (SHJ-20C, Nanjing Giant Machinery, Nanjing, China) was used. To ensure a homogenous dispersion of 3% GnPs into TPU, the TPU and GnPs were stir-mixed prior to melt compounding. The mixture was then fed into the extruder. The recommended starting extrusion temperature of TPU is 138 °C. To solve the die swelling problems and increase the dispersion of GnPs in the matrix, the barrel temperatures were set at 170 °C/180 °C/190 °C/185 °C/180 °C/170 °C, in order. The filament was subsequently cooled and then wound onto a reel.

2.4. Dog-Bone Specimens Printed by FDM

The preparation of the filament and the FDM printing process are shown in Figure 1. Initially, the FDM parts were modeled on CAD software according to the standard of ISO-527-2-2012 type-1BA with 5 mm thickness and exported in STL format. The STL file was sliced and transformed into G-code file by slicing software (Ultimaker Cura Version 4.2, Shenzhen, China). The dog-bone FDM parts were prepared by a commercial desktop FDM 3D printer (4Max Pro2.0, Shenzhen Anycubic, Shenzhen, China). In order to focus on the effect of GnPs, 3D printing process parameters such as layer thickness, feed rate, fill pattern, fill percentage or temperature were fixed for all the samples. Table 1 shows the values of these parameters. After printing, the supports were removed, and no further post-processing process was required.

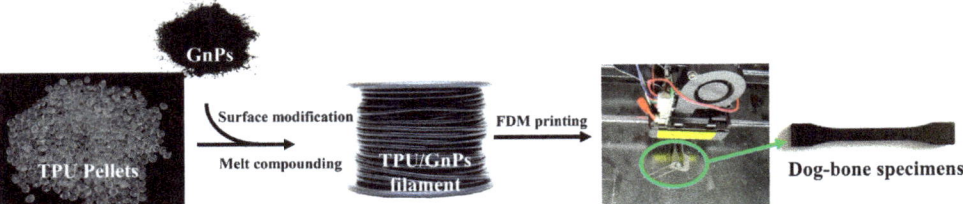

Figure 1. Preparation and FDM printing of specimens.

Table 1. 3D printer parameters.

Parameters	Value
Material	TPU-GnPs
Nozzle diameter	0.4 mm
Layer thickness	0.1 mm
Printing speed	40 mm/s
Nozzle temperature	190 °C
Bed Temperature	60 °C
Top/Bottom solid layers	1.2 mm
Outline/perimeters shell	1.2 mm
Internal fill pattern	Mesh
External fill pattern	Rectilinear
Internal fill percentage	20%
Filament diameter	1.75 mm

2.5. Characterization

2.5.1. Fourier Transfer Infrared (FTIR) Spectroscopy

Fourier transfer infrared (FTIR) spectra of modified GnPs were collected in the wavenumber range of 400–4000 cm^{-1} with a VERTEX70 (Bruker, Ettlingen, Germany) spectrometer. The samples were prepared by KBr-disk method [25,26].

2.5.2. Rheological Properties

A capillary rheometer (MLW-400B, Changchun Intelligent Instrument and Equipment, Changchun, China) was used to measure the rheological properties, having a die of 1 mm diameter and 10 mm length. The apparent viscosity of the blends was determined in the shear rates ranging from 100 to 1500 s^{-1} at 190 °C [27,28].

The melt flow index (MFI) [29,30] was carried out through the XNR-400C melt flow indexer (Jinhe Instruments, Chengde, China) at 190 °C by applying a 3.24 kg load to extrude the molten polymers. A 100 g rod was used a plunger. The MFI values were generated from at least five determinations.

2.5.3. Electrical Conductivity Measurement

The electrical conductivity was measured by using a resistance tester (AT683, Applent Instruments, Changzhou, China). The electrical conductivity (σ, S/m) was calculated using the following Equation (1) [31]:

$$\sigma = 1/\rho = L/RS \tag{1}$$

where ρ is the resistivity and R is the resistance. S and L are the length and cross-sectional area of the filament.

2.5.4. Dimensional Accuracy

Dimensions of the 3D-printed dog-bone specimens in length direction were measured using a digital Vernier caliper and compared with CAD dimensions. The dimensional deviations in length were calculated for different TPU/GnPs-composite-based dog-bone specimens and compared with CAD dimensions.

2.5.5. Surface Roughness

The average surface roughness (Ra) of the top surface of FDM parts is measured with a surface roughness tester (TR150A, Timech, Beijing, China) at 1 mm/s tracing speed, tracing length at 6 mm.

2.5.6. Tensile Properties

Tensile strength and elongation at break were conducted using a universal testing machine, YF-900 (Yuanfeng, Yangzhou, China) based on the ISO 527-1-2021, at a crosshead speed of 200 mm/min, and the sample length between benchmarks was 50 mm. At least five specimens were tested for each composite, and the medial values were reported.

2.5.7. Morphological Properties

The fracture surface morphologies of composites were investigated by field emission scanning electron microscopy (FE-SEM, FEI Nova NanoSEM 450, Brno, Czech Republic). The GnPs and facture surfaces of various TPU/GnPs filaments were sputtered with a thin layer of aurum [32].

3. Results

3.1. GnPs Characterization

FTIR was applied to analyze changes in chemical groups on the surface of GnPs before and after surface modification. Figure 2a presents the FTIR spectra of the GnP, PGnPs and CGnPs. The FTIR spectra of GnPs without any treatment shows bands at 2921–2765 cm^{-1} resulting from the C–H stretching. The amount 1151–1074 cm^{-1} corresponds to the C–O bending, and the C–H bending assigned at 1398, 860, 771 cm^{-1} is observed. Comparing pristine GnPs and PGnPs spectra, the low-temperature plasma treatment grafts' various functional polar groups on the GnPs surface, including O–H, N–H or NH_2 stretching ranging from 3056 to 3689 cm^{-1}, C=O stretching at 1747 cm^{-1} and C–N or C–O stretching at 1000–1294 cm^{-1} wavenumber, the existence of the characteristic bands confirmed the functional polar groups that contain oxygen or nitrogen on the surface of GnPs after plasma treatment (Figure 2b). The polar groups on PGnPs surface will benefit the homogenous dispersion of GnPs in the matrix and the interfacial adhesion between filler and macromolecules. Noticeably, in the FTIR spectra of CGnPs, the additional peaks at 3451, 1380 cm^{-1} and 1741 cm^{-1} are assigned to O–H bending and C=O stretching of the carboxylic groups. The peaks at 3451, 1639 and 1461 cm^{-1} ascribe to the N–H bending and stretching of the amine group, while the broad brands at 1226–1025 cm^{-1} are the uptake for C–N. From the results of the FTIR spectra, we conclude that the modification of GnPs by ADA was successful (Figure 2b) [33–39]. The organophilic absorption between the aliphatic chain of ADD and the TPU matrix leads to uniform dispersion of particles in the TPU matrix.

3.2. Rheological Properties

Rheological properties can confirm the molecular entanglement and molecular relaxation of the polymer composites and guideline of the final construction of FDM parts [40,41]. Figure 3a,b demonstrate the relationship between shear stress and shear viscosity on the steady shear rate of the different TPU/GnPs composites. All composites display the non-Newtonian characteristics; the shear viscosity of all TPU/GnPs composites decrease sharply with increasing shear rate, showing shear thinning behavior. As revealed in Figure 3a,b, at low shear rate, in the range of 100–800 s^{-1}, the shear stresses and shear viscosities of TPU/PGnPs and TPU/CGnPs are higher than those of TPU/GnPs. It is considered that PGnPs and CGnPs impede the chain mobility of TPU, during which their stresses and viscosities increase. However, the addition of IGnPs in TPU results in an obvious reduction in the shear stress and shear viscosity. This change indicates that the zwitterionic surfactant ADA is more of a plasticizer in the TPU/IGnPs composites than a surface modifier for GnPs during in situ treatment. Hence, the TPU/IGnPs filament shows a smoother surface than other filaments (Figure 3e). Over the entire range of shear rates investigated, all the non-Newtonian fluids followed a power-law relationship. At higher shear rates, the shear stress and shear viscosity of all the composites are revealed comparable regardless of the incorporation of fillers [41]. Figure 3c shows log–log plots of shear viscosity versus shear rate and the power-law index (n) for all TPU/GnPs composites at 190 °C. Comparing with TPU/GnPs composites ($n = 0.39$), both TPU/PGnPs ($n = 0.22$) and TPU/CGnPs ($n = 0.31$) composites have lower power-law index values, indicating that the surface modification of GnPs surface increases the interaction between TPU polymer chains and GnPs. The apparent viscosity of polymer melts is inversely proportional to their melt flow index (MFI). As shown in Figure 3d, the MFI of TPU/IGnPs increases slightly, while the TPU/PGnPs and TPU/CGnPs MFI values are smaller than that of TPU/GnPs. The variation is consistent with the shear viscosity. In summary, the four composites exhibit different viscosities (also different interactions between filler and polymer chains) in the order of TPU/IGnPs < TPU/GnPs < TPU/PGnPs < TPU/CGnPs [42,43].

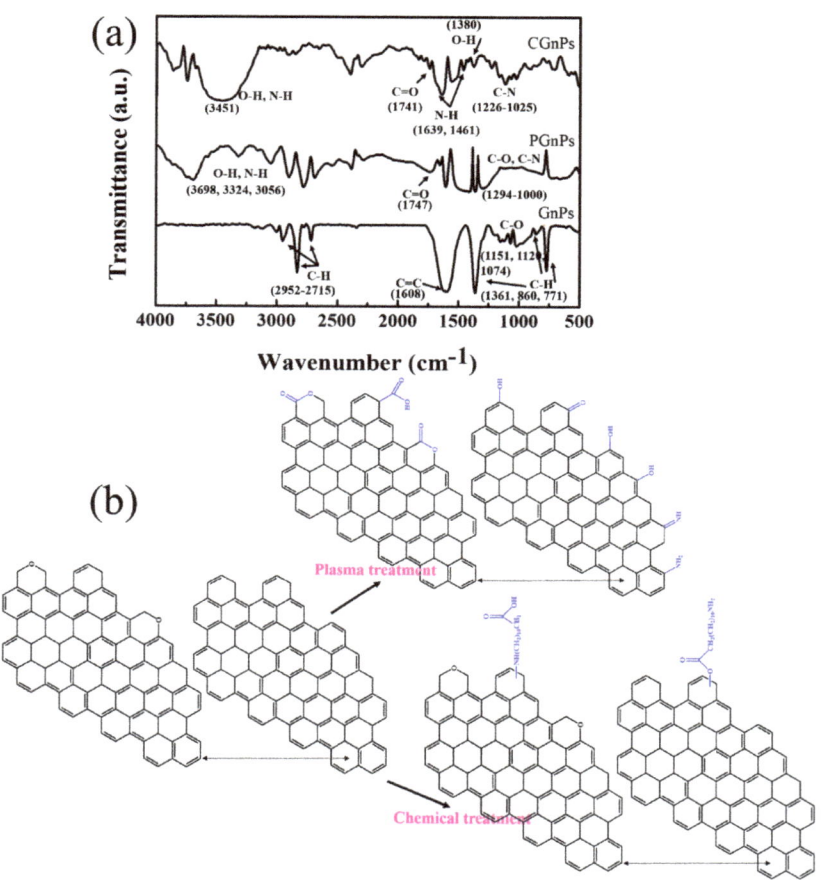

Figure 2. (**a**) FTIR spectra of pristine GnPs, PGnPs and CGnPs; (**b**) structure of pristine GnPs before and after surface modification.

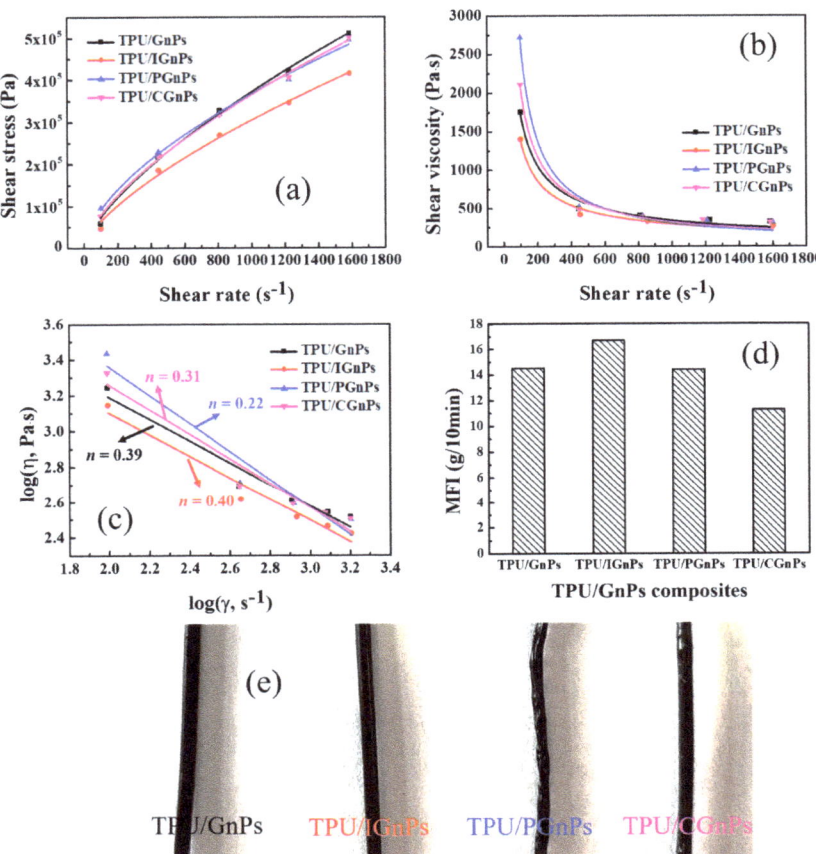

Figure 3. (a) Shear stress, (b) shear viscosity, (c) the bilogarithmic plots of viscosity versus shear rate, (d) MFI of various TPU/GnPs composites and (e) various TPU/GnPs-based filaments.

3.3. Effect of GnPs on Electrical Conductivity of TPU/GnPs Composites

Figure 4 depicts the electrical conductivity of various TPU/GnPs composites. It is observed that the electrical conductivity was associated with surface modification of GnPs. However, these differences were not very significant or diverse in the same order of magnitude. The highest conductivity is achieved by TPU/CGnPs; the electrical conductivity increased by 196%. The functionalization of GnPs helps in the uniform dispersion of GnPs flakes throughout the polymer matrix, forming an inter-connected network for electrical conduction. As a result, electrical properties will be increased [44,45].

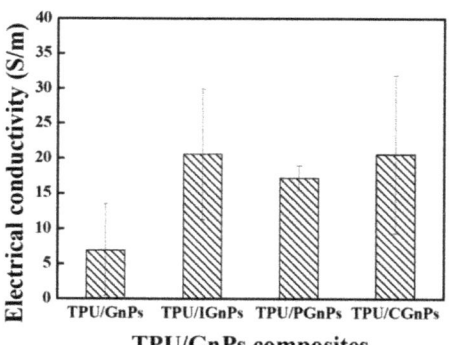

Figure 4. Electrical conductivity of various TPU/GnPs composites.

3.4. Mechanical Properties of FDM Parts

The change in functional groups on the surface of GnPs can differ polymer–filler interactions in the respective composites. To confirm the attractive polymer–filler interaction in TPU and GnPs experimentally, the mechanical properties of the various TPU/GnPs composites were examined and compared, and the tensile strength and elongation at break of the different TPU/GnPs composites are presented in Figure 5a,b. A serious relation between the surface treatment of GnPs and mechanical properties of composites can clearly be observed. IGnPs reduced the tensile strength of the composites, but when using PGnPs and CGnPs, the increase was 22% and 23%, respectively, which could improve the structural application of the composites. This result indicates that the polymer–filler interactions between TPU and PGnPs or CGnPs are stronger than those between TPU and IGnPs, which agree with the results of the rheological properties of composites. Meanwhile, the elongation at break values of the surface-modified GnPs exhibits slightly more greatly than the virgin GnPs composites. An application of CGnPs significantly increases the tensile strength and elongation at the break of composites, i.e., up to 69.79 MPa and 645%, respectively [37].

3.5. Dimensional Accuracy

Table 2 summarizes the mean values and standard deviations of dimensional variation in length between the FDM-printed parts, the CAD dimensions and the top surface roughness of the prototypes. Specimens showed values of positive differences, meaning that the length value exceed the CAD files. TPU/CGnPs composites show the most dimensional accuracy, while TPU/PGnPs show the least dimensional accuracy. More specifically, TPU/CGnPs and TPU/PGnPs show a mean error of 1.96% and 2.58%, respectively, while TPU/GnPs and TPU/IGnPs depict a mean error of 2.52% and 2.12%, respectively. In short, surface modification of GnPs can affect the dimensional accuracy of TPU/GnPs composites specimens; moreover, TPU/CGnPs samples show the best dimensional performance, probably because of the high viscosity of TPU/CGnPs. High viscosity allows for an increase in the content of spaces between paths [46–50]. Table 2 also indicates that homogenous distribution of modified GnPs in TPU helps in enhancement of the surface roughness. However, all specimens have high roughness values due to the layer-by-layer deposition process of the FDM technique.

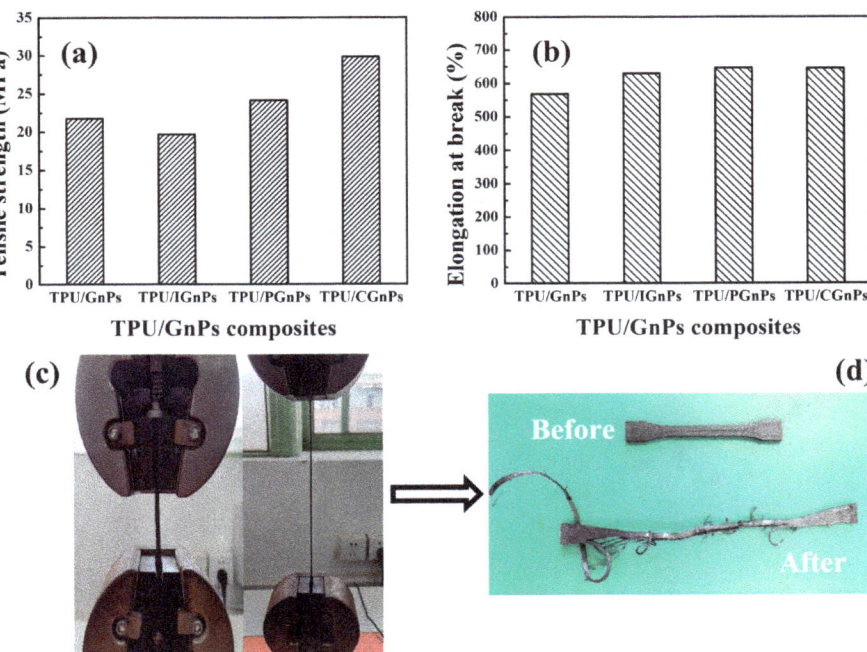

Figure 5. (**a**) Tensile strength and (**b**) elongation at break of various TPU/GnPs composites, (**c**) the deformed FDM parts at various stages and (**d**) the FDM-printed specimens before and after break.

Table 2. Mean value and standard deviation of the dimensional deviation and surface roughness.

Property	Samples	Mean Value	Standard Deviation
Dimensional deviation (%)	TPU/GnPs	2.52	0.015
	TPU/IGnPs	2.12	0.108
	TPU/PGnPs	2.58	0.077
	TPU/CGnPs	1.96	0.081
Surface roughness (µm)	TPU/GnPs	1.98	0.384
	TPU/IGnPs	1.91	0.233
	TPU/PGnPs	1.78	0.265
	TPU/CGnPs	1.82	0.252

3.6. Morphologies of Various TPU/GnPs Composites

Homogenous dispersion and distribution of the GnPs in the TPU matrix is a vertical factor for the property enhancement of the 3D-printed nanocomposites. The morphologies of pristine GnPs and the fracture surfaces of the various TPU/GnPs filaments characterized by FE-SEM are displayed in Figure 6. As Figure 6a shows, the GnPs is quite thin with smooth surface, and certain GnPs are aggregated and corrugated with a diameter of a few micrometers. The fracture surfaces of the TPU/GnPs (Figure 6b) and TPU/IGnPs (Figure 6c) filaments are significantly different from the fracture surfaces of TPU/GnPs (Figure 6d) and TPU/IGnPs (Figure 6e) filaments. Figure 6b,c shows rough and tortuous pathways, and the surfaces of the composites containing platelets projecting outside from the surfaces, meaning a weak adhesion to the matrix, and GnPs particles agglomerated into big agglomerates segregate in the TPU matrix. In contrast, the facture surfaces of the TPU/CGnPs filament

(Figure 6d) are quite smooth and flat; the CGnPs seem to be well-embedded in the TPU matrix, and bundle formations gradually disappear for the composites, an indication of strong interfacial interactions between the TPU and CGnPs, which are in good agreement with the high mechanical properties of TPU/CGnPs composites [51,52].

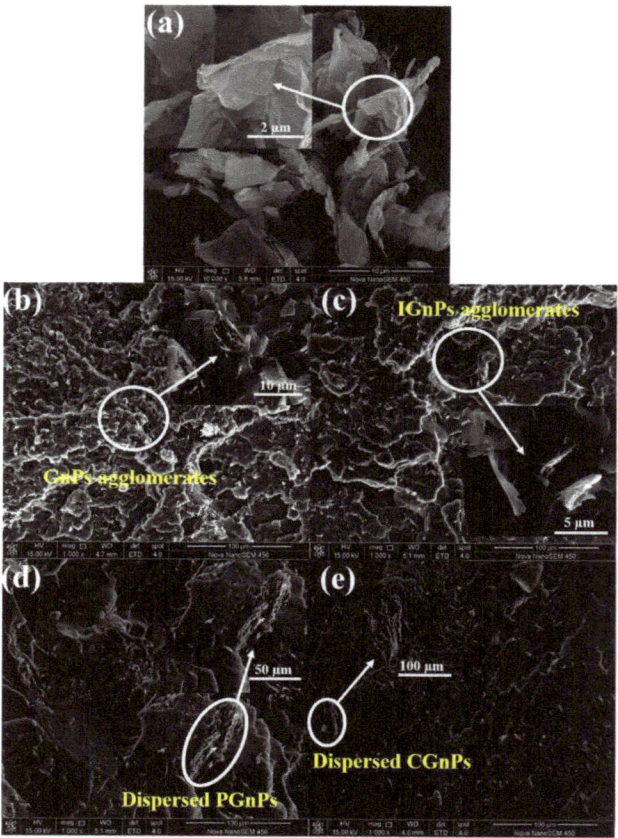

Figure 6. FE-SEM images of (**a**) GnPs and the fracture surfaces of (**b**) TPU/GnPs, (**c**) TPU/IGnPs, (**d**) TPU/PGnPs and (**e**) TPU/CGnPs filaments.

4. Conclusions

In this work, the three different surface treatments for GnPs have been compared, and their effects on the rheological properties of TPU/GnPs composites, as well as the mechanical performance, dimensional accuracy and surface roughness of FDM-printed TPU/GnPs specimens, have been analyzed. Furthermore, microscope imaging has provided insight into the reasons behind the observed changes in mechanical performance. The introduction of IGnPs decreased the viscosity of TPU/IGnPs composites. There were no significant rheological differences between TPU/PGnPs and TPU/CGnPs composites, but TPU/CGnPs composites exhibited higher electrical conductivity. TPU/CGnPs FDM parts showed a significant improvement of the dimensional accuracy and mechanical properties over the other three materials, mainly because the stronger interfacial interactions between TPU and CGnPs. However, the PGnPs is more inclined to reduce the dimensional accuracy of FDM parts. The TPU/IGnP composites exhibited comparable dimensional accuracy compared with the TPU/CGnPs composites. However, it is also noted that the TPU/IGnPS composites demonstrated the lowest tensile strength among the four types of composites. All specimens present similar but high surface roughness due to the nature of FDM techniques; post-processing is necessary for further application. The results suggest that GnPs modified with ADA is more effective in improving the multifunctional properties of TPU.

Author Contributions: X.Z.: conceptualization, methodology, investigation, formal analysis and writing—original draft. J.X.: supervision and funding acquisition. J.K.: writing—review and editing; L.C.: supervision, funding acquisition and writing—review and editing. All authors have read and agreed to the published version of the manuscript.

Funding: This work was financed by the National Natural Science Foundation of China (NSFC) (No. 52063021) and the PhD Research Startup Foundation of Nanchang Hangkong University (EA202001381).

Institutional Review Board Statement: Not applicable.

Data Availability Statement: The data presented in this study are available on request from the corresponding author.

Conflicts of Interest: The authors declare no conflict of interest.

References

1. Caminero, M.; Chacón, J.M.; García-Plaza, E.; Núñez, P.J.; Reverte, J.M.; Becar, J.P. Additive manufacturing of PLA-based composites using fused filament fabrication: Effect of graphene nanoplatelet reinforcement on mechanical properties, dimensional accuracy and texture. *Polymers* **2019**, *11*, 799. [CrossRef] [PubMed]
2. Zhang, H.; Zhang, K.; Li, A.; Wan, L.; Robert, C.; Brádaigh, C.M.; Yang, D. 3D printing of continuous carbon fibre reinforced powder-based epoxy composites. *Compos. Commun.* **2022**, *33*, 101239. [CrossRef]
3. Peng, X.; Zhang, M.; Guo, Z.; Sang, L.; Hou, W. Investigation of processing parameters on tensile performance for FDM-printed carbon fiber reinforced polyamide 6 composites. *Compos. Commun.* **2020**, *22*, 100478. [CrossRef]
4. Kozior, T.; Mamun, A.; Trabelsi, M.; Sabantina, L.; Ehrmann, A. Quality of the Surface Texture and Mechanical Properties of FDM Printed Samples after Thermal and Chemical Treatment. *Stroj. Vestn. J. Mech. Eng.* **2020**, *66*, 105–113. [CrossRef]
5. Garg, A.; Bhattacharya, A.; Batish, A. Chemical vapor treatment of ABS parts built by FDM: Analysis of surface finish and mechanical strength. *Int. J. Adv. Manuf. Technol.* **2017**, *89*, 2175–2191. [CrossRef]
6. Gao, S.; Liu, R.; Xin, H.; Liang, H.; Wang, Y.; Jia, J. The Surface Characteristics, Microstructure and Mechanical Properties of PEEK Printed by Fused Deposition Modeling with Different Raster Angles. *Polymers* **2021**, *14*, 77. [CrossRef]
7. Chohan, J.S.; Singh, R. Pre and post processing techniques to improve surface characteristics of FDM parts: A state of art review and future applications. *Rapid Prototyp. J.* **2017**, *23*, 495–513. [CrossRef]
8. Arif, M.; Alhashmi, H.; Varadarajan, K.; Koo, J.H.; Hart, A.; Kumar, S. Multifunctional performance of carbon nanotubes and graphene nanoplatelets reinforced PEEK composites enabled via FFF additive manufacturing. *Compos. Part B Eng.* **2020**, *184*, 107625. [CrossRef]
9. Ivanov, E.; Kotsilkova, R.; Xia, H.; Chen, Y.; Donato, R.K.; Donato, K.; Godoy, A.P.; Di Maio, R.; Silvestre, C.; Cimmino, S.; et al. PLA/Graphene/MWCNT composites with improved electrical and thermal properties suitable for FDM 3D printing applications. *Appl. Sci.* **2019**, *9*, 1209. [CrossRef]
10. Madhad, H.V.; Mishra, N.S.; Patel, S.B.; Panchal, S.S.; Gandhi, R.A.; Vasava, D.V. Graphene/graphene nanoplatelets reinforced polyamide nanocomposites: A review. *High Perform. Polym.* **2021**, *33*, 981–997. [CrossRef]
11. Mohan, D.; Sajab, M.S.; Bakarudin, S.B.; Roslan, R.; Kaco, H. 3D Printed Polyurethane Reinforced Graphene Nanoplatelets. *Mater. Sci. Forum* **2021**, *1025*, 47–52. [CrossRef]
12. Cha, J.; Kim, J.; Ryu, S.; Hong, S.H. Comparison to mechanical properties of epoxy nanocomposites reinforced by functionalized carbon nanotubes and graphene nanoplatelets. *Compos. Part B Eng.* **2019**, *162*, 283–288. [CrossRef]
13. Su, X.; Wang, R.; Li, X.; Araby, S.; Kuan, H.-C.; Naeem, M.; Ma, J. A comparative study of polymer nanocomposites containing multi-walled carbon nanotubes and graphene nanoplatelets. *Nano Mater. Sci.* **2022**, *4*, 185–204. [CrossRef]
14. Masarra, N.-A.; Batistella, M.; Quantin, J.-C.; Regazzi, A.; Pucci, M.F.; El Hage, R.; Lopez-Cuesta, J.-M. Fabrication of PLA/PCL/Graphene Nanoplatelet (GNP) Electrically Conductive Circuit Using the Fused Filament Fabrication (FFF) 3D Printing Technique. *Materials* **2022**, *15*, 762. [CrossRef] [PubMed]
15. Misra, S.K.; Ostadhossein, F.; Babu, R.; Kus, J.; Tankasala, D.; Sutrisno, A.; Walsh, K.A.; Bromfield, C.R.; Pan, D. 3D-printed multidrug-eluting stent from graphene-nanoplatelet-doped biodegradable polymer composite. *Adv. Healthc. Mater.* **2017**, *6*, 1700008. [CrossRef] [PubMed]
16. Jing, J.; Chen, Y.; Shi, S.; Yang, L.; Lambin, P. Facile and scalable fabrication of highly thermal conductive polyethylene/graphene nanocomposites by combining solid-state shear milling and FDM 3D-printing aligning methods. *Chem. Eng. J.* **2020**, *402*, 126218. [CrossRef]

17. Li, Z.; Li, B.; Chen, B.; Zhang, J.; Li, Y. 3D printed graphene/polyurethane wearable pressure sensor for motion fitness monitoring. *Nanotechnology* **2021**, *32*, 395503. [CrossRef]
18. Keramati, M.; Ghasemi, I.; Karrabi, M.; Azizi, H.; Sabzi, M. Incorporation of surface modified graphene nanoplatelets for development of shape memory PLA nanocomposite. *Fibers Polym.* **2016**, *17*, 1062–1068. [CrossRef]
19. Karatas, E.; Gul, O.; Karsli, N.G.; Yilmaz, T. Synergetic effect of graphene nanoplatelet, carbon fiber and coupling agent addition on the tribological, mechanical and thermal properties of polyamide 6, 6 composites. *Compos. Part B Eng.* **2019**, *163*, 730–739. [CrossRef]
20. Zhao, Z.; Teng, K.; Li, N.; Li, X.; Xu, Z.; Chen, L.; Niu, J.; Fu, H.; Zhao, L.; Liu, Y. Mechanical, thermal and interfacial performances of carbon fiber reinforced composites flavored by carbon nanotube in matrix/interface. *Compos. Struct.* **2017**, *159*, 761–772. [CrossRef]
21. Zang, C.G.; Zhu, X.D.; Jiao, Q.J. Enhanced mechanical and electrical properties of nylon-6 composite by using carbon fiber/graphene multiscale structure as additive. *J. Appl. Polym. Sci.* **2015**, *132*, n/a. [CrossRef]
22. Keramati, M.; Ghasemi, I.; Karrabi, M.; Azizi, H.; Sabzi, M. Dispersion of graphene nanoplatelets in polylactic acid with the aid of a zwitterionic surfactant: Evaluation of the shape memory behavior. *Polym. Plast. Technol.* **2016**, *55*, 1039–1047. [CrossRef]
23. Choi, J.T.; Dao, T.D.; Oh, K.M.; Lee, H.-I.; Jeong, H.M.; Kim, B.K. Shape memory polyurethane nanocomposites with functionalized graphene. *Smart Mater. Struct.* **2012**, *21*, 075017. [CrossRef]
24. Cao, Y.; Feng, J.; Wu, P. Alkyl-functionalized graphene nanosheets with improved lipophilicity. *Carbon* **2010**, *48*, 1683–1685. [CrossRef]
25. Wang, W.; Wang, X.; Pan, Y.; Liew, K.M.; Mohamed, O.A.; Song, L.; Hu, Y. Synthesis of phosphorylated graphene oxide based multilayer coating: Self-assembly method and application for improving the fire safety of cotton fabrics. *Ind. Eng. Chem. Res.* **2017**, *56*, 6664–6670. [CrossRef]
26. Zeng, M.; Wang, J.; Li, R.; Liu, J.; Chen, W.; Xu, Q.; Gu, Y. The curing behavior and thermal property of graphene oxide/benzoxazine nanocomposites. *Polymer* **2013**, *54*, 3107–3116. [CrossRef]
27. Li, L.; Li, B. Rheology, morphology and mechanical property relationship of non-halogen flame retarded glass fibre reinforced polyamide 66. *Polym. Polym. Compos.* **2011**, *19*, 603–610. [CrossRef]
28. Zadhoush, A.; Reyhani, R.; Naeimirad, M. Evaluation of surface modification impact on PP/MWCNT nanocomposites by rheological and mechanical characterization, assisted with morphological image processing. *Polym. Compos.* **2019**, *40*, E501–E510. [CrossRef]
29. Baimark, Y.; Srihanam, P. Influence of chain extender on thermal properties and melt flow index of stereocomplex PLA. *Polym. Test.* **2015**, *45*, 52–57. [CrossRef]
30. Zhu, D.; Ren, Y.; Liao, G.; Jiang, S.; Liu, F.; Guo, J.; Xu, G. Thermal and mechanical properties of polyamide 12/graphene nanoplatelets nanocomposites and parts fabricated by fused deposition modeling. *J. Appl. Polym. Sci.* **2017**, *134*, 45332. [CrossRef]
31. Jun, Y.S.; Hyun, B.G.; Hamidinejad, M.; Habibpour, S.; Yu, A.; Park, C.B. Maintaining electrical conductivity of microcellular MWCNT/TPU composites after deformation. *Compos. Part B Eng.* **2021**, *223*, 109113. [CrossRef]
32. Wang, Q.; Wang, Y.; Meng, Q.; Wang, T.; Guo, W.; Wu, G.; You, L. Preparation of high antistatic HDPE/polyaniline encapsulated graphene nanoplatelet composites by solution blending. *RSC Adv.* **2017**, *7*, 2796–2803. [CrossRef]
33. Albadarin, A.B.; Yang, Z.; Mangwandi, C.; Glocheux, Y.; Walker, G.; Ahmad, M. Experimental design and batch experiments for optimization of Cr (VI) removal from aqueous solutions by hydrous cerium oxide nanoparticles. *Chem. Eng. Sci. Des.* **2014**, *92*, 1354–1362. [CrossRef]
34. Rai, V.; Mukherjee, R.; Routray, A.; Ghosh, A.K.; Roy, S.; Ghosh, B.P.; Mandal, P.B.; Bose, S.; Chakraborty, C. Serum-based diagnostic prediction of oral submucous fibrosis using FTIR spectrometry. *Spectrochim. Acta A Mol. Biomol. Spectrosc.* **2018**, *189*, 322–329. [CrossRef]
35. Kumar, A.R.; Selvaraj, S.; Jayaprakash, K.; Gunasekaran, S.; Kumaresan, S.; Devanathan, J.; Selvam, K.; Ramadass, L.; Mani, M.; Rajkumar, P. Multi-spectroscopic (FT-IR, FT-Raman, 1H NMR and 13C NMR) investigations on syringaldehyde. *J. Mol. Struct.* **2021**, *1229*, 129490. [CrossRef]
36. Rashidi, A.; Shahidi, S.; Ghoranneviss, M.; Dalalsharifi, S.; Wiener, J. Effect of plasma on the zeta potential of cotton fabrics. *Plasma Sci. Technol.* **2013**, *15*, 455. [CrossRef]
37. Borooj, M.B.; Shoushtari, A.M.; Sabet, E.N.; Haji, A. Influence of oxygen plasma treatment parameters on the properties of carbon fiber. *J. Adhes. Sci. Technol.* **2016**, *30*, 2372–2382. [CrossRef]
38. Viji, S.; Anbazhagi, M.; Ponpandian, N.; Mangalaraj, D.; Jeyanthi, S.; Santhanam, P.; Devi, A.S.; Viswanathan, C. Diatom-based label-free optical biosensor for biomolecules. *Appl. Biochem. Biotech.* **2014**, *174*, 1166–1173. [CrossRef]
39. Rinawati, L.; Nugraha, R.E.; Munifa, R.M.I.; Chasanah, U.; Wahyuningsih, S.; Ramelan, A.H. Increasing the effectiveness of pesticides based urea nanofertilizer encapsulatednanosilica with addition of rice husk TiO_2 additive substances. *Chem. Pharm. Res.* **2015**, *7*, 85–89.
40. Kozior, T. Rheological properties of polyamide pa 2200 in sls technology. *Teh. VJesn.* **2020**, *27*, 1092–1100. [CrossRef]
41. Thumsorn, S.; Prasong, W.; Kurose, T.; Ishigami, A.; Kobayashi, Y.; Ito, H. Rheological Behavior and Dynamic Mechanical Properties for Interpretation of Layer Adhesion in FDM 3D Printing. *Polymers* **2022**, *14*, 2721. [CrossRef]
42. Li, J.; Li, Z.; Chen, H.; Yang, L.; Zheng, H.; Shang, Y.; Yu, D.; Christiansen, J.D.; Jiang, S. A qualitative analysis of particle-induced viscosity reduction in polymeric composites. *J. Mater. Sci.* **2016**, *51*, 3080–3096. [CrossRef]

43. Lim, B.; Poh, C.; Voon, C.H.; Salmah, H. Rheological and thermal study of chitosan filled thermoplastic elastomer composites. *Appl. Mech. Mater.* **2015**, *754–755*, 34–38. [CrossRef]
44. Rahaman, M.; Theravalappil, R.; Bhandari, S.; Nayak, L.; Bhagabati, P. Electrical conductivity of polymer-graphene composites. In *Polymer Nanocomposites Containing Graphene*; Elsevier: Cambridge, UK, 2022; pp. 107–139. [CrossRef]
45. Kim, H.; Kobayashi, S.; AbdurRahim, M.A.; Zhang, M.J.; Khusainova, A.; Hillmyer, M.A.; Abdala, A.A.; Macosko, C.W. Graphene/polyethylene nanocomposites: Effect of polyethylene functionalization and blending methods. *Polymer* **2011**, *52*, 1837–1846. [CrossRef]
46. Bilkar, D.; Keshavamurthy, R.; Tambrallimath, V. Influence of carbon nanofiber reinforcement on mechanical properties of polymer composites developed by FDM. *Mater. Today Proc.* **2021**, *46*, 4559–4562. [CrossRef]
47. Chohan, J.S.; Singh, R.; Boparai, K.S.; Penna, R.; Fraternali, F. Dimensional accuracy analysis of coupled fused deposition modeling and vapour smoothing operations for biomedical applications. *Compos. B Eng.* **2017**, *117*, 138–149. [CrossRef]
48. Butt, J.; Bhaskar, R.; Mohaghegh, V. Investigating the Influence of Material Extrusion Rates and Line Widths on FFF-Printed Graphene-Enhanced PLA. *J. Manuf. Mater. Process.* **2022**, *6*, 57. [CrossRef]
49. Gao, X.; Zhang, D.; Qi, S.; Wen, X.; Su, Y. Mechanical properties of 3D parts fabricated by fused deposition modeling: Effect of various fillers in polylactide. *J. Appl. Polym. Sci.* **2019**, *136*, 47824. [CrossRef]
50. Benwood, C.; Anstey, A.; Andrzejewski, J.; Misra, M.; Mohanty, A.K. Improving the impact strength and heat resistance of 3D printed models: Structure, property, and processing correlations during fused deposition modeling (FDM) of poly (lactic acid). *ACS Omega* **2018**, *3*, 4400–4411. [CrossRef]
51. Zhang, G.; Wang, F.; Dai, J.; Huang, Z. Effect of functionalization of graphene nanoplatelets on the mechanical and thermal properties of silicone rubber composites. *Materials* **2016**, *9*, 92. [CrossRef]
52. Sezer, H.K.; Eren, O. FDM 3D printing of MWCNT re-inforced ABS nano-composite parts with enhanced mechanical and electrical properties. *J. Manuf. Process.* **2019**, *37*, 339–347. [CrossRef]

Article

Improved Electrical Signal of Non-Poled 3D Printed Zinc Oxide-Polyvinylidene Fluoride Nanocomposites

Sharmad Joshi [1], Enrique Gazmin [2], Jayden Glover [2], Nathan Weeks [2], Fazeel Khan [1], Scott Iacono [2] and Giancarlo Corti [1,*]

[1] Department of Mechanical and Manufacturing Engineering, Miami University, Oxford, OH 45056, USA
[2] Department of Chemistry & Chemistry Research Center, United States Air Force Academy, Colorado Springs, CO 80840, USA
* Correspondence: corticlg@miamioh.edu

Abstract: Polyvinylidene fluoride (PVDF) presents highly useful piezo and pyro electric properties but they are predicated upon the processing methods and the ensuing volume fraction of the β-phase. Production of PVDF with higher β-phase content for additive manufacturing (AM) is particularly desirable because it can enable the creation of custom parts with enhanced properties. Necessary steps from compounding to the testing of a 3D printed piezo sensitive sensor are presented in this paper. AM process variables and the influence of zinc oxide (ZnO) nanofiller on crystallinity, viscosity, and electromechanical properties of PVDF, have been explored. Fourier-transform infrared spectroscopy (FTIR) measurements confirm that a high cooling rate (HCR) of 30 $°C \cdot min^{-1}$ promotes the conversion of the α-into the β-phase, reaching a maximum of 80% conversion with 7.5–12.5% ZnO content. These processing conditions increase the elastic modulus up to 40%, while maintaining the ultimate strength, $\approx$46 MPa. Furthermore, HCR 10% ZnO-PVDF produces four times higher volts per Newton when compared to low cooling rate, 5 $°C \cdot min^{-1}$, pristine PVDF. A piezoelectric biomedical sensor application has been presented using HCR and ZnO nanofiller. This technique also reduces the need for post-poling which can reduce manufacturing time and cost.

Keywords: zinc oxide; PVDF; 3D printing; cooling rates; poling

Citation: Joshi, S.; Gazmin, E.; Glover, J.; Weeks, N.; Khan, F.; Iacono, S.; Corti, G. Improved Electrical Signal of Non-Poled 3D Printed Zinc Oxide-Polyvinylidene Fluoride Nanocomposites. *Polymers* 2022, 14, 4312. https://doi.org/10.3390/polym14204312

Academic Editor: Matthew Blacklock

Received: 10 September 2022
Accepted: 3 October 2022
Published: 13 October 2022

Publisher's Note: MDPI stays neutral with regard to jurisdictional claims in published maps and institutional affiliations.

Copyright: © 2022 by the authors. Licensee MDPI, Basel, Switzerland. This article is an open access article distributed under the terms and conditions of the Creative Commons Attribution (CC BY) license (https://creativecommons.org/licenses/by/4.0/).

1. Introduction

Polyvinylidene fluoride (PVDF) is a semi-crystalline polymer with a monomer unit of CH_2–CF_2 [1]. It has gained considerable attention in recent years due to its excellent piezoelectric and pyroelectric properties [2–5]. PVDF has a piezoelectric constant d33 of 20×10^{-12} CN^{-1} and a voltage constant g33 of 160×10^{-3} mVN^{-1} [6]. PVDF polymer shows less acoustic impedance and lower density compared to piezoelectric crystals [7,8]. PVDF has been used in sensors and actuators because it can convert mechanical deformations into coupled electric signals [4,9–11]. PVDF has also been employed in spin-valve devices because of its inherent properties such as thermal stability, short switching time, and large polarization. PVDF has also been used in capacitors, piezoelectric nanogenerators, [12–16], and nerve tissue engineering in biomedical applications [11]. From the five major crystalline polymorphs of PVDF, namely α, β, γ, δ and ϵ, β-phase PVDF is of singular interest because the all-trans planar zigzag structure of molecules in this phase induces a net dipole moment that results in the aforementioned piezoelectric properties [17]. Thus, a variety of methods have been reported in the scientific literature to obtain β-phase PVDF. However, to the authors' knowledge, there is minimal research available on mechanically compounding and 3D printing PVDF in the β-phase without poling.

Traditionally, the α to β transformation takes place when the α-phase is mechanically drawn or stretched [18] at temperatures less than 100 °C, with stretch ratios ranging from 1 to 5 [19]. Solvent casting techniques such as electrospinning and spin coating have been widely used to obtain thin-film composites or a raw material to fabricate filaments for Fused Deposition Manufacturing (FDM) [20–25]. A variety of nanofillers have also been used to nucleate β-phase crystals such as barium titanate, zinc oxide (ZnO) and titanium dioxide (TiO_2) [26–29]. Certain PVDF copolymers such as P(VDF-TrFE) possess piezoelectric properties, as previously reported in the literature [23,30,31]. Dodds et al. found increased residual polarization with the addition of ZnO composites and speculated greater film piezoelectricity. Hu et al. also showed encouraging results demonstrating remnant polarization and β-phase crystallization with comparatively less amount of the nano-filler (graphene oxide or carbon nanotubes) [24]. Kim et al., Xiong et al., and Haddadi et al. used this process to obtain PVDF/silica composites where a marked transition from α to β phase was observed in all three cases, but a mixed result was seen in an attempt to enhance the mechanical properties of pure PVDF films [21,22,32].

Castanet et al. extruded a ZnO-PVDF nanocomposite and reported increments in β phase proportional to the amount of ZnO [27]. Kim et al. and Kennedy et al. employed a DMF solvent-cast approach to mix the nano-filler with PVDF before extruding the mix into FDM or Fused Filament Fabrication (FFF) or 3D print filaments. The 3D printed nanocomposite of Kim et al. used $BaTiO_3$ as filler, and it was thermally poled after printing the samples [33,34]. Kennedy et al. used multiwalled-carbon nanotube (MWCNT) as a filler, and the 3D printed to be used as chemresistors rather than a piezoelectric sensor [35]. Porter et al. 3D printed PVDF films and demonstrated measurable piezoelectricity after a corona poling was applied [36]. Recently, a rapid precipitation 3D printing technique where a DMF-PVDF solution is 3D printed into a reservoir containing a non-solvent demonstrated a 78% improvement on the β-phase fraction after hot-pressing and poling the 3D printed samples [37]. Most of these procedures have only been utilized for obtaining thin films, limiting the capabilities of 3D printers, which is a flexible technique that can be employed to fabricate complex bulk specimens. Minimum research has been done on 3D printing bulk samples of piezoelectric plastics such as PVDF. Moreover, the few studies that have 3D printed PVDF composite thin films have also required the need of additional techniques such as thermal poling or solvent evaporation to obtain a higher concentration of β-phase PVDF or are limited to the mechanical and printability properties of PVDF [10,25,33–35].

The research presented in this paper distinguishes itself from prior works by demonstrating how the flexibility of 3D printing can be harnessed to create PVDF parts directly in the β-phase. This effectively eliminates restrictions on sample/part shape and avoids time-cost intensive post processes, opening the possibility to 3D print sensors or bistable morphing generators [38]. Specifically, ZnO filler was used to promote β-phase PVDF formation in situ during the printing of the polymer composite material without poling. A simple and scalable manufacturing procedure (Figure 1) was envisioned by melt compounding PVDF and ZnO nanoparticles into a composite polymer filament to be used in commercially available FDM 3D printers. The extruded filaments were analyzed by differential scanning calorimetry (DSC), thermogravimetric analysis (TGA), Fourier transform infrared (FTIR) spectroscopy, and rheometry. Furthermore, this study expands on the effect that cooling rates have on the formation of the β-phase, without poling or stretching, to 3D print a ZnO-PVDF nanocomposite with an increased percentage of β phase content. Based on the material characterization results, two different 3D printing parameters were identified to directly affect the percentage of β-phase in the final printed specimen. Furthermore, the electrical and mechanical characterization methodology devised in this study demonstrates significant potential for creating 3D printed ZnO-PVDF sensors that are of custom design and could be integrated into shoe insoles to monitor human motion and posture. The various steps from the compounding of the ZnO-PVDF polymer to the creation and testing of the shoe insole sensor are illustrated in Figure 1 to explain the significance of the work from the materials and process design to application development.

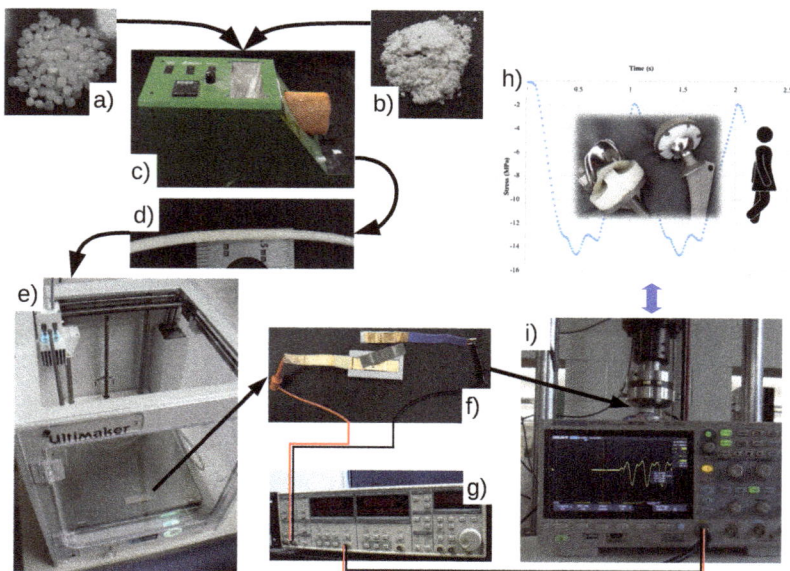

Figure 1. Materials, fabrication, and testing: (**a**) PVDF, (**b**) ZnO, (**c**) compounding and extrusion of the ZnO-PVDF nanocomposite into (**d**) 2.8 mm filaments were vacuum dried at 40 °C for 40 min, (**e**) 3D printing of test specimens, (**f**) electrically insulated test support with copper electrodes where the sample painted with conductive pain will rest, (**g**) lock-in amplifier connected directly to the test support electrodes, (**h**) walking profile with a 2 mm amplitude programmed into an MTS testing frame, and (**i**) measuring voltage generated during testing with an oscilloscope.

2. Materials and Methods

2.1. Materials

PVDF composite filaments were extruded with PVDF from ARKEMA Kynar® 740 with a molecular weight of 250,000 gmol^{-1} [3], melt flow rate ASTM D1238 of 1.5–3 kg, a density of 0.961 gcm^{-3}, and melting point of 165–172 °C. Zinc oxide (30-1405, CAS 1314-13-2) with an average size of 20 nm and a density of 5.606 gcm^{-3} was purchased from Strem Chemicals. The conductive nickel paint (841AR Super Shield Nickel Conductive Coating) with a surface resistance of 0.7 Ωsq^{-1} for a thickness of 0.05 mm was purchased from MG Chemicals, Burlington, ON, Canada.

2.2. Extrusion

ZnO nanopowders were mixed in isopropyl alcohol and sonicated for 15 min, then PVDF pellets were added, and the mixture was sonicated for another 15 min. This mixture was placed in a vacuum oven at 40 °C to dry. The mixture was then compounded in an filament extruder (EX-2 from Filabot, Barre, VT, USA) with a 3 mm nozzle and placed 285 mm above the air path, which was set at 100% fan speed, to cool the extruded filament to ensure its roundness. The mixture was compounded for several passes to improve the uniform distribution of the ZnO nanoparticles throughout the whole filament. The first two passes were extruded at 170 °C at the maximum extruder speed and half speed of the spooler. Under these conditions, a 1.5 mm filament was achieved. The temperature was raised to 210 °C for the next three passes while maintaining the extruder and spooler speeds. In between each pass, the filament was mechanically pelletized and placed back into the extruder. The final pass was performed at an extruder temperature of 170 °C. The extruder and spooler speeds were decreased to ≈50% and ≈0.9 turns, respectively, to achieve a diameter of ≈2.5 mm required by the Ultimaker3 3D printer. At 170 °C, the extruder did not have enough torque to produce a uniform filament for 20% concentration of ZnO; thus the velocity of the extruder was reduced to accommodate this limitation.

2.3. 3D Printing

An Ultimaker3 3D printer (Ultimaker, Utrecht, The Netherlands) with a closed environment was used to print all samples. The mechanical properties of the PVDF nanocomposites were tested on four types of specimens. A 20 mm diameter by 1 mm thick rheometer specimens, 50 × 6 × 1 mm DMA specimen, ASTM D790 50.8 × 12.7 × 1.6 mm three-point bending specimen, and ASTM D638 Type V tensile specimen for tensile tests. 3D printer g-code was generated using Cura with 100% infill following the conditions shown in Table 1. Three samples of each specimen type were printed for both cooling profiles, High and Low Cooling Rates (HCR and LCR), and also for each zinc oxide percentage. The LCR profile was achieved by turning off the printer cooling fan, increasing the nozzle and bed temperatures, and maintaining a closed-door during the print. Due to the enclosed environment and the higher temperatures, there is a smaller temperature differential between the melted plastic and the environment. Instead, the HCR profile had a lower bed, and nozzle temperature, an open door, and the printer cooling fan was also on to maximum speed, thus forcing cooler room air on the printed specimen. Although PVDF does not absorbe water, all filaments were dried for 40 min in a vacuum dry oven at a 40 °C.

Table 1. 3D printing parameters.

Parameter	Units	LCR	HCR
Initial layer height	mm	0.27	0.27
Layer height	mm	0.1	0.1
Line width	mm	0.35	0.35
Infill direction	°	[0, 0] & [45, 135]	
Infill percentage	%	100	100
Printing temperature	°C	250	225
Build plate temperature	°C	100	50
Printing speed	mm s^{-1}	20	20
Cooling fan	%	0	100

2.4. SEM Imaging and EDS

A field emission scanning electron microscope (FE-SEM) (Supra 35 VP FEG from Zeiss, Jena Germany) with an X-ray microanalysis system (XEDS) (Quantax 100 from Bruker, Billerica, MA, USA) was employed to analyze the fracture of the specimens. All specimens were coated with 5 nm of gold, and the images were taken at 5 KV to minimize charging with a working distance of 4 mm. Further characterization used mapping XEDS, and these spectra were collected using an accelerating voltage of 20 keV over a span of 300 live seconds at a 10k× magnification and 10 mm of working distance. The most prominent elements were fluorine (F), and zinc (Zn) which were used to identify the agglomeration of the ZnO nanoparticles.

2.5. Thermo-Gravimetric Analysis—TGA

The thermal stability of the ZnO-PVDF and PVDF-TiO$_2$ composite and the filler concentration were measured via a thermogravimetric analyzer (TGA) (TGAQ500 from TA Instruments, New Castle, DE, USA). All samples were heated in a nitrogen (N$_2$) atmosphere up to 850 °C at a heating rate of 10 °C·min^{-1}. Three random samples from the ends and the middle section of each filament were tested and analyzed using the TA Universal Analysis software (TA Instruments, New Castle, DE, USA).

2.6. Differential Scanning Calorimetry—DSC

Differential scanning calorimetry (DSC) was employed to identify the melting (T_m), crystallization (T_c), and glass transition (T_g) temperature of the samples. Two cooling rates were tested to identify the cooling rate's effect on the crystallization of PVDF nanocomposites, a LCR of 5 °C·min^{-1}, and a HCR of 30 °C·min^{-1}. T_m was always measured at a heating rate of 5 °C·min^{-1}. Considering that higher heating rates provide a better resolution to identify the T_g point, a 10 °C·min^{-1} was selected for these experiments. All DSC data were collected on the third run. Three samples from the ends of each filament were characterized using DSC TAQ100 (TA Instruments, New Castle, DE, USA). The collected data were analyzed with TA Universal Analysis software (TA Instruments, New Castle, DE, USA).

2.7. Fourier Transform Infrared Spectroscopy—FTIR

The polymeric nanocomposite material was 3D printed into specimens following the printing procedure stated earlier. The samples were characterized using a Nicolet™ iS™ 5 FTIR Spectrometer (Thermo Fisher Scientific, Waltham, MA, USA). FTIR measurements were collected at one cm^{-1} with 32 scans. All DSC samples and all printed DMA specimens were analyzed by FTIR.

2.8. Rheometry

The rheology data was collected on a HAAKE RheoWin 4.82.00 with a MARS 40 measuring device (Thermo Fisher Scientific, Waltham, MA, USA). A PP20 Adapter-01180765 was used for measuring the geometry. The PDVF-ZnO samples in Table S1 were all analyzed on the rheometer at three temperatures, 200 °C, 225 °C and 250 °C. The instrument was heated to the target temperature and then the sample was loaded into the test chamber. The instrument then allowed the sample to heat up for 120 s. Following the temperature adjustment, the instrument then began collecting data and ran for 100 s. During the 100 s^{-1} test interval, the instrument measured steady viscosity (Pa·s) vs. the shear rate (s^{-1}).

2.9. DMA

The width and thicknesses of all DMA specimens were measured at three different points, and the average dimension was used in the stress and strain calculations. A RS3 DMA (TA Instruments, New Castle, DE, USA) was used to measure the storage modulus (E′), loss modulus (E″), and tan(δ) (E″/E′). During testing, specimens were allowed to stabilize at room temperature of 23 °C by introducing an isothermal segment of 2 min before a ramp-up cycle up to 150 °C at a rate of 5 °C·min^{-1}. Data were collected at an interval of 10 s.

2.10. Tensile and Flexion

Tensile tests were conducted according to the ASTM D638 standard type V for tensile properties of reinforced plastics using [39]. Similar to the DMA test, the widths and thicknesses of all the samples were measured at three different points, and the average of these dimensions was used for the calculation of tensile strength and modulus of the specimen. The crosshead speed was set at 1 mm·mm^{-1}·min^{-1} according to the standard [39]. The specimen was allowed to reach the rupture point. The load and the extension and of each specimen were recorded every 50 ms. Three samples for each printing condition and material combination were tested.

Flexural tests were conducted according to the ASTM D790 standard. In this case, a 3 point bend test fixture was fitted in the grippers of 30 kN Instron. Subsequently, the specimen was tested by placing them centrally on the supports. The crosshead speed was set at 1 mm·min^{-1} according to the standard [40]. Similar to the tensile test, the dimensions were taken at three different points and the average was used for the calculation of flexural modulus. The load and the extension of each specimen were recorded every 20 ms. Three samples for each printing condition and material combination were tested. Both tensile and flexural tests were performed on a Universal test frame (series 5960 Instron, Norwood, MA, USA).

2.11. Electrical Response

The electrical response of the 3D printed PVDF and ZnO-PVDF nanocomposite was measure in a three-point bending test under a cyclic strain of 0.2–0.4% at 6 Hz initially and later at 0.8 and 4 Hz to match the walking gait profile of an average person [41]. The ASTM D790 3-point bending specimens were coated with a conductive nickel paint with an approximate thickness of 0.5 mm. The RS3 DMA (TA Instruments, New Castle, DE, USA) was employed to strain the sample for 10 s. A 3-point bending plastic support with a 25 mm span (Figure 1) fitted with copper electrodes was mounted on the DMA. The copper electrodes were connected with a BNC cable to a lock-in amplifier (SR810 from Stanford Research Systems, Sunnyvale, CA, USA) and the output signal was measured with an oscilloscope (SDS1104X-E from Siglent, Shenzhen China). The applied force was calculated using the measured dimensions and the average modulus of elasticity of the samples. The simulated walking gait profile was programmed into a hydraulic MTS load frame (100 kN Landmark from MTS, Eden Prairie, MN, USA) and use the same electrical connections as previous tests. Two samples samples for each printing condition and material combination were tested.

3. Results and Discussion

During the filament extrusion provides the first insights on the behavior concentration of ZnO nanopowder

3.1. Materials Characterization

Since 3D printing PVDF requires a nozzle temperature of 250 °C and PVDF decomposes in toxic fumes, the thermal stability of the PVDF nanocomposite was first evaluated. Thermogravimetric analysis of ZnO-PVDF (Figure 2) shows a thermal stability increase as a function of the ZnO content. The decomposition temperature increases up to 25 °C higher than that of pristine PVDF for a 20% content of ZnO. This increased thermal stability is attributed to the interaction between the polymer and ZnO nanoparticles. TGA data also determined the amount of ZnO present in each compound filament. Table S1 shows the average amount of ZnO remaining after the calcination of PVDF, while there was a loss of ZnO during compounding, TGA measurements show a low standard deviation and error on the ZnO weight left after PVDF's calcination, suggesting a uniform extruded filament.

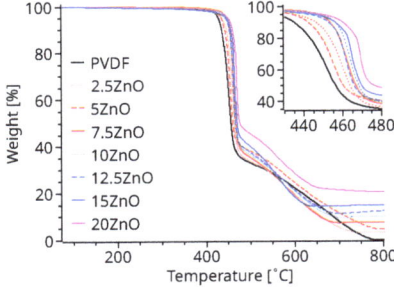

Figure 2. TGA measurements (inset shows a magnification in the area of interest).

The glass transition temperature (Figure S1), measured using a differential scanning calorimetry, increased as a function of the percentage of ZnO content of the samples. In agreement with existing literature, smaller changes were noticed for samples with less than 7.5% of ZnO, which presented T_g at −40.1 °C and 30.9 °C for pristine PVDF and 7.5% of ZnO, respectively. 20% ZnO samples showed the highest T_g at −7.5 °C (Figure S1). The increasing T_g can be explained by the lower molecular mobility of the PVDF chains induced by the ZnO nanoparticles' and the crystallinity of ZnO [42].

The crystallinity of PVDF and ZnO-PVDF nanocomposite samples (Figure 3) was calculated using Equation (1) with a melting enthalpy of 100% PVDF H_f^* equal to 104.6 Jg^{-1}, and heat flow ΔH_f equal to the area under the crystallization peak (Figure S2) [43–45]. The overall crystallization time (Figure 3) was calculated by subtracting the time at the crystallization peak temperature from the on-set temperature-time (Figure S2). The effect of the cooling rate on the crystallinity and time of crystallization was analyzed by cooling the samples at two different rates: a LCR of 5 °C·min^{-1} and a HCR of 30 °C·min^{-1}.

$$X_c = \frac{\Delta H_f}{\Delta H_f^*} \times 100\%. \tag{1}$$

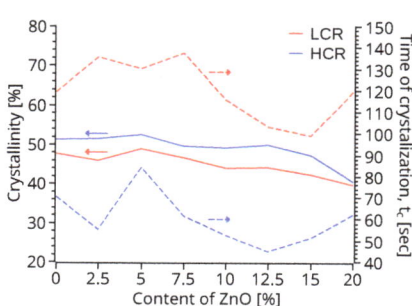

Figure 3. Comparison of the crystallinity percentage and the time of crystallization, t_c as function of the ZnO sample content and cooling rate.

High cooling rate samples produced a slightly larger percentage of crystallization (Figure 3) than the low cooling rate samples, while with HCR, the degree of crystallization is relatively stable until 15% of ZnO; thereafter, the crystallinity decreases. This behavior is also observable on LCR, where the crystallinity of the samples drops for contents of ZnO larger than 12.5%. This reduction of crystallinity could be attributed to the reduction of nucleation sites due to the agglomeration of the ZnO nanoparticles as shown on the mapping XEDS (Figure S3) due to the increased content of ZnO as previously reported by Castanet et al. [27]. ZnO nanoparticles have a significant contribution, mainly in the range of 10–15% of ZnO, on the reduction of the time to crystallization (Figure S1), ≈35% and ≈12% for the HCR and LCR, respectively, with respect to pristine PVDF. The reduction of the time to crystallization can be attributed to the increased nucleation sites created by the presence of the ZnO nanoparticles. The ZnO nanofiller also increased the temperature of crystallization (Figure S2), reaching up to ≈7 °C higher for both cooling rates. Samples with 10–15% ZnO also depict a narrower crystalization peak compared to pristine PVDF, suggesting a narrow crystallite size distribution [46,47].

Following the crystallization cycles, both high and low cooling rate samples were melted at 5 °C·min^{-1}. DSC melting cycles following LCR exhibited a lower temperature shoulder or a decrease of the melting temperature, with respect to pristine PVDF, 170.3 °C, for all ZnO-PVDF samples above 5% of ZnO (Figure 4). Samples with 7.5% of ZnO presented the lowest melting temperature at 167.9 °C for LCR samples. Instead, HCR produced a double melting peak or a shoulder of the main melting temperature peak for all samples. This lower-temperature melting peak decreased with the increase of ZnO content until reaching a minimum with 15% of ZnO content at 164.4 °C, which is the melting temperature of β-phase of PVDF. Sencadas et al., Harstad et al. and Soin et al. reported the formation of a shoulder and a second peak at 165 °C to be proportional to the increment the PVDF β-phase [48–50]. Literature also indicates that the γ-phase have a 175 °C $< T_m <$ 175 °C, the α-phase have a 172 °C $< T_m <$ 180 °C, and the β-phase has a 165 °C $< T_m <$ 172 °C [12,14,16,50,51]. Both HCR and LCR samples, present no visisble peaks or shoulders at temperatures above 175 °C, suggesting a minimum presence of the γ phase on the ZnO-PVDF nanocomposite [51].

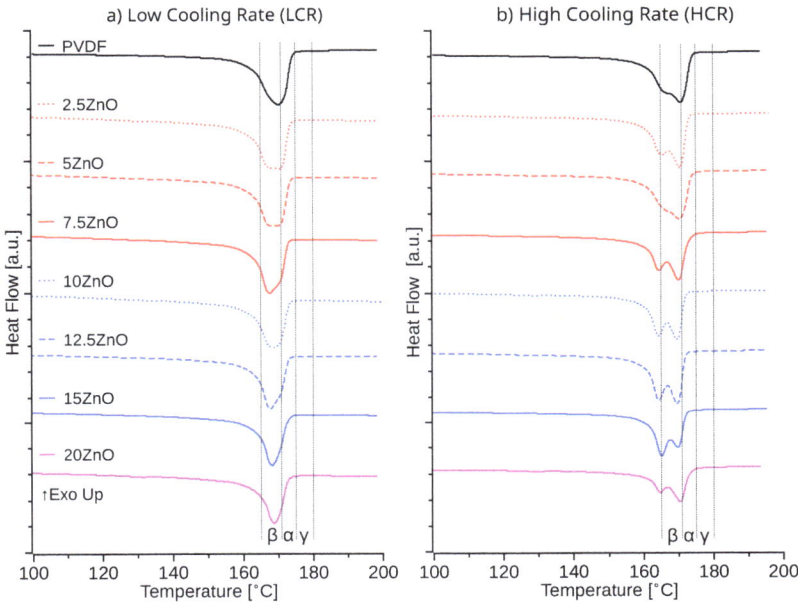

Figure 4. DSC melting temperature for (**a**) 5 °C·min^{-1} LCR, and (**b**) 30 °C·min^{-1} HCR.

FTIR spectra (Figure S4) were collected from the bottom of the DSC samples after cooling the samples at HCR and LCR on the third DSC cycle. There are no appreciable shoulders on 1234 cm^{-1}, suggesting the lack or minimum content of the γ-phase. The 840 cm^{-1}, which represents the β phase and the γ-phase, decreases its transmittance as a function of the ZnO reaching a minimum transmission with 10–12.5% of ZnO, for both HCR and LCR samples. Similarly, the 1275 cm^{-1} peak, which represents the β phase solely, decreases its transmittance as a function of the amount of ZnO reaching a minimum transmission with 10% of ZnO, while the α phase 760 cm^{-1} increase for both HCR and LCR samples. These findings are in agreement with existing literature [14,37,51,52]. Moreover, FTIR result corroborate the findings from DSC melting temperatures. Based on these findings, two FDM or 3D printing profiles were designed, Table 1, to mimic as close as possible both DSC, high and low, cooling rates. It is important to mention that the 3D printed LCR profile was limited by the 3D printing environmental control, which produced a cooling rate higher than 5 °C·min^{-1}. Specimens for rheometry, DMA, tensile and flexural tests were 3D printed for subsequent analysis of the PVDF and ZnO-PVDF nanocomposite mechanical properties.

FTIR measurements (Figure S5) were also performed on the upper surface of all the 3D printed DMA specimens. These spectra exhibit a reduction of the transmittance on the 400–500 cm^{-1} band assigned to the characteristic stretching mode of Zn-O bond proportional to the ZnO percentage in the sample. In addition (Figure 5), the α peak, 760 cm^{-1}, increases its transmittance as a function of the ZnO percentage, reaching a maximum at 10–12.5% of ZnO, while the $\beta - \gamma$ peak, 840 cm^{-1}, and the β peak, 1275 cm^{-1}, decrease their transmittance proportionally to the percentage of ZnO. A minimum transmittance for the peak of the 840 cm^{-1}, and 1275 cm^{-1} peaks was found at concentrations of 10–12.5% of ZnO. Similarly to the DSC samples, there is no apparent change to the 1234 cm^{-1} γ peak. When comparing the changes in transmittance, there is also a reduction proportional to the ZnO percentage with a maximum of 80% reduction (Figure 5) of the α minus β ratio with respect to the pristine PVDF for the 10% ZnO samples. Due to the smaller difference in cooling rates of the 3D printed samples compared to the DSC samples, there are no statistical differences between the two printing profiles.

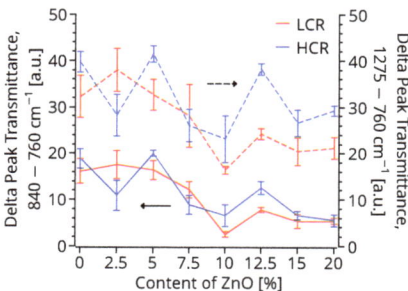

Figure 5. Comparison of transmittance changes for the α and β peaks as a function of the ZnO sample content and cooling rate.

3.2. Rheological and Mechanical Properties

Lower shear viscosity and elasticity have been related to ease of 3D printing and quality [53,54]. The viscosity curves (Figure 6) obtained from the rheometer show a broad plateau of constant shear viscosity, e.g., below shear rates of 10 s^{-1} for pristine PVDF. Above shear rates of 10 s^{-1}, there is a significant reduction, e.g., one order of magnitude, of the shear viscosity as a function of the temperature. There are no significant differences between the HCR and LCR samples using pristine PVDF. However, ZnO-PVDF nanocomposite samples exhibit increased shear stress (Supporting Information Figure S5) and shear viscosity (Figure 6) at temperatures below 200 °C. Suggesting, a reinforcing effect due to the ZnO nanoparticles. This effect also increased pressure on the extruder for ZnO concentration above 15%, resulting in a reduction of the extrusion speed to produce a uniform filament. However, at higher temperatures, above 225 °C, ZnO produced a reduction of the shear viscosity of ≈678 and ≈827 Pa·s for the LCR and HCR, respectively, at lower shear rates of ≈2 s^{-1} than the pristine PVDF. The thinning effect of the ZnO is more significant with HCR of PVDF nanocomposites, reaching a minimum shear viscosity of ≈68 Pa·s, and shear stress of ≈680 Pa, at 250 °C and 10 s^{-1}, for a ZnO concentration of 7.5–12.5%. The ZnO thinning effect can be attributed to the ZnO nanoparticle size being smaller than the polymer's RMS radius of gyration as seen before by Mackay et al. and Kairn et al. [55,56]. However, as the concentration of ZnO increases above 12.5%, the shear viscosity (Figure 6) and shear stress of PVDF (Figure S6) began to increase. The filler-like behavior is attributed to the large concentration of ZnO and agglomeration of the nanoparticles (Supporting Information Figure S3). Similar thinning behavior, for <12.5% ZnO, and filler-like behavior, for >12.5% ZnO, is present in both the LCR and HCR samples for temperatures above 225 °C. Due to the thinning effect of ZnO, PVDF-ZnO nanocomposites have significant lower viscosities, ≈10 Pa·s, than pristine PVDF, ≈100 Pa·s, at higher shear rates, e.g., $\geq 100 \text{ s}^{-1}$. Thus, making PVDF-ZnO nanocomposite a better option for FDM 3D printers since typical shear rates at the nozzle are in the range of 10^2–10^3 s^{-1} depending on the printing filament speeds [54].

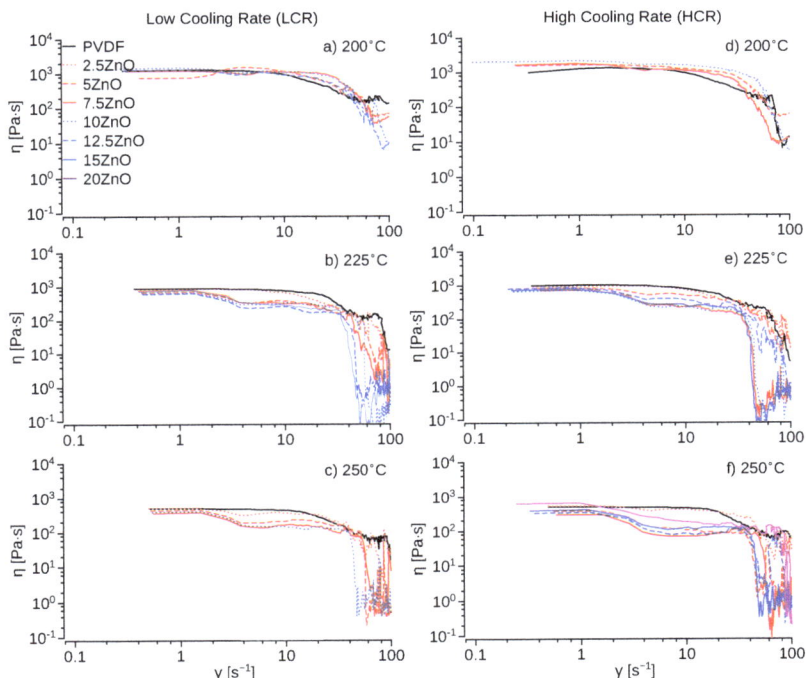

Figure 6. Shear viscosity η versus shear rate γ at different temperature profiles for low and high cooling rate 3D printed samples.

DMA specimens of both 3D printing profiles show a storage modulus (Figure 7), an order of magnitude higher than the loss modulus (Figure S7). Hence, material elasticity remains dominant at lower temperatures. In the high-temperature region, the difference in the magnitudes of the two moduli reduces to an extent, yet the loss modulus remains lower than the storage modulus. In the LCR samples, E' (Figure 7a) and E" (Figure S7a) decrease with the increasing percentage of ZnO. A similar trend was observed in the material viscosity (Figure 6). The HCR specimens show a proportional elastic modulus (Figure 7b) and loss modulus (Figure S7b) to the percentage of ZnO. The stiffening effect could be due to the lower time to crystallization (Figure 3), good dispersion of the nanoparticles in the matrix, and the high surface area of the ZnO particles as reported by Cao et al. [57]. The increment in the storage modulus (Figure 7b) indicates that HCR condition with ZnO-PVDF nanocomposite material creates higher resilience to high temperatures. The opposite trend observed in LCR specimens (Figure 7a) indicates that lower cooling rates hinder the stiffening effect of ZnO nanoparticles. Similar results were obtained by Flyagina et al. [58]. The larger crystallites may explain the softening effect that slow cooling has on ZnO-PVDF nanocomposite due to the longer time to crystallization of ZnO-PVDF nanocomposite (Figure 3).

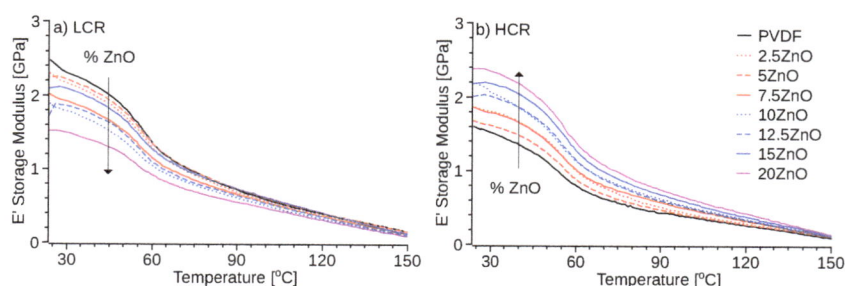

Figure 7. Storage modulus (E') of DMA specimens 3D printed with: (**a**) low cooling rate (LCR), and (**b**) high cooling rate (HCR).

Tensile and flexural tests were also carried out following ASTM D638 and D790 for both cooling rates specimens [39,40]. Two infill configurations were studied; an infill orientation [0°, 0°], and an infill orientation of [45°, 135°] with respect of the tensile force. Pristine PVDF has a natural elongation, which decreases with an increased percentage of ZnO filler (Supporting Information Figure S8a,b) when printed at LCR regardless of the printing infill direction. HCR printed samples produced a higher elongation than LCR for pristine PVDF at low concentrations of ZnO, e.g., <5%, (Supporting Information Figure S8c,d) regardless of the infill printing orientation. SEM images of the tested specimens also corroborate these findings. LCR presented minimum deformation at the breaking point, while HCR showed (Figure 8) a plastic deformation before rupture. PVDF and PVDF-ZnO nanocomposite produced high-quality prints (Figure 9) with no appreciable difference between HCR and LCR printing profiles.

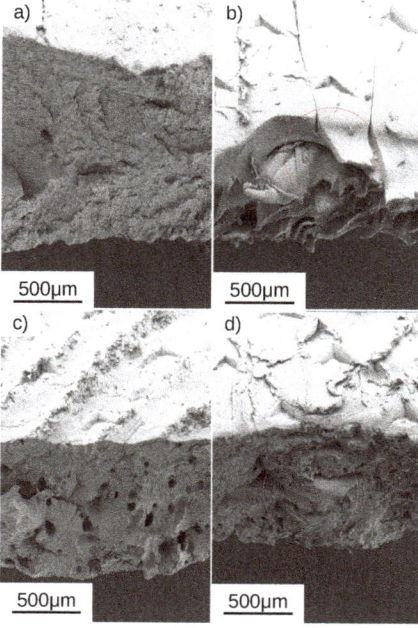

Figure 8. SEM micrographs of representative tensile specimens 3D printed with a [45°, 135°] infill direction: (**a**) LCR pristine PVDF, (**b**) HCR pristine PVDF, (**c**) LCR of 10% ZnO-PVDF, and (**d**) HCR of 10% ZnO-PVDF. The red circles shows elongation characteristic of the ductile failure of the HCR sample.

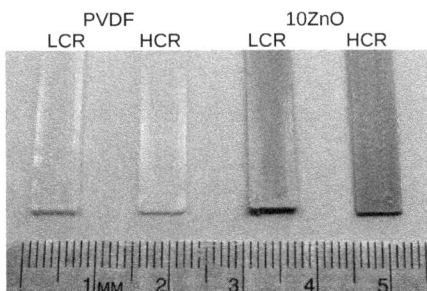

Figure 9. Optical image comparing DMA specimens 3D printed with a [45°, 135°] with HCR and LCR printing profiles.

Further analysis of the tensile specimens showed an increase of ≈25% in the elastic modulus (Figure 10b,d) between pristine PVDF and the 20% of ZnO HCR sample with a [45°, 135°] infill. The modulus for specimens printed with [0°, 0°] infill orientation and HCR (Figure 10a,c) also increases proportionally to the percentage ZnO in the samples, to a maximum of 0.9 GPa, approximately 30% larger than the modulus of pristine PVDF. The lower modulus of the [0°, 0°] infill direction samples is attributed to a higher degree of anisotropy and defects of the [0°, 0°] printed samples. Three-point flexion tests of the HCR samples also display a similar increase of the elastic modulus (Figure 10c,d) with the increasing percentage of ZnO nanofiller, with a maximum elastic modulus of 2.0–2.3 GPa, which is ≈40% greater than that of pristine PVDF. The elastic modulus of the LCR samples printed with [0°, 0°] infill direction shows no statistically significant difference between the pristine PVDF and ZnO-PVDF nanocomposite specimens. The elastic modulus of LCR samples printed with a [45°,135°] infill pattern decreased as the content of ZnO is increased in the ZnO-PVDF nanocomposite. 3D printed specimens, regardless of the infill direction, with HCR produced a significant increase in the elastic modulus (Figure 10b,d) as a function of the ZnO content. The elastic modulus increase could be attributed to the lower time to crystallization (Figure 3), increased β phase formation (Figure 5), and the stiffening due to the nanofiller as reported previously in the literature [27,59].

The strength of the 3D printed specimens (Figure 10e,f) with HCR did not show any statistically significant difference between pristine PVDF samples and ZnO-PVDF nanocomposite for either of the infill directions [0°, 0°] and [45°, 135°]. LCR produced stronger pristine PVDF specimens, but the strength reduces with increased content of ZnO. These results were confirmed by testing the filament used to print the specimens (Figure S9). The extruded filament was cooling down to room temperature using fans. Thus, its cooling rate is comparable to the HCR rate of the 3D printed samples. The filament displayed no statistical difference in strength, except for the 20% of ZnO. The filament composite displayed an increase of the elastic modulus proportional to the content of ZnO with a maximum elastic modulus of ≈2.1 GPa, an increase of 30% with respect to the pristine PVDF. This result is in agreement with the strength and elastic modulus of the specimens 3D printed with HCR and a [45°, 135°] infill pattern.

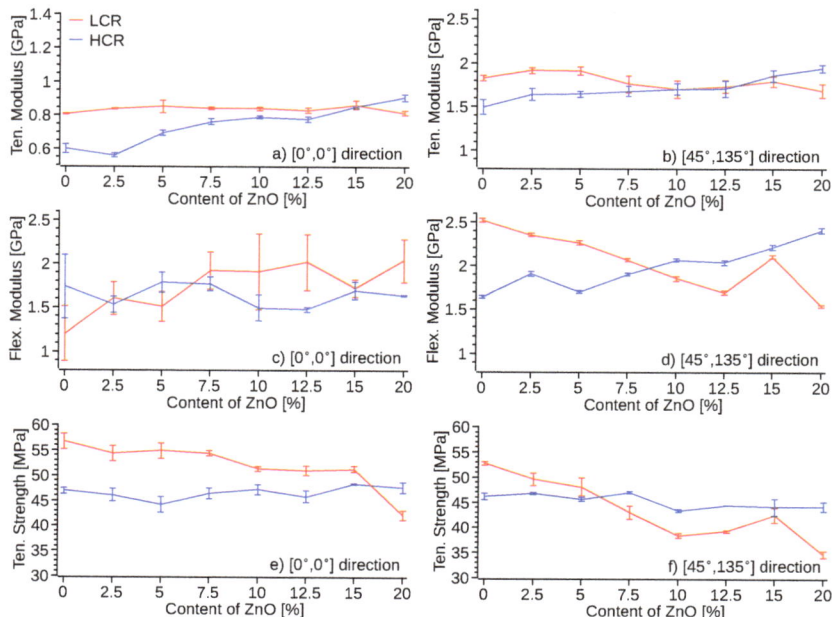

Figure 10. Average tensile and flexural test results of specimens 3D printed with HCR (blue) and LCR (red). Elastic modulus from tensile test (**a**) [0°, 0°] infill direction, (**b**) [45°, 135°] infill direction, flexural modulus from three-point bending (**c**) [0°, 0°] infill direction, (**d**) [45°, 135°] infill direction, tensile strength of 3D printed specimens with (**e**) [0°, 0°] infill direction, and (**f**) [45°, 135°] infill direction.

3.3. Electrical Response

The electrical response (Video S1) of the 3D printed specimens was also measured. The three DMA samples, with infill direction of [45°, 135°], were coated with conductive nickel paint with an average resistance of 2.2 ± 0.2 Ω at 25 mm. These samples were subjected to 6 Hz cyclic loading, a 0.2% and 0.4% strain amplitude with a span of 25 mm. When the locking amplifier was set to match the driven frequency of the DMA (Figures S10 and S11), the signal saturate at the peak voltage, reaching a maximum after two seconds of oscillating input. The driving force of the DMA was calculated as a function of the measured elastic modulus of the samples (Figure 10d) and the input strain. The measured voltage was related to the force and plotted as a function of the content of ZnO (Figure 11). Samples with 7.5–15% ZnO produced a two fold increment of the voltage per unit force with respect of the pristine PVDF samples. Furthermore, the HCR samples produced ≈70% increment of the generated voltage compare to the LCR and approximately a four times higher voltage per unit force (Figure 11) for HCR 10% ZnO-PVDF with respect of LCR pristine PVDF. These results correlate with the FTIR (Figure 5) and DSC (Figure 4) measurements, which show that samples with 7.5–15% ZnO had a larger percentage of β phase. ZnO nanoparticles may have also improve the charge motion to the electrodes, in accordance with Zhang et al. findings, and thus increasing the voltage signal [60].

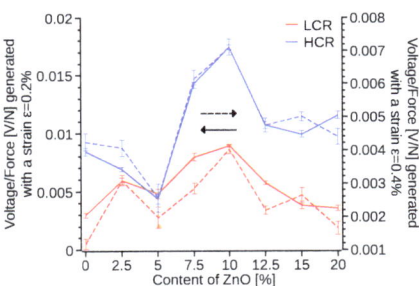

Figure 11. Average volts per newton generated by the 3D printed PVDF and ZnO-PVDF nanocomposite samples as function of the percentage of ZnO for low (red) and high (blue) cooling rates.

Experimental testing also revealed a frequency dependency of the signal generated by the test samples. As the setting frequency on the locking amplifier increased above the driven frequency, the signal of the PVDF-nano composite samples became sinusoidal and its frequency increased proportional to the locking amplifier's frequency set point. When the locking amplifier matching frequency is set to two times the driven frequency, then the specimen signal match (Figure 12a,b) the input force frequency. Setting the locking amplifier at a higher matching frequency increases the signal frequency and decreases the signal amplitude (Figure 12c,d), while the frequency response remains constant, ZnO increases (Figure 11) the signal amplitude, with 10% of ZnO content producing the maximum volts per newton of applied force, ≈0.0175 V/N. Similarly, HCR samples (Figures 11 and 12) increase the voltage to force signal by approximately two times regardless of the ZnO content, with the exception of the 5% of ZnO. 5ZnO also presents the lowest (Figure 5) alpha to beta transmittance difference, and its melting point (Figure 4) is similar to pristine PVDF, suggesting a lower percentage of β-phase formation. This effect is attributed to the higher concentration and agglomeration of ZnO (Figure S3).

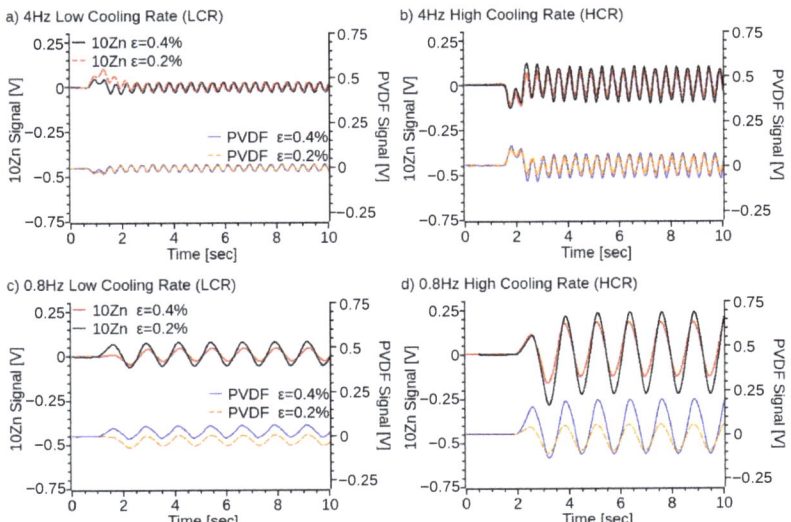

Figure 12. Voltage signal generated by the 3D printed PVDF and ZnO-PVDF in response to a 0.2% and 0.4% sinusoidal strain at a frequency of: (**a**) 4 Hz for LCR samples, (**b**) 4 Hz for HCR samples, (**c**) 0.8 Hz for LCR samples, and (**d**) 0.8 Hz for HCR samples.

Since the 10ZnO specimens produce a better electrical response, they were also tested by applying a strain that simulates a walking profile (Figure 13) of an average person (Video S2) [41,61]. The walking profile was set to a maximum deflection amplitude of 2 mm, while the MTS test frame measured the force, and the sample signal was amplified and measured with the same setup from previous experiments. The 10Zn samples, HCR and LCR, were capable of reproducing the input signal (Figure 13) with higher accuracy than the signal of pristine PVDF samples (Video S2). Similar to the frequency tests on the DMA samples (Figure 12) and due to the superimposed frequencies of the walking profile, there is a frequency mismatch with respect to the applied wave. The HCR specimens for the PVDF and 10Zn specimens produced ≈2 times more volts per newton compared to the LCR specimens. The 10Zn also produced a ≈30% and ≈20% increase in V/N than pristine PVDF samples for HCR and LCR specimens, respectively. The 10Zn samples, HCR and LCR, were capable of reproducing the input signal (Figure 13) with higher accuracy than the pristine PVDF samples (Video S2). The increased signal of the 10Zn samples with respect to pristine PVDF is similar to the observed V/N at a constant driving frequency (Figure 12). Two 10ZnO samples were tested (Figure S12) under the same condition producing an equivalent electrical response to the applied load. Similarly, the two PVDF specimens produced an equivalent response. These results suggest that the 3D printed ZnO-PVDF composite material is repeatable and correlates to the locking amplifier's matching frequency. Thus, the need for a frequency response characterization of the sensor, which is outside of the scope of the current project.

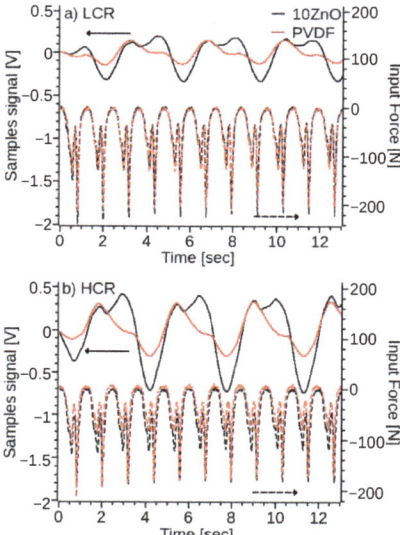

Figure 13. Voltage signal generated in response to a walking profile simulated on an MTS tensile frame for by pristine PVDF and 10%ZnO PVDF nanocomposite samples created with: (**a**) LCR and (**b**) HCR 3D printing profiles.

4. Conclusions

PVDF and PVDF metal oxide nanocomposites were compounded and extruded into filaments for FDM or FFF 3D printers. The filaments properties were characterized by analyzing thermal stability, crystallinity, and crystal phase composition by using TGA, DSC, and FTIR. Furthermore, the filaments' viscosity was studied at temperatures close to the nozzle temperature to 3D print PVDF. The mechanical properties of 3D printed parts were also characterized along with the electrical output to a mechanical strain. TGA analyses exhibited an increase of ≈25 °C of the decomposition temperature with respect to pristine

PVDF. DSC measurements also showed an increase in the glass transition temperature as a function of the ZnO content. Although DSC did not show a significant increase in crystallinity of the nanocomposite filament as a function of the ZnO content, it did exhibit a ≈35% and ≈12% reduction in the time of crystallinity for a HCR of 30 °C·min^{-1} and LCR of 5 °C·min^{-1}, respectively, for samples with a 12.5–15% ZnO. Samples with 10–15% of ZnO also depict a narrower crystalization peak compared to pristine PVDF, suggesting a narrow crystallite size distribution. DSC melting cycles following a HCR of 30 °C·min^{-1} produced a double melting peak. The lower temperature melting peak increase with the ZnO content until reaching a minimum with 15% at 164.4 °C, which is typically attributed to PVDF β-phase. FTIR corroborated the DSC measurements and showed a decrease of the 840 cm^{-1} and 1275 cm^{-1} peaks, which represent the β phase, reaching a minimum transmission with 10–12.5% of ZnO content. The α-phase 760 cm^{-1} increased as a function of the amount of ZnO, reaching an 80% reduction of the α/β ratio with 10–12.5% of ZnO. The ZnO-PVDF nanocomposite filament's shear viscosity reduced as a function of the ZnO percentage reaching a minimum at 7.5–12.5% of ZnO, due to the size of the ZnO nanoparticles. Overall the HCR ZnO-PVDF produces up to ≈40% increase of the elastic modulus while maintaining the same ultimate strength of the pristine PVDF. Furthermore, concentrations of ZnO up to 5% increased the toughness of the ZnO-PVDF nanocomposite. Printing infill direction does not affect HCR samples, but it increases the PVDF and ZnO-PVDF nanocomposite strength when 3D printed with an LCR. FTIR measurements indicate that 7.5–12.5% of ZnO promote up to 80% conversion of the α into β phase, in agreement with the DSC measurement. Finally, the electrical response measured by applying a cyclic strain of 6 Hz generated up to 70% higher voltage per unit force for samples with 7.5–12.5% of ZnO compared to pristine PVDF. High cooling rate printed samples generate an electrical signal up two times large for the same amount of force compared to the LCR specimens. Thus, an overall four times higher voltage per unit force for HCR 10% ZnO-PVDF with respect to LCR pristine PVDF. Furthermore, 10%ZnO-PVDF printed were capable of reproducing a walking profile more accurately than pristine PVDF, and HCR 10% ZnO-PVDF samples also generated a higher volt signal. Thus, suggesting that under the proper printing conditions, the ZnO can be used to promote β phase of PVDF nanocomposite material by directly 3D print the desired part without post poling and samples thicker than a thin film. The advantages of the proposed technique for creating parts with PVDF include improved mechanical properties, higher thermal decomposition, increased electrical response and elimination of extra poling steps. Optimization of the process has helped identify the ideal cooling rates and filler content.

Supplementary Materials: The following supporting information can be downloaded at: https://www.mdpi.com/article/10.3390/polym14204312/s1.

Author Contributions: S.J.: Conceptualization of this study, Acquisition of Data, Methodology, Writing. E.G.: Acquisition of Data. J.G.: Acquisition of Data. N.W.: Acquisition of Data. F.K.: Acquisition of Data, Writing, Review & Editing. S.I.: Data curation, Writing, Review & Editing. G.C.: Data curation, Supervision, Writing—Original draft preparation. All authors have read and agreed to the published version of the manuscript.

Funding: G.C. was partially supported by the Air Force Research Laboratory Summer Faculty Fellowship Program and the Miami University Faculty Grants to Promote Research (GPR), and the. S.T.I. recognizes the Air Force Office of Scientific Research (AFOSR) and the Joint Insensitive Munitions Program (JIMTP) for funding. N.J.W. acknowledges funding for this work made available by AFOSR, sponsored by the U.S. Air Force Academy, Chemistry Research Center under the cooperative agreement FA7000-19-2-006 (NJW Scientific Ltd. Colorado Springs USA).

Institutional Review Board Statement: Not applicable.

Informed Consent Statement: Not applicable.

Data Availability Statement: Additional data will be made available on reasonable request to the authors.

Acknowledgments: The authors thank the Center for Advanced Microscopy and Imaging at Miami University for their assistance. G.C. also thanks the Air Force Research Laboratory Summer Faculty Fellowship Program and the Miami University Faculty Grants to Promote Research (GPR) for their support.

Conflicts of Interest: The authors have no conflict of interest/competing interests to declare that are relevant to the content of this article.

References

1. Moore, A.L. *Fluoroelastomers Handbook: The Definitive User's Guide and Databook*; William Andrew Pub: Norwich, NY, USA, 2006.
2. Eun, J.H.; Sung, S.M.; Kim, M.S.; Choi, B.K.; Lee, J.S. Effect of Mwcnt Content on the Mechanical and Piezoelectric Properties of Pvdf Nanofibers. *Mater. Des.* **2021**, *206*, 109785. [CrossRef]
3. He, Z.; Rault, F.; Lewandowski, M.; Mohsenzadeh, E.; Salaün, F. Electrospun Pvdf Nanofibers for Piezoelectric Applications: A Review of the Influence of Electrospinning Parameters on the β Phase and Crystallinity Enhancement. *Polymers* **2021**, *13*, 174. [CrossRef] [PubMed]
4. Wu, Y.; Ma, Y.; Zheng, H.; Ramakrishna, S. Piezoelectric Materials for Flexible and Wearable Electronics: A Review. *Mater. Des.* **2021**, *211*, 110164. [CrossRef]
5. Santos, B.P.S.; Arias, J.J.R.; Jorge, F.E.; Értola Pereira de Deus Santos, R.; da Silva Fernandes, B.; da Silva Candido, L.; de Carvalho Peres, A.C.; Chaves, E.G.; de Fátima Vieira Marques, M. Preparation, Characterization and Permeability Evaluation of poly(vinylidene fluoride) Composites with Zno Particles for Flexible Pipelines. *Polym. Test.* **2021**, *94*, 107064. [CrossRef]
6. Uchino, K., 3.24 Piezoelectro Composites. In *Comprehensive Composite Materials II*; Elsevier: Amsterdam, The Netherlands, 2018; pp. 613–624. [CrossRef]
7. Foster, F.; Harasiewicz, K.; Sherar, M. A history of medical and biological imaging with polyvinylidene fluoride (PVDF) transducers. *IEEE Trans. Ultrason. Ferroelectr. Freq. Control* **2000**, *47*, 1363–1371. [CrossRef]
8. Barrau, S.; Ferri, A.; Costa, A.D.; Defebvin, J.; Leroy, S.; Desfeux, R.; Lefebvre, J.M. Nanoscale Investigations of α- and γ-Crystal Phases in Pvdf-Based Nanocomposites. *ACS Appl. Mater. Interfaces* **2018**, *10*, 13092–13099. [CrossRef]
9. Kashfi, M.; Fakhri, P.; Amini, B.; Yavari, N.; Rashidi, B.; Kong, L.; Bagherzadeh, R. A Novel Approach to Determining Piezoelectric Properties of Nanogenerators Based on Pvdf Nanofibers Using Iterative Finite Element Simulation for Walking Energy Harvesting. *J. Ind. Text.* **2020**, *51*, 531S–553S. [CrossRef]
10. Bodkhe, S.; Turcot, G.; Gosselin, F.P.; Therriault, D. One-Step Solvent Evaporation-Assisted 3D Printing of Piezoelectric PVDF Nanocomposite Structures. *ACS Appl. Mater. Interfaces* **2017**, *9*, 20833–20842. [CrossRef]
11. Ramesh, D.; D'Souza, N. Experimental and Computational Investigation of PVDF-BaTiO$_3$ Interface for Impact Sensing and Energy Harvesting Applications. *SN Appl. Sci.* **2020**, *2*, 1129. [CrossRef]
12. Zandesh, G.; Gheibi, A.; Bafqi, M.S.S.; Bagherzadeh, R.; Ghoorchian, M.; Latifi, M. Piezoelectric Electrospun Nanofibrous Energy Harvesting Devices: Influence of the Electrodes Position and Finite Variation of Dimensions. *J. Ind. Text.* **2016**, *47*, 348–362. [CrossRef]
13. Meng, N.; Ren, X.; Santagiuliana, G.; Ventura, L.; Zhang, H.; Wu, J.; Yan, H.; Reece, M.J.; Bilotti, E. Ultrahigh β-phase Content poly(vinylidene fluoride) with Relaxor-Like Ferroelectricity for High Energy Density Capacitors. *Nat. Commun.* **2019**, *10*, 4535. [CrossRef] [PubMed]
14. Cai, X.; Lei, T.; Sun, D.; Lin, L. A critical analysis of the α, β and γ phases in poly(vinylidene fluoride) using FTIR. *RSC Adv.* **2017**, *7*, 15382–15389. [CrossRef]
15. Maurya, D.; Peddigari, M.; Kang, M.G.; Geng, L.D.; Sharpes, N.; Annapureddy, V.; Palneedi, H.; Sriramdas, R.; Yan, Y.; Song, H.C.; et al. Lead-Free Piezoelectric Materials and Composites for High Power Density Energy Harvesting. *J. Mater. Res.* **2018**, *33*, 2235–2263. [CrossRef]
16. Gheibi, A.; Bagherzadeh, R.; Merati, A.A.; Latifi, M. Electrical Power Generation From Piezoelectric Electrospun Nanofibers Membranes: Electrospinning Parameters Optimization and Effect of Membranes Thickness on Output Electrical Voltage. *J. Polym. Res.* **2014**, *21*, 571. [CrossRef]
17. Ruan, L.; Yao, X.; Chang, Y.; Zhou, L.; Qin, G.; Zhang, X. Properties and Applications of the β Phase Poly(vinylidene fluoride). *Polymers* **2018**, *10*, 228. [CrossRef]
18. Kawai, H. The Piezoelectricity of Poly (vinylidene Fluoride). *Jpn. J. Appl. Phys.* **1969**, *8*, 975–976. [CrossRef]
19. Sencadas, V.; Moreira, M.V.; Lanceros-Méndez, S.; Pouzada, A.S.; Gregório Filho, R. α- to β Transformation on PVDF Films Obtained by Uniaxial Stretch. *Mater. Sci. Forum* **2006**, *514–516*, 872–876. [CrossRef]
20. Fortunato, M.; Chandraiahgari, C.; Bellis, G.D.; Ballirano, P.; Sarto, F.; Tamburrano, A.; Sarto, M. Piezoelectric Effect and Electroactive Phase Nucleation in Self-Standing Films of Unpoled Pvdf Nanocomposite Films. *Nanomaterials* **2018**, *8*, 743. [CrossRef]
21. Xiong, X.; Li, Q.; Zhang, X.C.; Wang, L.; Guo, Z.X.; Yu, J. Poly(vinylidene fluoride)/silica nanocomposite membranes by electrospinning. *J. Appl. Polym. Sci.* **2013**, *129*, 1089–1095. [CrossRef]
22. Haddadi, S.A.; Ghaderi, S.; Amini, M.; Ramazani, S.A. Mechanical and piezoelectric characterizations of electrospun PVDF-nanosilica fibrous scaffolds for biomedical applications. *Mater. Today Proc.* **2018**, *5*, 15710–15716. [CrossRef]

23. Dodds, J.S.; Meyers, F.N.; Loh, K.J. Piezoelectric Characterization of PVDF-TrFE Thin Films Enhanced with ZnO Nanoparticles. *IEEE Sens. J.* **2012**, *12*, 1889–1890. [CrossRef]
24. Hu, Y.C.; Hsu, W.L.; Wang, Y.T.; Ho, C.T.; Chang, P.Z. Enhance the Pyroelectricity of Polyvinylidene Fluoride by Graphene-Oxide Doping. *Sensors* **2014**, *14*, 6877–6890. [CrossRef] [PubMed]
25. Kumar, R.; Singh, R.; Singh, M.; Kumar, P. On Zno Nano Particle Reinforced Pvdf Composite Materials for 3d Printing of Biomedical Sensors. *J. Manuf. Process* **2020**, *60*, 268–282. [CrossRef]
26. An, N.; Liu, H.; Ding, Y.; Zhang, M.; Tang, Y. Preparation and electroactive properties of a PVDF/nano-TiO_2 composite film. *Appl. Surf. Sci.* **2011**, *257*, 3831–3835. [CrossRef]
27. Castanet, E.; Thamish, M.A.; Hameed, N.; Krajewski, A.; Dumée, L.F.; Magniez, K. Zinc Oxide PVDF Nano-Composites-Tuning Interfaces toward Enhanced Mechanical Properties and UV Protection: Zinc Oxide PVDF Nano-Composites. *Adv. Eng. Mater.* **2017**, *19*, 1600611. [CrossRef]
28. Muralidhar, C.; Pillai, P.K.C. Dielectric behaviour of barium titanate ($BaTiO_3$)/polyvinylidene fluoride (PVDF) composite. *J. Mater. Sci. Lett.* **1987**, *6*, 346–348. [CrossRef]
29. Kalani, S.; Kohandani, R.; Bagherzadeh, R. Flexible Electrospun Pvdf-$Batio_3$ Hybrid Structure Pressure Sensor With Enhanced Efficiency. *RSC Adv.* **2020**, *10*, 35090–35098. [CrossRef]
30. Wu, L.; Jing, M.; Liu, Y.; Ning, H.; Liu, X.; Liu, S.; Lin, L.; Hu, N.; Liu, L. Power generation by PVDF-TrFE/graphene nanocomposite films. *Compos. Part Eng.* **2019**, *164*, 703–709. [CrossRef]
31. Li, Y.; Feng, W.; Meng, L.; Tse, K.M.; Li, Z.; Huang, L.; Su, Z.; Guo, S. Investigation on In-Situ Sprayed, Annealed and Corona Poled Pvdf-Trfe Coatings for Guided Wave-Based Structural Health Monitoring: From Crystallization To Piezoelectricity. *Mater. Des.* **2021**, *199*, 109415. [CrossRef]
32. Kim, Y.J.; Ahn, C.H.; Choi, M.O. Effect of thermal treatment on the characteristics of electrospun PVDF-silica composite nanofibrous membrane. *Eur. Polym. J.* **2010**, *46*, 1957–1965. [CrossRef]
33. Kim, H.; Fernando, T.; Li, M.; Lin, Y.; Tseng, T.L.B. Fabrication and characterization of 3D printed $BaTiO_3$/PVDF nanocomposites. *J. Compos. Mater.* **2018**, *52*, 197–206. [CrossRef]
34. Kim, H.; Torres, F.; Wu, Y.; Villagran, D.; Lin, Y.; Tseng, T.L. Integrated 3D printing and corona poling process of PVDF piezoelectric films for pressure sensor application. *Smart Mater. Struct.* **2017**, *26*, 085027. [CrossRef]
35. Kennedy, Z.C.; Christ, J.F.; Evans, K.A.; Arey, B.W.; Sweet, L.E.; Warner, M.G.; Erikson, R.L.; Barrett, C.A. 3D-printed poly(vinylidene fluoride)/carbon nanotube composites as a tunable, low-cost chemical vapour sensing platform. *Nanoscale* **2017**, *9*, 5458–5466. [CrossRef] [PubMed]
36. Porter, D.A.; Hoang, T.V.; Berfield, T.A. Effects of in situ poling and process parameters on fused filament fabrication printed PVDF sheet mechanical and electrical properties. *Addit. Manuf.* **2017**, *13*, 81–92. [CrossRef]
37. Tu, R.; Sprague, E.; Sodano, H.A. Precipitation-Printed High-β Phase Poly(vinylidene fluoride) for Energy Harvesting. *ACS Appl. Mater. Interfaces* **2020**, *12*, 58072–58081. [CrossRef]
38. Elsheikh, A. Bistable Morphing Composites for Energy-Harvesting Applications. *Polymers* **2022**, *14*, 1893. [CrossRef]
39. D20 Committee. *ASTM D638—14 Test Method for Tensile Properties of Plastics*; Technical report; ASTM International: West Conshohocken, PA, USA, 2014. [CrossRef]
40. D20 Committee. *ASTM D790—17 Test Methods for Flexural Properties of Unreinforced and Reinforced Plastics and Electrical Insulating Materials*; Technical report; ASTM International: West Conshohocken, PA, USA, 2017. [CrossRef]
41. Institute, J.W. Orthoload Loading of Orthopaedic Implant. Available online: https://orthoload.com/ (accessed on 5 September 2020).
42. Indolia, A.P.; Gaur, M.S. Investigation of structural and thermal characteristics of PVDF/ZnO nanocomposites. *J. Therm. Anal. Calorim.* **2013**, *113*, 821–830. [CrossRef]
43. Buckley, J.; Cebe, P.; Cherdack, D.; Crawford, J.; Ince, B.S.; Jenkins, M.; Pan, J.J.; Reveley, M.; Washington, N.; Wolchover, N. Nanocomposites of poly(vinylidene fluoride) with organically modified silicate. *Polymer* **2006**, *47*, 2411–2422. [CrossRef]
44. Bafqi, M.S.S.; Bagherzadeh, R.; Latifi, M. Nanofiber Alignment Tuning: An Engineering Design Tool in Fabricating Wearable Power Harvesting Devices. *J. Ind. Text.* **2016**, *47*, 535–550. [CrossRef]
45. Bafqi, M.S.S.; Bagherzadeh, R.; Latifi, M. Fabrication of Composite Pvdf-Zno Nanofiber Mats By Electrospinning for Energy Scavenging Application with Enhanced Efficiency. *J. Polym. Res.* **2015**, *22*, 130. [CrossRef]
46. Zanrosso, C.D.; Piazza, D.; Lansarin, M.A. Pvdf/zno Composite Films for Photocatalysis: A Comparative Study of Solution Mixing and Melt Blending Methods. *Polym. Eng. Sci.* **2020**, *60*, 1146–1157. [CrossRef]
47. Tsonos, C.; Zois, H.; Kanapitsas, A.; Soin, N.; Siores, E.; Peppas, G.; Pyrgioti, E.; Sanida, A.; Stavropoulos, S.; Psarras, G. Polyvinylidene Fluoride/magnetite Nanocomposites: Dielectric and Thermal Response. *J. Phys. Chem. Solids* **2019**, *129*, 378–386. [CrossRef]
48. Sencadas, V.; Lanceros-Méndez, S.; Mano, J. Characterization of poled and non-poled β-PVDF films using thermal analysis techniques. *Thermochim. Acta* **2004**, *424*, 201–207. [CrossRef]
49. Harstad, S.; D'Souza, N.; Soin, N.; El-Gendy, A.A.; Gupta, S.; Pecharsky, V.K.; Shah, T.; Siores, E.; Hadimani, R.L. Enhancement of β-phase in PVDF films embedded with ferromagnetic Gd_5Si_4 nanoparticles for piezoelectric energy harvesting. *AIP Adv.* **2017**, *7*, 056411. [CrossRef]

50. Soin, N.; Boyer, D.; Prashanthi, K.; Sharma, S.; Narasimulu, A.A.; Luo, J.; Shah, T.H.; Siores, E.; Thundat, T. Exclusive self-aligned β-phase PVDF films with abnormal piezoelectric coefficient prepared via phase inversion. *Chem. Commun.* **2015**, *51*, 8257–8260. [CrossRef]
51. Martins, P.; Lopes, A.; Lanceros-Mendez, S. Electroactive phases of poly(vinylidene fluoride): Determination, processing and applications. *Prog. Polym. Sci.* **2014**, *39*, 683–706. [CrossRef]
52. Fakhri, P.; Amini, B.; Bagherzadeh, R.; Kashfi, M.; Latifi, M.; Yavari, N.; Kani, S.A.; Kong, L. Flexible Hybrid Structure Piezoelectric Nanogenerator Based on Zno Nanorod/pvdf Nanofibers With Improved Output. *RSC Adv.* **2019**, *9*, 10117–10123. [CrossRef]
53. Khaliq, M.H.; Gomes, R.; Fernandes, C.; Nóbrega, J.; Carneiro, O.S.; Ferrás, L.L. On the Use of High Viscosity Polymers in the Fused Filament Fabrication Process. *Rapid Prototyp. J.* **2017**, *23*, 727–735. [CrossRef]
54. Sanchez, L.C.; Beatrice, C.A.G.; Lotti, C.; Marini, J.; Bettini, S.H.P.; Costa, L.C. Rheological Approach for an Additive Manufacturing Printer Based on Material Extrusion. *Int. J. Adv. Manuf. Technol.* **2019**, *105*, 2403–2414. [CrossRef]
55. Mackay, M.E. The importance of rheological behavior in the additive manufacturing technique material extrusion. *J. Rheol.* **2018**, *62*, 1549–1561. [CrossRef]
56. Kairn, T.; Daivis, P.J.; Ivanov, I.; Bhattacharya, S.N. Molecular-Dynamics Simulation of Model Polymer Nanocomposite Rheology and Comparison with Experiment. *J. Chem. Phys.* **2005**, *123*, 194905. [CrossRef] [PubMed]
57. Cao, Y.; Deng, Q.; Liu, Z.; Shen, D.; Wang, T.; Huang, Q.; Du, S.; Jiang, N.; Lin, C.T.; Yu, J. Enhanced thermal properties of poly(vinylidene fluoride) composites with ultrathin nanosheets of MXene. *RSC Adv.* **2017**, *7*, 20494–20501. [CrossRef]
58. Flyagina, I.S.; Mahdi, E.M.; Titov, K.; Tan, J.C. Thermo-mechanical properties of mixed-matrix membranes encompassing zeolitic imidazolate framework-90 and polyvinylidine difluoride: ZIF-90/PVDF nanocomposites. *APL Mater.* **2017**, *5*, 086104. [CrossRef]
59. Suresh, G.; Jatav, S.; Mallikarjunachari, G.; Rao, M.S.R.; Ghosh, P.; Satapathy, D.K. Influence of Microstructure on the Nanomechanical Properties of Polymorphic Phases of Poly(vinylidene fluoride). *J. Phys. Chem. B* **2018**, *122*, 8591–8600. [CrossRef] [PubMed]
60. Zhang, X.; Huang, X.; Li, C.; Jiang, H. Dye-Sensitized Solar Cell With Energy Storage Function Through PVDF/ZnO Nanocomposite Counter Electrode. *Adv. Mater.* **2013**, *25*, 4093–4096. [CrossRef] [PubMed]
61. Damm, P.; Kutzner, I.; Bergmann, G.; Rohlmann, A.; Schmidt, H. Comparison of in Vivo Measured Loads in Knee, Hip and Spinal Implants During Level Walking. *J. Biomech.* **2017**, *51*, 128–132. [CrossRef]

Article

Development and Processing of New Composite Materials Based on High-Performance Semicrystalline Polyimide for Fused Filament Fabrication (FFF) and Their Biocompatibility

Igor Polyakov [1,*], Gleb Vaganov [2], Andrey Didenko [2], Elena Ivan'kova [2], Elena Popova [2], Yuliya Nashchekina [1,3], Vladimir Elokhovskiy [2], Valentin Svetlichnyi [2] and Vladimir Yudin [1,2]

1. Institute of Biomedical Systems and Biotechnology, Peter the Great St. Petersburg Polytechnic University, Polytechnicheskaya St. 29, St. Petersburg 195251, Russia
2. Institute of Macromolecular Compounds of Russian Academy of Sciences, V.O., Bol'shoy Pr. 31, St. Petersburg 199004, Russia
3. Institute of Cytology, Russian Academy of Sciences, St. Petersburg 194064, Russia
* Correspondence: polyakov-iv@yandex.ru

Citation: Polyakov, I.; Vaganov, G.; Didenko, A.; Ivan'kova, E.; Popova, E.; Nashchekina, Y.; Elokhovskiy, V.; Svetlichnyi, V.; Yudin, V. Development and Processing of New Composite Materials Based on High-Performance Semicrystalline Polyimide for Fused Filament Fabrication (FFF) and Their Biocompatibility. *Polymers* **2022**, *14*, 3803. https://doi.org/10.3390/polym14183803

Academic Editor: Matthew Blacklock

Received: 1 September 2022
Accepted: 9 September 2022
Published: 11 September 2022

Publisher's Note: MDPI stays neutral with regard to jurisdictional claims in published maps and institutional affiliations.

Copyright: © 2022 by the authors. Licensee MDPI, Basel, Switzerland. This article is an open access article distributed under the terms and conditions of the Creative Commons Attribution (CC BY) license (https://creativecommons.org/licenses/by/4.0/).

Abstract: Samples of composite materials based on high-performance semicrystalline polyimide R-BAPB (based on the dianhydride R: 1,3-bis-(3',4,-dicarboxyphenoxy)benzene and diamine BAPB: 4,4'-bis-(4''-aminophenoxy)diphenyl)) filled with carbon nanofibers and micron-sized discrete carbon fibers were obtained by FFF printing for the first time. The viscosity of melts of the composites based on R-BAPB, thermal, mechanical characteristics of the obtained composite samples, their internal structure, and biocompatibility were studied. Simultaneously with FFF printing, samples were obtained by injection molding. The optimal concentrations of carbon fillers in polyimide R-BAPB for their further use in FFF printing were determined. The effect of the incorporation of carbon fillers on the porosity of the printed samples was investigated. It was shown that the incorporation of carbon nanofibers reduces the porosity of the printed samples, which leads to an increase in deformation at break. Modification of polyimide with discrete carbon fibers increases the strength and Young's modulus sufficiently but decreases the deformation at break. The cytotoxicity analysis showed that the obtained composite materials are bioinert.

Keywords: polyimide; FFF-printing; additive technologies; nanocomposites; carbon fiber; biocompatibility

1. Introduction

Fused filament fabrication is the most popular approach among the additive technologies [1,2]. The frequency of application of the FFF method can be explained by the low cost of printing equipment, easy maintenance, and, at the same time, the possibility of obtaining high-quality parts of complex configuration. One of the main problems of FFF technology for functional application is the low mechanical properties of the obtained products in comparison with traditional production methods, such as extrusion or injection molding. This is due to the unavoidable presence of pores because of the layered deposition of the material [3,4], insufficient adhesion between the layers, as well as the tendency of thermoplastic polymers to shrink while cooling during printing [5]. To improve the mechanical characteristics of these products, there are two main solutions: use of new materials or modification of existing ones [6,7]. Currently, FFF printing most often uses such polymers as ABS (acrylonitrile butadiene styrene) plastic and polylactide, as well as polyamides with low mechanical characteristics (their strength does not exceed 60 MPa). To increase the mechanical properties of polymer samples processed by FFF, it is necessary to involve high-performance thermoplastic polymers for this type of printing. Materials obtained from high-performance thermoplastics are distinguished not only by high strength characteristics, but also have increased thermal, fire, and frost resistance, high fracture toughness, chemical resistance to aggressive liquids, and some other special properties. One of the representatives of this class of thermoplastics are polyimides. Products from this class of thermoplastics primarily ensure the stable operation of products obtained from them at high temperatures [8]. Polyimides are used in the production of structural materials, coatings in the aerospace industry, and the bases of flexible printed circuit boards, and they are also promising structural materials for biomedical purposes, due to such operational characteristics as heat resistance, chemical stability, hydrophobicity, and bioinertness [9].

One of the promising thermoplastic polyimides is the semicrystalline polyimide R-BAPB based on the dianhydride R: 1,3-bis-(3′,4,-dicarboxyphenoxy)benzene and diamine BAPB: 4,4′-bis-(4″-aminophenoxy)diphenyl) (Figure 1).

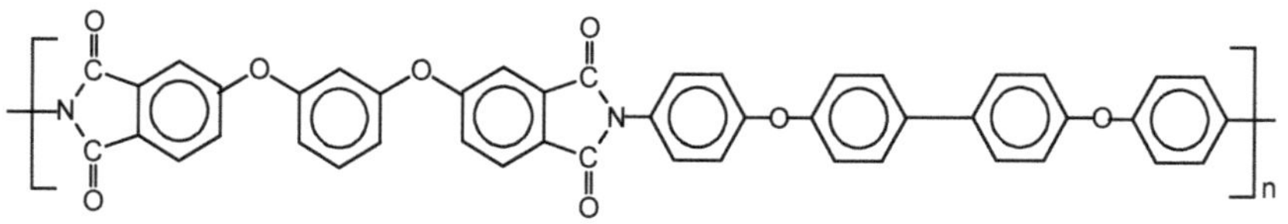

Figure 1. Chemical structure of the polyimide R-BAPB.

The glass transition temperature of R-BAPB is about 205 °C and the melting point is about 318 °C [10]. A main feature of polyimide R-BAPB is the ability to control crystallization and recrystallization at low melt viscosity (up to 1000 Pa·s at 360 °C), which contributes to its processing by traditional methods suitable for thermoplastics (injection molding, extrusion, hot pressing, etc.). The ability to crystallize from the melt allows one to increase the actual temperature of operation, wear resistance, chemical resistance, and several other operational characteristics of products. To increase the crystallization rate, it is possible to add some nucleants, for example, various carbon nanoparticles [11].

Along with the use of high-performance plastics, an effective way to improve the 3D-printing process and the characteristics of the resulting products is the introduction of various fillers into the polymer material [12,13]. Composite materials are used to achieve the desired mechanical and functional properties by improving the matrix material by adding particles, fibers, or nanomaterials [14–16].

Over the last decade, polymer nanocomposites have received much attention, both in fundamental and applied research, due to the possibility of controlling the operational properties of lightweight materials based on them by adding nanofillers. The reason for this is that the nanofillers have a significantly higher ratio of surface area to volume compared to micro- and macrofillers [17]. In particular, the incorporation of carbon nanotubes and nanofibers has a positive effect on the mechanical characteristics of products, increasing the Young's modulus and tensile strength [18].

The study [19] compared the behavior of amorphous polyetherimide (PEI) (ODPA-P3) with semicrystalline (BPDA-P3) with the incorporation of single-walled carbon nanotubes (SWCNT; 0.1–4.4 vol.%). Tensile tests showed that the yield strength was the same for both polyimides when loading SWCNT up to 0.3 vol.%. Above this concentration, the yield strength for the BPDA-P3 nanocomposites remained constant, whereas for the ODPA-P3 nanocomposites, it increased from 80 up to 126 MPa (1.2 vol.%). In our previous work [20], we showed the influence of carbon nanofibers such as VGCF (vapor grown carbon fiber) on the mechanical properties of polyethyrimide samples processed by FFF. The incorporation of 1 wt.% VGCF into the polyethyrimide increased the strength and Young's modulus of the printed samples and these values approached those values of the PEI samples processed by injection molding. In [21], it was noted that with an increase in the mass fraction of the carbon nanofibers incorporated into ABS plastic, the tensile strength of FFF-printed samples also increased. Incorporation of 2 wt.% carbon nanofibers into the ABS-matrix resulted in a 40% increase in strength compared to pure ABS. Similarly, an increase of 26.1% in strength was observed for ABS with 1 wt.% carbon nanofibers. The authors in [22] achieved an increase in strength and Young's modulus for printed ABS samples filled with

VGCF by 39% and 60%, respectively, compared with ABS without filler. The article [23] reports that the tensile strength of the printed samples increased almost 3 times with the addition of 7 wt.% of multi-walled carbon nanotubes in ABS.

Mostly, the improvement in strength with the incorporation of nanoparticles into a volume of polymer is explained by the structure organization of the particles in the matrix, and their additional orientation during printing that provides a small reinforcing effect. The increase in the modulus is facilitated by an increase in the stiffness of products as the particles limit the mobility of macromolecular chains. In addition, well-dispersed fillers have more possibilities for physical or chemical bonding with the macromolecular chains thanks to their large surface area and high surface energy. As a result, when cracks begin to form, an effective stress transfer occurs between the polymer and the nanofillers.

If one talks about microfillers, short fibers with a high degree of anisometry are most often used to modify the polymer matrix of materials for FFF [24]. Glass fibers and carbon fibers are most often used to modify the polymer matrix of materials for FFF [25,26]. Carbon fibers are in some way unique among the reinforcing fibers. The share of their use in the production of the composite materials is constantly growing, which is explained by the high level of their properties. Carbon fibers have the largest Young's modulus among other fibers and exceed all heat-resistant fibers in specific parameters.

The data in [27] showed that printed PEEK samples with reinforced carbon fiber had significantly better strengths than the pure PEEK in the tensile and bending tests. The tensile strength of PEEK with 5 wt.% of carbon fibers (CF) was 101 MPa. The results in [28] suggest that the addition of carbon fiber or glass fiber to PEEK (polyetheretherketone) can significantly increase the tensile and bending strength, but, at the same time, reduce the deformation at break. The fiber content of 5 wt.% increased the tensile strength up to 94 MPa. The study [29] showed that printed samples based on ABS with a carbon fiber content of 5 wt.% had the highest average tensile strength, and the sample with a carbon fiber content of 7.5 wt.% had the highest average value of Young's modulus. The tensile strength and Young's modulus of the samples with a carbon fiber content of 5% and 7.5% increased by 22.5% and 30.5%, respectively. At the same time, compared with the sample made of pure plastic, the composite sample with a carbon fiber content of 5 wt.% had greater bending strength, flexural modulus, and fracture toughness. Products made on the basis of high-performance polymers by the FFF can be used in a variety of fields, for example, in the aerospace and automobile industry for the production of durable and lightweight products of complex shape such as air ducts, turbine parts, and aerodynamics elements. Aurora Flight Sciences, specializing in unmanned aerial vehicles (UAVs), in collaboration with Stratasys, created a UAV from ULTEM using 80% FFF-printed parts [30]. Another study reported on nanosatellites for space applications printed from PEEK. Moreover, researchers created nuclear shielding materials. Their study found that FFF-printing of PEEK composites allows the creation of cheaper and lightweight screens that provide greater protection from low-energy gamma rays than from higher-energy ones [31]. However, one of the most relevant areas of application of high-performance polymers in additive technologies is medicine. Currently, stainless steel, titanium, and their alloys are most often used to produce endoprostheses because of their good corrosion resistance, high mechanical properties, and good biocompatibility. However, the obvious hidden dangers are the harmfulness of the released metal ions and the radiopacity of metal alloys in vivo. Another serious problem is the mismatch of the Young's modulus between the metal and the surrounding bone tissue, which can cause stress screening after surgery, leading to bone resorption [32]. However, these risks can be avoided by replacing metals with strong biocompatible polymers. In recent years, polyetheretherketone (PEEK) being a highly efficient semicrystalline thermoplastic structural polymer has been recognized as a suitable replacement for metal implants, mainly because the elastic modulus of PEEK (3–4 GPa) is much closer to the modulus of human cortical bone (6–30 GPa), which is much lower than the Young's modulus of titanium and its alloys (more than 100 GPa) [33]. The limiting factor in the use of polymers is still the sufficiently low strength characteristics (not exceeding 110 MPa) compared to metals. In this regard, active research has been conducted in recent years to improve these parameters.

The aim of the study [34] was to compare the cytotoxicity of polyetheretherketone (PEEK) and polyetherketoneketone (PEKK) with conventional materials for dental implants and abutments, namely titanium alloy (Ti-6Al-4V) and tetragonal polycrystal of zirconium dioxide stabilized with yttrium oxide (Y-TZP). According to the analysis of cellular cytotoxicity and expression of proinflammatory cytokine genes, there were no differences between the materials. The PEEK surface, where fibroblast culture showed the best metabolic activity of cells, looks like the more promising material for the implant abutment. In [35], a series of PEEK composites with different carbon fiber (CF) contents (25 wt.% 30 wt.%, 35 wt.%, and 40 wt.%) was successfully obtained by injection molding. Evaluation of mechanical properties showed that the CF-PEEK composites have a higher bending strength, compressive strength, and hardness than pure PEEK, but lower impact strength. The modulus of elasticity of the CF-PEEK composites was much closer to human bones than to metals. In [36], the structural changes of implants fixed on a sintered polyamide skull model was investigated under the influence of mechanical stress in four simplified models. In a simplified model with a quasistatic load, both implants withstood forces exceeding those capable of causing skull fractures.

In one of our previous works [37], the cytotoxicity of samples based on polyetherimide modified with nanofibers such as VGCF was investigated using the MTT test (tetrazolium dye MTT, which is chemically 3-(4,5-dimethylthiazol-2-yl)-2,5-diphenyltetrazolium bromide). The samples made of both pure polyimide and with the addition of 1 wt.% VGCF did not have a negative effect on human fibroblast culture that may indicate their bioinertness. These materials possess good cell adhesion to the surface and have suitable conditions for cell proliferative activity.

Based on the foregoing, the purpose of the present work is to obtain new composite materials for FFF printing based on high-performance semicrystalline polyimide such as R-BAPB modified with nano- and microsized carbon fillers, as well as to study the possibility of using these materials for medical purposes, investigating their cytotoxicity.

2. Materials and Methods

A polyimide R-BAPB was synthesized in the form of a powder based on R (1,3-bis(3′,4-dicarboxyphenoxy) dianhydridebenzene), T_m ~164 °C (Techhimprom LLC, Yaroslavl, Russia), and diamine BAPB (4,4′-bis(4″-aminophenoxy)biphenyl), T_m ~198–199 °C (VWR International, Radnor, PA, USA). Phthalic anhydride, T_m ~131–134 °C (Sigma-Aldrich Co., LLC, St. Louis, MO, USA), was used as a chain growth limiter for the polycondensation reaction. Acetic anhydride, benzene, and triethylamine were supplied by Sigma-Aldrich Co., LLC.

The following fillers were used:

- VGCF—carbon nanofibers obtained by gas phase deposition (Pyrograf®-III, Cedarville, OH, USA) with an outer diameter of ~100 nm and a length of 20 to 200 μm.
- CF—discrete carbon fibers (Umatex, Moscow, Russia) with a diameter of ~7 μm and a length of ~7 mm.

2.1. Preparation of R-BAPB and R-BAPB-Based Composite Materials Modified with Carbon Nanoparticles

Synthesis of R-BAPB was carried out by chemical imidization. A more detailed description of the synthesis is presented in our early works [10]. To obtain the nanocomposite, the required number of VGCF particles was injected into the resulting polyamide acid solution. As a result of synthesis, pure R-BAPB powder was obtained, as well as powders with different concentrations of the carbon nanofibers VGCF—0.5%, 1%, 3%, and 5% (weight percentages). The synthesized polyimide had an average molecular weight of M_w ~20,000 Da [38], determined by the light scattering method. Earlier, the structure of polyimide was proven by IR (infrared) spectroscopy [39].

2.2. Preparation of the Composite Materials Based on R-BAPB Modified with Micron-Sized Discrete Carbon Fibers

To obtain a composite with discrete carbon fibers, polyimide powder obtained by the method described above was mixed with the carbon fibers in proportions corresponding to the concentrations studied. Before mixing, both powder and fibers were dried in a vacuum thermostat at a temperature of 150 °C for 1 day. Mixing was carried out in a twin-screw microextruder "DSM Xplore MC5" (Xplore, Sittard, The Netherlands) at a temperature of 370 °C and a screw rotation speed of 50 rpm for 10 min to ensure a more uniform dispersion of the fiber in the melt. As a result, strands of the composite material with carbon fiber concentrations of 10%, 20%, and 30% (weight percentages) were obtained.

2.3. Granulation of the Obtained Composite Materials

To achieve the most uniform loading of the material for obtaining filaments for FFF printing, all composites were subjected to granulation. Strands were prepared from pure R-BAPB, and the nanocomposites based on them according to the procedure described in paragraph 2.2. After that, all the strands obtained, including composites with the discrete carbon fibers, were crushed in a laboratory mill to granules~1–3 mm long and 1 mm in diameter.

2.4. Study of the Viscosity of the Obtained Materials

The viscosity of the polymer melt was studied on the Physica MCR301 rheometric unit (Anton Paar, Graz, Austria) in the CP25-2 cone-plane measuring system (diameter 25 mm, angle 2°, gap between the cone and the plane 0.05 mm) at a temperature of 360 °C. The test was carried out in an oscillating mode in the frequency range from 100 rad/s down to 1 rad/s.

2.5. Thermal Analysis of the Samples

Thermal analysis of the samples was carried out using the method of differential scanning calorimetry (DSC) at the "DSC 204 F1 Phoenix" device (NETZSCH, Selb, Germany) in an inert medium (argon), in the temperature range from 25 °C up to 350 °C at a heating rate of 10 K/min. The degree of crystallinity of the R-BAPB samples was estimated from the enthalpy of melting ΔH_m calculated earlier for R-BAPB with a degree of crystallinity of 100%, which was equal to 90 J/g [10]. A total of 3 samples were tested for each material (the deviation for each parameter did not exceed 0.5%).

To study the temperature of thermal degradation, thermogravimetric analysis (TGA) was used, using the TG 209 F1 Iris device (NETZSCH, Selb, Germany). The sample was heated in an inert medium (argon), ranging from a temperature of 30 °C to 800 °C at a speed of 10 K/min.

2.6. Obtaining Samples by Injection Molding

To study the mechanical properties and internal structure, as well as to determine the optimal concentrations of the fillers for the subsequent production of the filaments, the samples were obtained from the granules of the studied materials by injection molding. Before molding, all pellets were dried in a vacuum thermostat at a temperature of 150 °C for 1 day. Injection molding was carried out using the DSM Xplore MC5 microextruder (Xplore, Sittard, The Netherlands) and the DSM Xplore IM5.5 micro-injector (Xplore, Sittard, The Netherlands). The samples were prepared according to the following parameters:

- Pure R-BAPB: extruder temperature 360 °C, screw speed 50 rpm, cylinder temperature 370 °C, mold form temperature 180 °C, and pressure 16 bar.
- R-BAPB with carbon nanofibers VGCF: extruder temperature 360 °C, screw speed 50 rpm, cylinder temperature 370 °C, mold form temperature 190 °C, and pressure 16 bar.
- R-BAPB with discrete carbon fiber: extruder temperature 370 °C, screw speed 50 rpm, cylinder temperature 380 °C, mold form temperature 190 °C, and pressure 16 bar.

The material cooled in the mold form for about 10 s. As a result, "dog-bone" samples with a width of 4 mm, a thickness of 2 mm, and a length of the working part of 25 mm were obtained by injection molding.

2.7. Filaments Production

Based on studies of the viscosity, and thermal and mechanical properties of the prepared materials, optimal concentrations for both types of the fillers were determined. First, the composite granules were dried in a vacuum thermostat at a temperature of 150 °C for 12 h. Next, the granules were loaded into the DSM Xplore MC5 microextruder (Xplore, Sittard, The Netherlands). After that, the filament was extracted from the melt using the coil of the receiving device. The filaments were obtained with the following parameters:

- Pure R-BAPB: screw speed 35 rpm, chamber temperature 360 °C, chamber force 200 N, and coil speed 250.
- R-BAPB with carbon nanofibers: screw speed 35 rpm, chamber temperature 360 °C, chamber force 250 N, and coil speed 250.
- R-BAPB with discrete carbon fiber: screw speed 50 rpm, chamber temperature 370 °C, chamber force 350 N, and coil speed 150.
- As a result, filaments were obtained from pure R-BAPB, R-BAPB with 1 wt.% VGCF, and R-BAPB with 20 wt.% discrete carbon fiber with a diameter of 1.6–1.85 mm.

2.8. FFF Printing

FFF printing was performed on an experimental setup for 3D-printing with high-temperature-resistant plastics. Samples were printed in "dog-bone" form with a width of 4 mm, a thickness of 2 mm, and a working length of 25 mm, which corresponds to the samples obtained by injection molding described in paragraph 2.6. The COMPASS-3D software (ASKON, Saint-Petersburg, Russia) was used to create the "dog-bone" model. The Cura v.4.13.0 software (Ultimaker, Utrecht, The Netherlands) was used as a slicer for setting printing parameters.

Samples from different materials were printed with the following parameters:

- Pure R-BAPB: nozzle diameter 0.4 mm, extruder temperature 365 °C, build platform temperature 180 °C, chamber temperature 150 °C, printing speed 50 mm/s, layer thickness 0.1 mm, raster angle ± 45°, and wall thickness 04 mm.
- R-BAPB + 1 wt.% VGCF: nozzle diameter 0.4 mm, extruder temperature 365 °C, build platform temperature 180 °C, chamber temperature 150 °C, printing speed 50 mm/s, layer thickness 0.1 mm, raster angle ± 45°, and wall thickness 0.4 mm.
- R-BAPB + 20 wt.% CF: nozzle diameter 0.4 mm, extruder temperature 380 °C, build platform temperature 180 °C, chamber temperature 150 °C, printing speed 50 mm/s, layer thickness 0.1 mm, raster angle ± 45°, and wall thickness 0.4 mm.

2.9. Investigation of the Mechanical Characteristics

The mechanical properties of the samples were studied at the ElectroPuls E1000 (Instron, Norwood, MA, USA). The samples were tested in the form of the "dog-bone", 4 mm wide and 2 mm thick, and a working part with a length of 25 mm at a speed of 1 mm/min. At least 5 samples from each material were tested to measure the mechanical properties. According to the test results, the Young's modulus, tensile strength, and deformation at break of each sample were determined.

2.10. Investigation of the Internal Structure of the Samples

The porosity of the samples was investigated using the pycnometric method. A sample of the material was placed in a capillary pycnometer and filled with 96% ethanol at 25 °C. Next, the volume occupied by the material was calculated as the difference with the volume of a pycnometer filled with ethanol. The porosity was calculated based on the differences between the theoretical and experimentally determined volume of the material according to the following formula:

$$P = \frac{V_t - \frac{m}{\rho}}{V_t} \times 100\% \qquad (1)$$

where P is porosity; V_t is the volume of the test sample determined by the pycnometric method; ρ is the density of the material; m is the mass of the sample measured on analytical scales.

Micrographs of the fracture surface of the block samples at various magnifications were obtained using a Supra-55 VP scanning electron microscope (SEM) (Carl Zeiss, Oberkochen, Germany). To obtain a high-quality fracture surface, the sample was broken in liquid nitrogen. The resulting cleavages of the samples were fixed with a special conductive glue on the microscope holders and a thin layer of platinum was sprayed.

2.11. Cytotoxicity Study of the Printed Samples

Human osteosarcoma MG63 cell lines were used to study cytotoxicity. The cells were cultured in a CO_2 incubator at 37 °C in a humidified atmosphere containing air and 5% CO_2 in an EMEM nutrient medium (Dulbecco's modified Eagle's medium; Gibco) containing 1% essential amino acids, 10% (by volume) thermally inactivated fetal bovine serum (FBS; HyClone Marlborough, MA, USA), 1% L-glutamine, 50 U/mL of penicillin, and 50 mcg/mL of streptomycin. For this experiment, the samples in the form of disks with a diameter of 11 mm printed by FFF were filled with 2 mL of complete nutrient medium and incubated for 1 and 3 days. To assess cytotoxicity, cells in the amount of 5.0×10^3 cells/100 mcL/well were plated in 96-well plates and cultivated for 24 h for their attachment. Then, 100 mcL of the medium was added and then incubated for 72 h. At the end of the incubation period, the medium was removed and 50 mcL/well of EMEM medium with MTT (0.1 mg/mL) was introduced. The cells were incubated in a CO_2 incubator for 2 h at 37 °C. After removal of the supraplastic fluid, formazane crystals formed by metabolically vital cells were dissolved in dimethyl sulfoxide (50 mcL/well) and the optical density was measured at 570 nm on a flatbed spectrophotometer. Ethical Statement: The MG-63 osteosarcoma cell lines were obtained from the Vertebrate Cell Culture Collection (Institute of Cytology RAS, St-Petersburg, Russia).

Cell adhesion to the surface of the samples was studied using a Supra-55 scanning electron microscope (Carl Zeiss, Oberkochen, Germany). Before placing the samples inside the microscope chamber, a thin conductive layer of platinum was sprayed onto their surface. The accelerating voltage was 3–5 kV.

3. Results and Discussion

3.1. Investigation of the Viscosity of Melts of Composites Based on R-BAPB

One of the most important parameters affecting the FFF printing process is the viscosity of the materials. If the viscosity is too high, the movement of the melt may be hindered, and its integrity at the outlet of the nozzle may also be disrupted. The introduction of VGCF particles into the polymer matrix based on R-BAPB leads to an increase in the viscosity with a decrease in angular frequency (Figure 2). The viscosity enhances with an increase in the VGCF concentration. At the angular frequencies of about 1, it increases quite significantly starting from 1 wt.% VGCF, which may be due to the high anisometry of the nanofibers. This fact indicates a good dispersion of the nanoparticles in the polymer matrix [40]. With deformations such as those occurring during the printing process, the melt viscosity is in the range suitable for high-quality printing for all concentrations except 5 wt.% VGCF. The introduction of the discrete carbon fibers leads to a noticeable increase in viscosity (Figure 3). With the introduction of 30 wt.% CF, the melt viscosity reaches 4000 Pa·s, which is too high for high-quality FFF printing as there is a pure flow and the appearance of breaks in the melt jet. However, for 20 wt.% CF, the value of the complex viscosity at 10 rad/s does not exceed 2000 Pa·s, which is an acceptable value for processing by both injection molding and FFF printing.

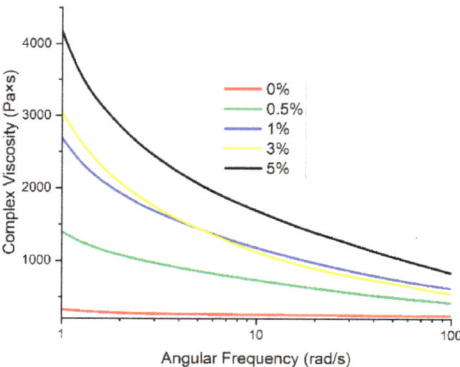

Figure 2. Dependence of the melt viscosity on angular frequency for R-BAPB modified with VGCF.

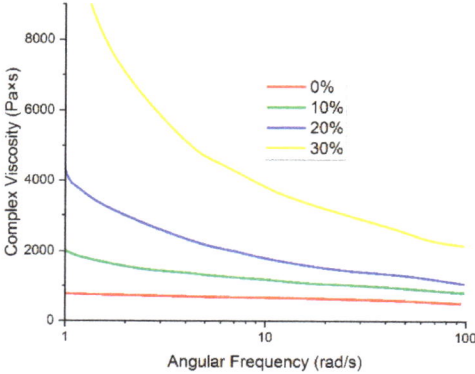

Figure 3. Dependence of melt viscosity on angular frequency for R-BAPB modified with CF.

3.2. Studies of Thermal Properties of Melts of the R-BAPB-Based Composites

According to the study of materials by the DSC method, the introduction of various concentrations of both VGCF and discrete CF practically does not affect the glass transition temperature (Table 1, Figure 4). At the same time, with an increase in the concentration of VGCF, there is a significant increase in the degree of crystallinity of the nanocomposites and the crystallization rate, which is observed on the DSC curve owing to the appearance of a crystallization peak (Figure 3). All this indicates that the carbon nanofibers act as crystallization centers for polyimide R-BAPB [11]. When the micron carbon fibers are added, a weak peak of crystallization appears during DSC heating. It is worth noting that the samples crystallize under sufficiently slow heating during the DSC experiment. When samples are obtained by injection molding and FFF printing, the composites do not have time to crystallize and remain in an amorphous state.

Table 1. The dependence of the thermal properties of the composites based on R-BAPB on the concentration of various fillers. (T_g is the glass transition temperature, T_m is the melting point, T_{cr} is the crystallization temperature, χ is the degree of crystallinity, τ_5 is the temperature at the loss of 5% of the sample mass; the deviation for each value did not exceed 0.5%).

Sample	T_g, °C	T_m, °C	T_{cr}, °C	χ, %	τ_5, °C
R-BAPB	201	326	305	2.7	526
R-BAPB + 0.5% VGCF	201	320	289	3.7	513
R-BAPB + 1% VGCF	200	320	287	13.3	511
R-BAPB + 3% VGCF	201	320	287	16.3	510
R-BAPB + 5% VGCF	200	322	282	23.5	513
R-BAPB + 10% CF	199	320	290	6.5	534
R-BAPB + 20% CF	199	324	289	6.3	535

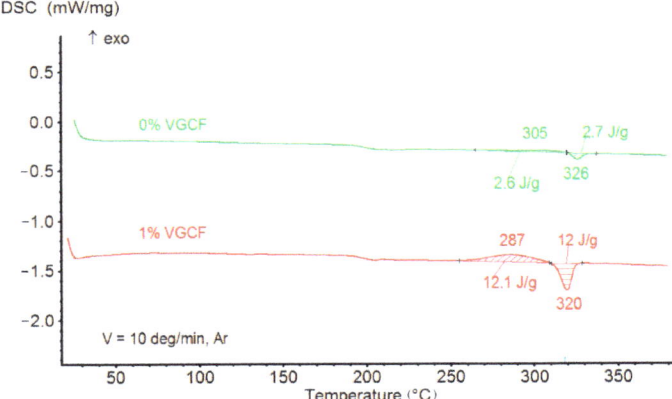

Figure 4. An example of the effect of VGCF on phase transitions in R-BAPB-based composites.

The temperature of thermal degradation of the materials was determined with the aid of the TGA method. The received data revealed that these materials have a single-stage mechanism of thermal destruction. At the same time, a sharp increase in the rate of the thermal destruction process is observed with the loss of 5% of the sample mass. The addition of different concentrations of the nanofibers has no essential effect on the temperature τ_5; with the introduction of the discrete carbon fibers, the thermal degradation temperature increases due to the greater thermal stability of the fibers themselves (Table 1).

3.3. Investigation of the Mechanical Characteristics and Internal Structure of the Molded and Printed Samples Made of the R-BAPB-Based Composites

Tensile tests of molded samples were carried out to determine the optimal concentration of the fillers for the preparation of the filament. Analysis of the data showed that with the incorporation of VGCF, the strength and Young's modulus slightly increase with an increase in VGCF concentration, but the deformation at break significantly decreases (Figure 5). The samples with concentrations of 0.5 wt.% and 1 wt.% VGCF retain about 50% of deformation regarding pure R-BAPB, in contrast to 3 wt.% and 5 wt.%. As the mechanical characteristics of the samples with 0.5 wt.% and 1 wt.% VGCF turned out to be quite close, the choice in favor of 1 wt.% VGCF was made due to a 3 times higher degree of crystallinity (Table 1), because of the presence of a larger amount of crystallization centers. This factor may be important in further studies of the crystallized samples obtained by FFF printing, as this composite will be easier to transfer to a crystallized state due to a higher crystallization rate.

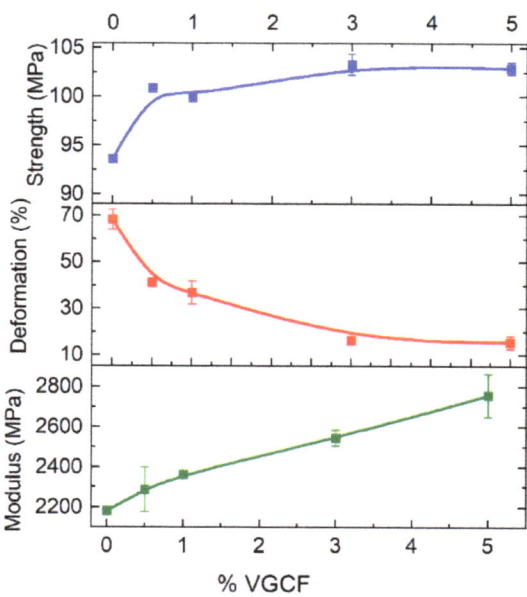

Figure 5. Dependence of strength, Young's modulus, and deformation at break of the R-BAPB-based molded samples on the concentration of VGCF.

The incorporation of the discrete CF leads to a visible increase in the strength and Young's modulus of the studied samples, while the deformation at break is reduced by more than 10 times starting from 10 wt.% CF (Figure 6). The sample with 30 wt.% CF has a similar strength and a larger modulus than that with 20 wt.%, but due to the high melt viscosity of this composite (Figure 3), a choice was made in favor of R-BAPB with 20 wt.% of CF.

Based on the collected data on the composite's viscosity, as well as on their thermal and mechanical characteristics, the filaments and, consequently, the samples were obtained by FFF printing from pure R-BAPB, and also filled with 1 wt.% VGCF and 20 wt.% CF. The samples were printed with a raster angle of ±45°, which is the optimal direction for FFF printing to achieve good mechanical properties for most types of tests and can also overlap the pores by changing the direction of the lines. The results of the tensile tests (Figure 7) showed that the strength and Young's modulus of the printed samples from pure R-BAPB are very close to the parameters of the molded samples (91 and 93 MPa, 2462 and 2179 MPa, respectively). At the same time, the deformation at break in the case of the printed samples does not exceed 10.5%, which is significantly lower than that of the injection-molded samples. This fact is explained by the presence of voids inside the sample caused by layer-by-layer deposition of materials (Figure 8b). FFF samples from R-BAPB filled with 1 wt.% VGCF have a strength and Young's modulus that are almost identical to the printed samples from pure R-BAPB. However, the deformation at break is almost 4 times higher compared to the unmodified printed R-BAPB sample and this value is 37%. This effect is probably caused by a decrease in porosity and a possible increase in adhesion between layers [41]. The SEM analysis revealed that the voids in the R-BAPB samples with 1 wt.% VGCF become smaller compared to those of pure R-BAPB (Figure 8c,d), and the boundaries between adjacent layers become blurred (Figure 8d). For pure R-BAPB, the fracture surface is sufficiently smooth, which is a typical characteristic of brittle fracture (Figure 8b) [42]. With the introduction of 1 wt.% VGCF, the texture of the fracture surface of the printed sample (Figure 8g) becomes rougher, i.e., plastic destruction takes place. All this suggests that VGCF effectively distorts the crack tip trajectory and increases the complexity of crack propagation [43], which leads to an increased value of deformation at break in the printed R-BAPB sample with 1 wt.% VGCF compared to the pure R-BAPB sample. The samples made of R-BAPB modified with the discrete CF, in turn, have low deformation, but at the same time, their strength increases by almost 50%, and the modulus is 2.5 times higher compared to pure R-BAPB. On the images of the fracture surface, it can be seen how the fibers line up along the direction of laying the polymer thread under 45°, which provides a reinforcing effect (Figure 8e).

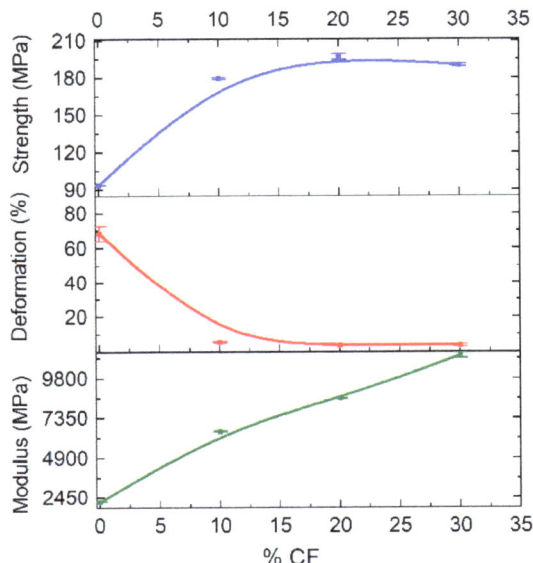

Figure 6. Dependence of strength, modulus of elasticity, and deformation at break of the R-BAPB-based molded samples on the concentration of CF.

For a more detailed study of the internal structure of the samples, their porosity was investigated. The study by the pycnometric method showed that for all samples obtained by injection molding, the porosity barely exceeds 2%, while for the samples obtained by FFF printing, the porosity for pure R-BAPB is ~4% (Table 2), which leads to a decrease in the deformation of the printed sample compared to molding by almost 7 times. The porosity of the sample with 1 wt.% VGCF is only 1.9%, while the deformation and strength properties are comparable to the samples obtained by injection molding, in particular, the printed sample has a high deformation at break of 37%. It can be concluded that the incorporation of VGCF reduces the number of pores in the sample that appear because of the layered deposition of the material, which leads to the fact that the deformation at break increases significantly. For 20 wt.% CF, the porosity of the printed sample is ~5%. In this case, the incorporation of the rigid carbon fibers results in a serious drop in deformation and an increase in Young's modulus. Summing up, in terms of strength characteristics for FFF-printing, composites based on R-BAPB with discrete carbon fiber exceed materials based on PEEK and Ultem with discrete reinforcing fibers. In turn, for composite materials with carbon nanoparticles, the deformation at break of printed samples exceeds all known analogs.

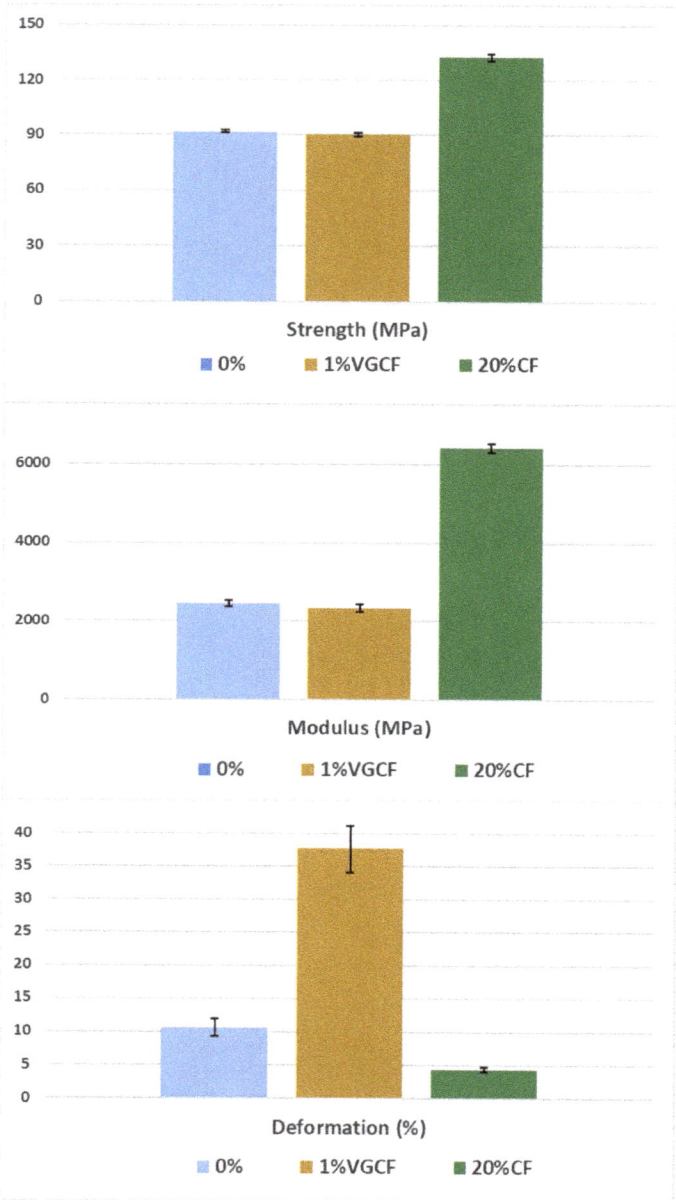

Figure 7. Mechanical characteristics of the FFF samples from R-BAPB and its composites.

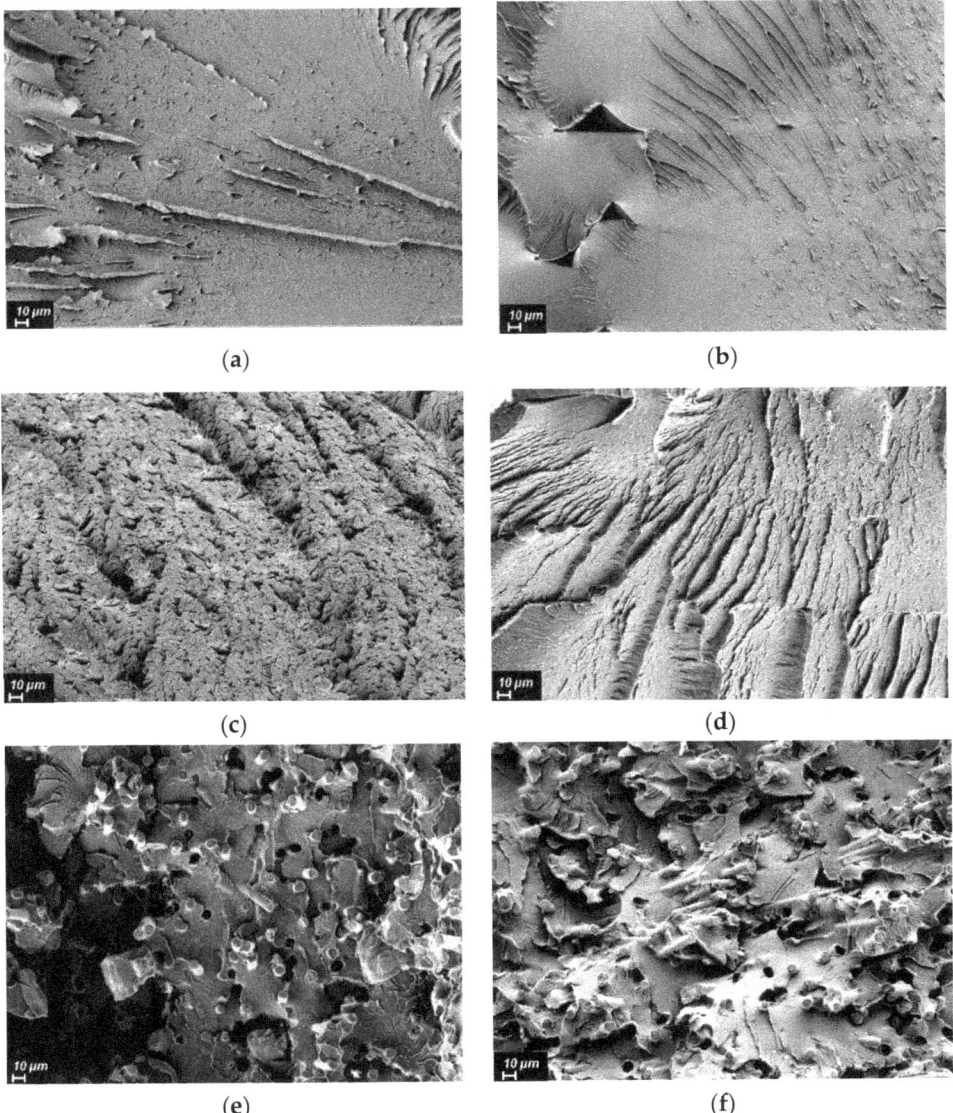

Figure 8. SEM images of the fracture surface of the studied samples (injection molding on the left; FFF printing on the right) from: (**a**,**b**)—pure R-BAPB; (**c**,**d**)—R-BAPB + 1% VGCF; (**e**,**f**)—R-BAPB + 20% CF.

Table 2. Comparison of porosity of the samples from R-BAPB-based composites produced by injection molding (IM) and FFF printing.

Sample	IM	FFF
R-BAPB	0.33 ± 0.07%	3.84 ± 0.13%
R-BAPB + 1% VGCF	1.04 ± 0.05%	1.90 ± 0.09%
R-BAPB + 20% CF	2.13 ± 0.11%	5.27 ± 0.15%

3.4. Investigation of Cytotoxicity of the Obtained Products

Materials that will be in direct contact with living tissue should not have a negative impact on the organism with which they will interact. Exposure to toxic materials can lead to irreversible damage or even the death of cells. Therefore, for use in medicine, any materials are first checked for the level of cytotoxicity through laboratory in vitro tests and analyses. Cells interacting with the test material should not change their normal cycle of functioning, and their proliferation should not be disrupted. In turn, the materials that are planned to be used for implantation should ensure good attachment and growth of cells on their surface.

The viability and proliferative activity of cells incubated on the surface of the material was studied using an MTT test. Incubation of cells on the printed samples from pure R-BAPB showed that this material does not have a pronounced cytotoxic effect on human osteosarcoma cell culture. At the same time, when comparing the optical densities of formazane solutions between the levels of proliferative activity of cells incubated on both pure R-BAPB and R-BAPB modified with various fillers, no statistically essential difference was revealed (Figure 9). In the photographs, where cell culture was recorded a day after plating (Figure 10), taken with a scanning electron microscope, it is clear that the cells are well spread out on the surface of the sample. This circumstance indicates that the material does not have a pronounced toxic effect and the surface properties of the samples are favorable for adhesion and proliferation of human osteosarcoma cells.

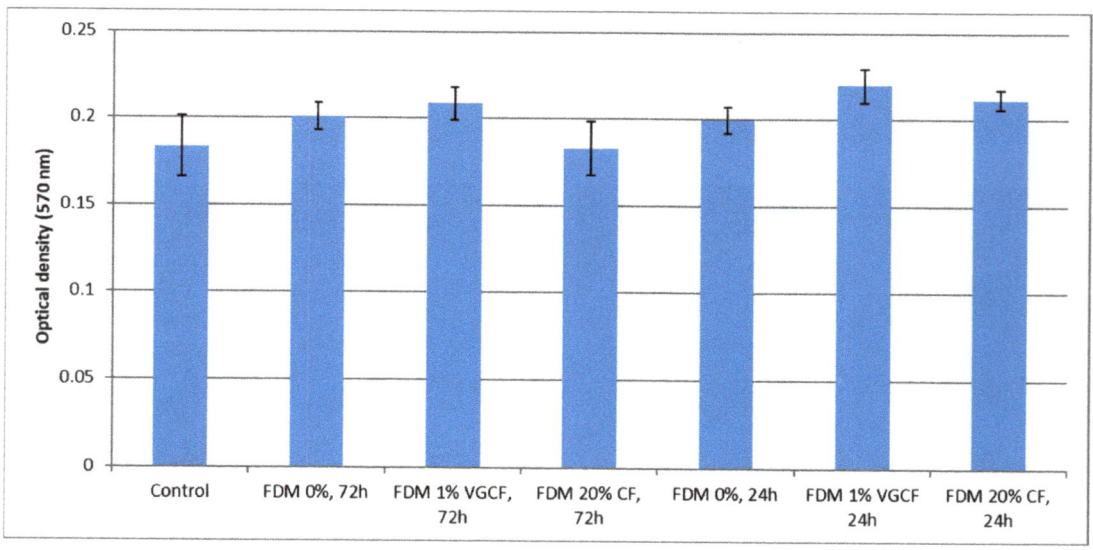

Figure 9. Comparative characteristics of the optical density of formazane solutions obtained from the studied samples.

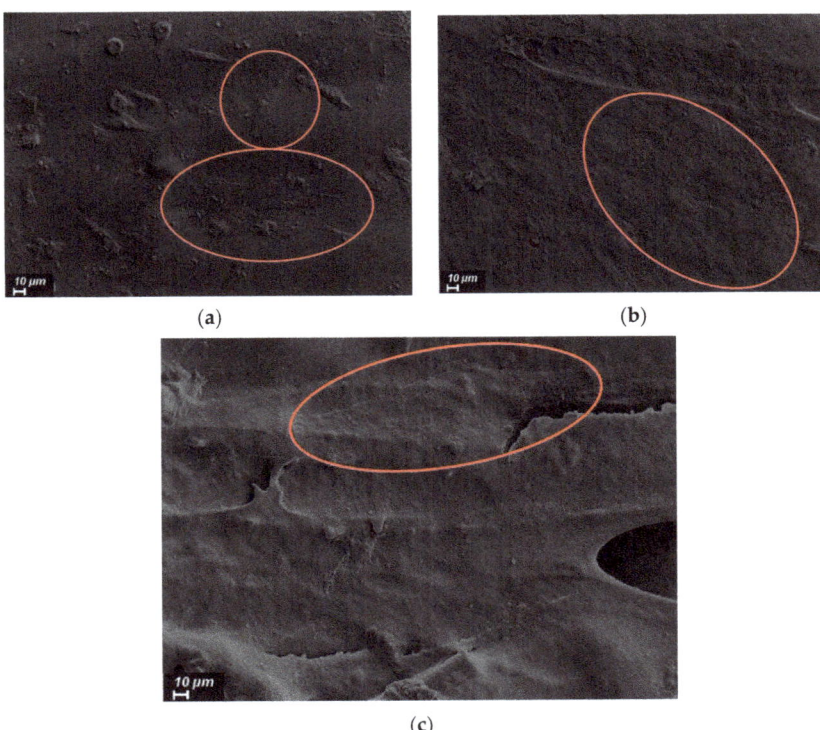

Figure 10. SEM images of cells 1 day after plating on the samples surface prepared by the FFF method from (**a**) pure R-BAPB, (**b**) R-BAPB + 1% VGCF, and (**c**) R-BAPB + 20% CF.

4. Conclusions

In this work, composite materials based on semicrystalline polyimide R-BAPB filled with the carbon nanofibers and micron-sized discrete carbon fibers were obtained.

Based on the study of the melt viscosity of the composites with R-BAPB, and thermal and mechanical properties, the optimal concentration of the carbon nanoparticles in polyimide R-BAPB was determined as 1 wt.%, and for the discrete carbon fibers—20 wt.%.

A study of the printed samples revealed that the introduction of 1 wt.% VGCF reduces the porosity of the printed samples by more than 2 times, which leads to an increase in deformation at break by more than 3 times compared to pure R-BAPB. When 20 wt.% of discrete carbon fibers are added to R-BAPB, there is a sharp increase in the strength of the printed samples to 135 MPa, and the modulus of elasticity to 6.2 GPa.

There was no cytotoxic effect of the polyimide composite materials on the culture of human osteosarcoma cells. Moreover, a good cell adhesion on the surface of the material was observed, indicating the bioinertness of the investigated composites.

Due to the high strength characteristics when incorporating discrete carbon fibers and high deformation when incorporating carbon nanofibers, the developed biocompatible composite materials for FFF can be widely used both in various industrial sectors and in medicine.

Author Contributions: Conceptualization, I.P. and G.V.; methodology, I.P. and G.V.; formal analysis, I.P.; investigation, I.P., G.V., E.I., E.P., Y.N. and V.E.; resources, A.D. and V.S.; data curation, I.P., G.V., E.I., E.P., Y.N. and V.E.; writing—original draft preparation, I.P. and G.V.; writing—review and editing, I.P., G.V., E.I. and V.Y.; visualization, I.P.; supervision, V.Y.; project administration, V.Y.; funding acquisition, V.Y. and I.P. All authors have read and agreed to the published version of the manuscript.

Funding: The reported study was funded by RFBR, project number 20-33-90145/20.

Institutional Review Board Statement: Not applicable.

Informed Consent Statement: Not applicable.

Data Availability Statement: Not applicable.

Conflicts of Interest: The authors declare no conflict of interest.

References

1. Wendel, B.; Rietzel, D.; Kühnlein, F.; Feulner, R.; Hülder, G.; Schmachtenberg, E. Additive Processing of Polymers. *Macromol. Mater. Eng.* **2008**, *293*, 799–809. [CrossRef]
2. Ngo, T.D.; Kashani, A.; Imbalzano, G.; Nguyen, K.T.Q.; Hui, D. Additive manufacturing (3D printing): A review of materials, methods, applications and challenges. *Compos. Part B Eng.* **2018**, *143*, 172–196. [CrossRef]
3. Tao, Y.; Kong, F.; Li, Z.; Zhang, J.; Zhao, X.; Yin, Q.; Xing, D.; Li, P. A review on voids of 3D printed parts by fused filament fabrication. *J. Mater. Res. Technol.* **2021**, *15*, 4860–4879. [CrossRef]
4. Wang, X.; Zhao, L.; Fuh, J.Y.H.; Lee, H.P. Effect of Porosity on Mechanical Properties of 3D Printed Polymers: Experiments and Micromechanical Modeling Based on X-ray Computed Tomography Analysis. *Polymers* **2019**, *11*, 1154. [CrossRef] [PubMed]
5. 3D Printing High-Strength Carbon Composites Using PEEK, PAEK | Designnews.Com. Available online: https://www.designnews.com/design-hardware-software/3d-printing-high-strength-carbon-composites-using-peek-paek (accessed on 30 August 2022).
6. Acquah, S.F.A.; Leonhardt, B.E.; Nowotarski, M.S.; Magi, J.M.; Chambliss, K.A.; S.Venzel, T.E.; Delekar, S.D.; Al-Hariri, L.A. Carbon Nanotubes and Graphene as Additives in 3D Printing. In *Carbon Nanotubes—Current Progress of Their Polymer Composites*; IntechOpen: London, UK, 2016. [CrossRef]
7. Penumakala, P.K.; Santo, J.; Thomas, A. A critical review on the fused deposition modeling of thermoplastic polymer composites. *Compos. Part B Eng.* **2020**, *201*, 108336. [CrossRef]
8. Ha, C.-S.; Anu, S.; Mathews, G. Polyimides and High Performance Organic Polymers. In *Advanced Functional Materials*; Springer: Berlin/Heidelberg, Germany, 2011; pp. 1–36. [CrossRef]
9. Richardson, R.R.; Miller, J.A.; Reichert, W.M. Polyimides as biomaterials: Preliminary biocompatibility testing. *Biomaterials* **1993**, *14*, 627–635. [CrossRef]
10. Vaganov, G.; Didenko, A.; Ivan'Kova, E.; Popova, E.; Yudin, V.; Elokhovskii, V.; Lasota, I. Development of new polyimide powder for selective laser sintering. *J. Mater. Res.* **2019**, *34*, 2895–2902. [CrossRef]
11. Vaganov, G.; Ivan'kova, E.; Didenko, A.; Popova, E.; Elokhovskii, V.; Yudin, V. Effect of Carbon Nanoparticles on the Structure and Properties of Melt-Extruded R-BAPB Polyimide Fibers. *Fibers Polym.* **2022**, *23*, 611–616. [CrossRef]
12. Pegoretti, A.; Dorigato, A. Polymer Composites: Reinforcing Fillers. In *Encyclopedia of Polymer Science and Technology*; Wiley: Hoboken, NJ, USA, 2019; pp. 1–72. [CrossRef]
13. Pratama, J.; Cahyono, S.I.; Suyitno, S.; Muflikhun, M.A.; Salim, U.A.; Mahardika, M.; Arifvianto, B. A Review on Reinforcement Methods for Polymeric Materials Processed Using Fused Filament Fabrication (FFF). *Polymers* **2021**, *13*, 4022. [CrossRef]
14. Stansbury, J.W.; Idacavage, M.J. 3D printing with polymers: Challenges among expanding options and opportunities. *Dent. Mater.* **2016**, *32*, 54–64. [CrossRef]
15. Christ, S.; Schnabel, M.; Vorndran, E.; Groll, J.; Gbureck, U. Fiber reinforcement during 3D printing. *Mater. Lett.* **2015**, *139*, 165–168. [CrossRef]
16. Harikrishnan, U.; Soundarapandian, S. Fused Deposition Modelling based Printing of Full Complement Bearings. *Procedia Manuf.* **2018**, *26*, 818–825. [CrossRef]
17. Mai, Y.W.; Yu, Z.Z. *Polymer Nanocomposites*; Elsevier Ltd.: Amsterdam, The Netherlands, 2006; ISBN 9781855739697.
18. Wang, X. Processing and Characterization of Multifunctional Thermoplastic Nanocomposite Films. Master's Thesis, University of Central Florida, Orlando, FL, USA, 2014.
19. Hegde, M.; Lafont, U.; Norder, B.; Samulski, E.T.; Rubinstein, M.; Dingemans, T.J. SWCNT induced crystallization in amorphous and semi-crystalline poly(etherimide)s: Morphology and thermo-mechanical properties. *Polymer* **2014**, *55*, 3746–3757. [CrossRef]
20. Polyakov, I.V.; Vaganov, G.V.; Yudin, V.E.; Ivan'kova, E.M.; Popova, E.N.; Elokhovskii, V.Y. Investigation of Properties of Nanocomposite Polyimide Samples Obtained by Fused Deposition Modeling. *Mech. Compos. Mater.* **2018**, *54*, 33–40. [CrossRef]
21. Bilkar, D.; Keshavamurthy, R.; Tambrallimath, V. Influence of carbon nanofiber reinforcement on mechanical properties of polymer composites developed by FDM. *Mater. Today Proc.* **2021**, *46*, 4559–4562. [CrossRef]

22. Shofner, M.L.; Lozano, K.; Rodríguez-Macías, F.J.; Barrera, E.V. Nanofiber-reinforced polymers prepared by fused deposition modeling. *J. Appl. Polym. Sci.* **2003**, *89*, 3081–3090. [CrossRef]
23. Sezer, H.K.; Eren, O. FDM 3D printing of MWCNT re-inforced ABS nano-composite parts with enhanced mechanical and electrical properties. *J. Manuf. Process.* **2019**, *37*, 339–347. [CrossRef]
24. Seymour, R.B. The Role of Fillers and Reinforcements in Plastics Technology. *Polym.-Plast. Technol. Eng.* **2006**, *7*, 49–79. [CrossRef]
25. Wang, X.; Jiang, M.; Zhou, Z.; Gou, J.; Hui, D. 3D printing of polymer matrix composites: A review and prospective. *Compos. Part B Eng.* **2017**, *110*, 442–458. [CrossRef]
26. Tekinalp, H.L.; Kunc, V.; Velez-Garcia, G.M.; Duty, C.E.; Love, L.J.; Naskar, A.K.; Blue, C.A.; Ozcan, S. Highly oriented carbon fiber–polymer composites via additive manufacturing. *Compos. Sci. Technol.* **2014**, *105*, 144–150. [CrossRef]
27. Han, X.; Yang, D.; Yang, C.; Spintzyk, S.; Scheideler, L.; Li, P.; Li, D.; Geis-Gerstorfer, J.; Rupp, F. Carbon Fiber Reinforced PEEK Composites Based on 3D-Printing Technology for Orthopedic and Dental Applications. *J. Clin. Med.* **2019**, *8*, 240. [CrossRef] [PubMed]
28. Wang, P.; Zou, B.; Ding, S.; Huang, C.; Shi, Z.; Ma, Y.; Yao, P. Preparation of short CF/GF reinforced PEEK composite filaments and their comprehensive properties evaluation for FDM-3D printing. *Compos. Part B Eng.* **2020**, *198*, 108175. [CrossRef]
29. Ning, F.; Cong, W.; Qiu, J.; Wei, J.; Wang, S. Additive manufacturing of carbon fiber reinforced thermoplastic composites using fused deposition modeling. *Compos. Part B Eng.* **2015**, *80*, 369–378. [CrossRef]
30. World's First Jet-Powered, 3D Printed UAV Tops 150 MPH—Stratasys. Available online: https://www.stratasys.com/en/resources/blog/aurora-uav-3d-printing/ (accessed on 7 September 2022).
31. Dua, R.; Rashad, Z.; Spears, J.; Dunn, G.; Maxwell, M. Applications of 3D-Printed PEEK via Fused Filament Fabrication: A Systematic Review. *Polymers* **2021**, *13*, 4046. [CrossRef]
32. Nune, K.C.; Misra, R.D.K.; Li, S.J.; Hao, Y.L.; Yang, R. Osteoblast cellular activity on low elastic modulus Ti–24Nb–4Zr–8Sn alloy. *Dent. Mater.* **2017**, *33*, 152–165. [CrossRef]
33. Deng, Y.; Yang, Y.; Ma, Y.; Fan, K.; Yang, W.; Yin, G. Nano-hydroxyapatite reinforced polyphenylene sulfide biocomposite with superior cytocompatibility and in vivo osteogenesis as a novel orthopedic implant. *RSC Adv.* **2017**, *7*, 559–573. [CrossRef]
34. Peng, T.Y.; Shih, Y.H.; Hsia, S.M.; Wang, T.H.; Li, P.J.; Lin, D.J.; Sun, K.T.; Chiu, K.C.; Shieh, T.M. In Vitro Assessment of the Cell Metabolic Activity, Cytotoxicity, Cell Attachment, and Inflammatory Reaction of Human Oral Fibroblasts on Polyetheretherketone (PEEK) Implant–Abutment. *Polymers* **2021**, *13*, 2995. [CrossRef]
35. Qin, W.; Li, Y.; Ma, J.; Liang, Q.; Tang, B. Mechanical properties and cytotoxicity of hierarchical carbon fiber-reinforced poly (ether-ether-ketone) composites used as implant materials. *J. Mech. Behav. Biomed. Mater.* **2019**, *89*, 227–233. [CrossRef]
36. Lethaus, B.; Safi, Y.; Ter Laak-Poort, M.; Kloss-Brandstätter, A.; Banki, F.; Robbenmenke, C.; Steinseifer, U.; Kessler, P. Cranioplasty with Customized Titanium and PEEK Implants in a Mechanical Stress Model. *J. Neurotrauma* **2012**, *29*, 1077–1083. [CrossRef]
37. Polyakov, I.V.; Vaganov, G.V.; Yudin, V.E.; Smirnova, N.V.; Ivan'kova, E.M.; Popova, E.N. Study of Polyetherimide and Its Nanocomposite 3D Printed Samples for Biomedical Application. *Polym. Sci. Ser. A* **2020**, *62*, 337–342. [CrossRef]
38. Vaganov, G.V.; Ivan'Kova, E.M.; Didenko, A.L.; Popova, E.N.; Elokhovskiy, V.Y.; Yudin, V.E. The study of heat-resistant fibers obtained from a melt of thermoplastic crystallizable polyimide. *J. Phys. Conf. Ser.* **2020**, *1697*, 012114. [CrossRef]
39. Vaganov, G.; Ivan'kova, E.; Didenko, A.; Popova, E.; Elokhovskiy, V.; Kasatkin, I.; Yudin, V. High-performance crystallized composite carbon nanoparticles/polyimide fibers. *J. Appl. Polym. Sci.* **2022**, *139*, e52748. [CrossRef]
40. Huang, Y.Y.; Ahir, S.V.; Terentjev, E.M. Dispersion rheology of carbon nanotubes in a polymer matrix. *Phys. Rev. B-Condens. Matter Mater. Phys.* **2006**, *73*, 125422. [CrossRef]
41. Chen, Q.; Zhang, Y.Y.; Huang, P.; Li, Y.Q.; Fu, S.Y. Improved bond strength, reduced porosity and enhanced mechanical properties of 3D-printed polyetherimide composites by carbon nanotubes. *Compos. Commun.* **2022**, *30*, 101083. [CrossRef]
42. Yue, Z.; Yu, M.Y.; Lan, X. Study on Properties of Carbon Nanotubes/Epoxy Resin Composite Prepared by In Situ Polymerization. *Adv. Mater. Res.* **2013**, *750–752*, 132–135. [CrossRef]
43. Zhou, Y.; Jeelani, M.I.; Jeelani, S. Development of photo micro-graph method to characterize dispersion of CNT in epoxy. *Mater. Sci. Eng. A* **2009**, *506*, 39–44. [CrossRef]

Article

Influence of Rapid Consolidation on Co-Extruded Additively Manufactured Composites

Chethan Savandaiah [1,2,*], Stefan Sieberer [3], Bernhard Plank [4], Julia Maurer [4], Georg Steinbichler [2] and Janak Sapkota [5]

1. Bio-Based Composites and Processes, Wood K Plus-Kompetenzzentrum Holz GmbH, 4040 Linz, Austria
2. Institute for Polymer Injection Moulding and Process Automation, Johannes Kepler University, 4040 Linz, Austria; georg.steinbichler@jku.at
3. Institute of Structural Lightweight Design, Johannes Kepler University, 4040 Linz, Austria; stefan.sieberer@jku.at
4. Research Group Computed Tomography, University of Applied Sciences Upper Austria, 4600 Wels, Austria; bernhard.plank@fh-wels.at (B.P.); julia.maurer@fh-wels.at (J.M.)
5. Research Centre of Applied Science and Technology, Tribhuvan University, Kirtipur 44600, Nepal; sapkota.janak@outlook.com
* Correspondence: k1606005@students.jku.at

Citation: Savandaiah, C.; Sieberer, S.; Plank, B.; Maurer, J.; Steinbichler, G.; Sapkota, J. Influence of Rapid Consolidation on Co-Extruded Additively Manufactured Composites. *Polymers* 2022, 14, 1838. https://doi.org/10.3390/polym14091838

Academic Editor: Matthew Blacklock

Received: 28 March 2022
Accepted: 25 April 2022
Published: 29 April 2022

Publisher's Note: MDPI stays neutral with regard to jurisdictional claims in published maps and institutional affiliations.

Copyright: © 2022 by the authors. Licensee MDPI, Basel, Switzerland. This article is an open access article distributed under the terms and conditions of the Creative Commons Attribution (CC BY) license (https://creativecommons.org/licenses/by/4.0/).

Abstract: Composite filament co-extrusion (CFC) additive manufacturing (AM) is a bi-matrix rapid fabrication technique that is used to produce highly customisable composite parts. By this method, pre-cured, thermoset-based composite carbon fibre (CCF) is simultaneously extruded along with thermoplastic (TP) binding melt as the matrix. Like additive manufacturing, CFC technology also has inherent challenges which include voids, defects and a reduction in CCF's volume in the fabricated parts. Nevertheless, CFC AM is an emerging composite processing technology, a highly customisable and user-oriented manufacturing unit. A new TP-based composites processing technique has the potential to be synergised with conventional processing techniques such as injection moulding to produce lightweight composite parts. Thus, CFC AM can be a credible technology to replace unsustainable subtractive manufacturing, if only the defects are minimised and processing reliability is achieved. The main objective of this research is to investigate and reduce internal voids and defects by utilising compression pressing as a rapid consolidation post-processing technique. Post-processing techniques are known to reduce the internal voids in AM-manufactured parts, depending on the TP matrices. Accordingly, the rapid consolidated neat polylactic acid (PLA) TP matrix showed the highest reduction in internal voids, approximately 92%. The PLA and polyamide 6 (PA6) binding matrix were reinforced with short carbon fibre (SCF) and long carbon fibre (LCF), respectively, to compensate for the CCF's fibre volume reduction. An increase in tensile strength (ca. 12%) and modulus (ca. 30%) was observed in SCF-filled PLA. Furthermore, an approximately 53% increase in tensile strength and a 76% increase in modulus for LCF-reinforced PA6 as the binding matrix was observed. Similar trends were observed in CFC and rapidly consolidated CFC specimens' flexural properties, resulting due to reduced internal voids.

Keywords: additive manufacturing; rapid consolidation; composites; carbon fibres

1. Introduction

Composite filament co-extrusion (CFC) additive manufacturing (AM) is a newly developed bi-matrix processing technology. CFC processing combines thermoplastic (TP) filament as a binding matrix and pre-cured thermoset (TS)-based composite carbon fibre (CCF) filament as a reinforcement. The CCF constituents are continuous carbon fibres infused with low viscous epoxy-based thermoset and heat cured to form a filament-like structure for CFC processing. The binding process of TP melt onto the CCF filament is analogous to crosshead extrusion processing of wire and cable coating. This enables the user to select off-the-shelf filled and unfilled binding TP matrix filaments, such as polylactic acid, polyolefin, polyamide, polyimide, et cetera. The carbon fibre volume fraction in CCF is approximately 57 vol.% to 62 vol.% [1–3]. However, when combined with a TP binding matrix the volume fraction is further reduced to approximately 18 vol.% to 35 vol.% depending on CCF layer height settings [1,2,4,5].

The incompatibility of TP matrix and carbon fibre in continuous fibre AM, often resulting due to high viscous TP, has been extensively reported by researchers [6–8]. The CCF filament matrix is an epoxy-based low viscous TS that conforms to better carbon fibre wetting, and the TS curing process results in the void and defect-free composite filament [1,9]. According to Azarov et al. [1] and Adumitroaie et al. [5], the CCF filament usage in CFC aims to overcome the perceived disadvantages of producing composite parts entirely from TP or TS. The detailed study showed that the CFC technology is aimed at fabricating lattice composites that are often difficult to fabricate by the existing automated technologies [2,9]. Furthermore, researchers concluded that CFC AM, as such, is not suitable to achieve significant improvement in the composite material properties because fabricating defect-free composite parts is challenging [1]. Also, few studies have shown filling the TP matrix [10] and reinforcing it with short [11] and long carbon fibre [12] can increase the mechanical and thermal properties of material-extruded (MEX) AM specimens [13]. Likewise, to compensate for the reduction in CCF fibre volume fraction and increase the flexural properties, the researchers used 20 wt. % short carbon fibre filled polyamide 6 as a binding matrix [14]. The short carbon fibre-filled polyamide 6 as a binding matrix had improved flexural properties compared to the unfilled TP matrix. Furthermore, the researchers found high process-induced voids within the CFC-fabricated specimens for both TP matrices, and reduction in a high degree of voids is elusive due to low compaction in situ consolidation during layered manufacturing.

Van de Werken et al. [15,16], used the hot isotactic pressing (HIP) technique to reduce the process-induced voids by 51% and, improved the flexural strength and interlaminar shear strength of TP-based continuous fibre additive manufactured coupons by 30%–45%. Similarly, Savandaiah et al. [17] performed rapid consolidation on highly anisotropic short and long carbon fibre-reinforced MEX specimens. The post-processing increased the thermal and mechanical properties of MEX specimens comparable to the injection-moulded specimens. The advantage of the rapid consolidation techniques such as compression pressing is the ability to achieve higher densification and reduction in voids between 50% to 75% at a reduced cycle time of less than 30 min [17]. However, in HIP the process duration varies between 1 h to 4 h depending on the type of thermoplastic used in TP-based continuous fibre AM.

Henceforth, the scope of the research was centred on the question of whether the rapid consolidation technique is suitable to achieve better composite material properties in CFC specimens. In this study, the researchers have investigated the influence of polylactic acid (PLA), and polyamide 6 (PA6) as a binding matrix on processing and mechanical properties. Similarly, short carbon fibre-filled PLA and long carbon fibre-reinforced PA6 were studied to quantify the influence of fibre reinforcement on CFC-printed specimens. The tensile and flexural specimens were assessed to evaluate the mechanical properties of CFC and post-processed CFC specimens. Computed tomography (CT) ensured the volume of process-induced voids in the CFC and reduction in post-consolidated specimens.

2. Materials and Methods

2.1. Materials

PLA homopolymer [18], was bought from Total Corbion, Gorinchem, Netherlands. The particulate carbon fibre without sizing, Tenax HT, was acquired from Teijin Carbon Europe GmbH, Wuppertal, Germany. Film extrusion grade PA6 [19] was purchased from DSM N. V, Geleen, Netherlands. The PLA and PA6 compounding, filament extrusion, and injection moulding (IM) is based on the procedure detailed in the previously published research work [14,17,20]. Short carbon fibre (20 wt.%, length 83 µm) filled PLA was filament-extruded in-house (1.75 mm) and a commercially available long carbon fibre (20 wt.%, length 218 μm) reinforced PA6 filament (1.75 mm) was supplied by Prirevo 3D solutions GmbH, Ried im Traunkreis, Austria. The CCF filament was purchased from Anisoprint SARL, Mondercange, Luxembourg. For simplification, the nomenclature of the samples is presented in Table 1, the thermal and mechanical test data are tabulated in Table 2, and the tensile test data of CCF is given in Table 3.

Table 1. Material nomenclature.

Material Description	Nomenclature
Neat PLA	N-PLA
Injection-moulded neat PLA	N-PLA-IM
Neat PLA material extruded with a raster angle of 0°	N-PLA-MEX-0°
Neat PLA as a binding matrix in CFC	N-PLA-CFC
Post-consolidated neat PLA as a binding matrix in CFC	N-PLA-CFC-PC
Short carbon fibre-filled PLA	S-PLA
Injection-moulded short carbon fibre-filled PLA	S-PLA-IM
Short carbon fibre-filled PLA material extruded with a raster angle of 0°	S-PLA-MEX-0°
Short carbon fibre-filled PLA as a binding matrix in CFC	S-PLA-CFC
Post-consolidated short carbon fibre-filled PLA as a binding matrix in CFC	S-PLA-CFC-PC
Neat PA6	N-PA6
Injection-moulded neat PA6	N-PA6-IM
Neat PA6 material extruded with a raster angle of 0°	N-PA6-MEX-0°
Neat PA6 as a binding matrix in CFC	N-PA6-CFC
Post-consolidated neat PA6 as a binding matrix in CFC	N-PA6-CFC-PC
Long carbon fibre-reinforced PA6	L-PA6
Injection-moulded long carbon fibre-reinforced PA6	L-PA6-IM
Long carbon fibre reinforced PA6 material extruded with a raster angle of 0°	L-PA6-MEX-0°
Long carbon fibre-reinforced PA6 as a binding matrix in CFC	L-PA6-CFC
Post-consolidated long carbon fibre-reinforced PA6 as a binding matrix in CFC	L-PA6-CFC-PC

Table 2. Injection-moulded material thermal test data. Melting temperature (MT), heat deflection temperature (HDT).

Properties	Test	Unit	N-PLA	S-PLA	N-PA6	L-PA6
MT	ISO-11357-1:2016	°C	170–175	170–175	220–225	220–225
HDT	EN ISO 75-HDT A	°C	58.0 ± 0.1	59.0 ± 0.2	48.0 ± 0.3	-- *
HDT	EN ISO 75-HDT C	°C	--	--	--	133.0 ± 3.0

* No maximum deflection at maximum temperature (testing machine limit at 200 °C).

Table 3. CCF tensile test data. For the tensile test, 100 mm of CCF was taken directly from the spool, tabs were glued onto each end for gripping and to reduce damage to CCF during clamping and testing.

Properties	Unit	Value
Diameter	mm	0.36
Tensile strength	MPa	2224 ± 283
Young's modulus	MPa	130000 ± 9000
Elongation at break	%	1.6 ± 0.2
Fibre volume fraction	%	57

2.2. Preparation of Printed Specimens

The samples for mechanical characterization were printed in composer A4 CFC printer (Figure 1), with two print heads, manufactured by Anisoprint SARL, Mondercange, Luxembourg. Conventional MEX print head to fabricate parts with unfilled and filled thermoplastic alike, and CFC print head to print CCF with thermoplastic melt as a binder.

The instruction sets for the CFC printer were prepared in a proprietary slicing software, AURA (ver. 1.27.3) and the configuration is shown in Figure 2 and summarized in Table 4. The print bed and printed samples were cooled down to nominal room temperature to avoid part distortion and to facilitate the removal. The standard tensile sample dimension $220 \times 15 \times 3$ mm^3 [21], and standard flexural sample dimension $155 \times 13 \times 4$ mm^3 [22] were set according to ASTM standards. The sample thickness for rapid consolidation of the flexural specimen was increased $155 \times 13 \times 4.2$ mm^3, i.e., a single bottom (0.1 mm) and a top layer (0.1 mm) of pure plastic printing was set, to improve the rapid consolidated surface quality and flexural property. Furthermore, after post-consolidation, the specimen thickness was 4 mm, without a major change in final fibre loading. Furthermore, before testing the printed samples were stocked in the standard control cabinet (relative humidity of 50% at 23 °C) for 72 h, as per ASTM D618-21 standard for specimen conditioning [23].

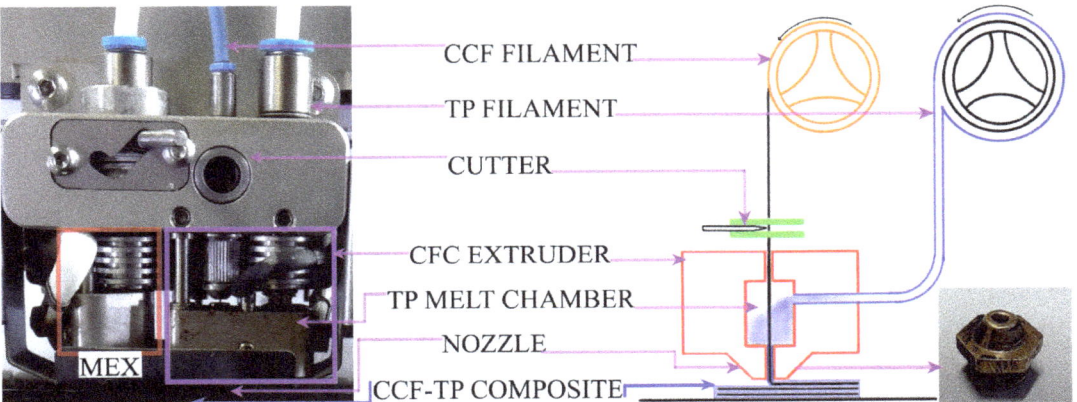

Figure 1. MEX and CFC print head photography and schematics of Composer A4 CFC print head. The photography shows two print heads, a typical MEX print head for plastic printing (red box) and a CFC print head (purple box) for composite printing.

Table 4. Important printing settings.

Parameter	Unit	PLA	PA6
CFC nozzle temperature	°C	225	255
MEX nozzle temperature	°C	220	250
CFC TP flow multiplier	--	0.95	1.05
CFC layer height	mm	0.36	0.36
CFC extrusion width	mm	0.75	0.75
MEX TP flow multiplier	--	0.90	1
MEX layer height	mm	0.12	0.12
MEX extrusion width co-efficient	--	1	1.05
Bed temperature	°C	80	95
TP perimeter count	--	2	2
Inner CCF perimeter count	--	1	1
CCF infill pattern	--	Solid	Solid
CCF infill angle	°	0	0
MEX print speed	mm·s^{-1}	60	60
CFC print speed	mm·s^{-1}	10	10

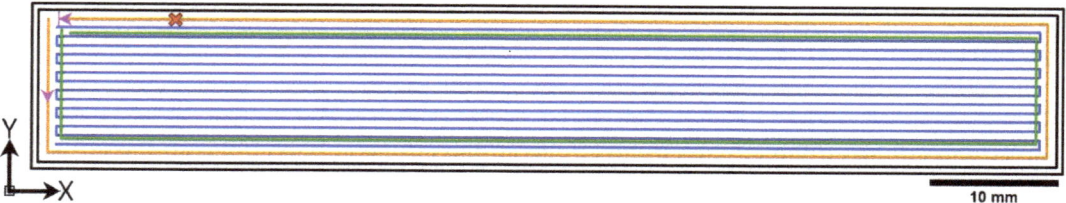

Figure 2. Graphical representation of part configuration generated via slicing software. Black coloured lines are outer thermoplastic perimeters, green coloured line represents micro infill thermoplastic filling by MEX print head, and orange and blue series lines are CCF reinforced perimeter and infill by CFC print head. The purple pointed arrow shows the direction of CCF printing and the pattern "x" marks the symbolic fibre cut operation.

2.3. Compression Press Moulding

The compression platen press, LabEcon from Fontijne platen presses and services BV, Delft, Netherlands, was used to rapidly consolidate the printed samples. The rapid consolidation of CFC specimens is based on the procedure detailed in the previously published work [17]. The heating and cooling cycle for each thermoplastic matrix is presented in Table 5. The pressed MEX samples were stored in a controlled cabinet (relative humidity of 50% at 23 °C) for 72 h, as per ASTM D618-21 standard for specimen conditioning [23].

Table 5. Compression pressing process conditions.

Setting	PLA	PA6
Set temperature	180 °C	220
1st cooling cycle	cool down to 70 °C at 50 °C·min^{-1}	cool down to 150 °C at 50 °C·min^{-1}
2nd cooling cycle	cool down to 23 °C at 5 °C·min^{-1}	cool down to 50 °C at 5 °C·min^{-1}

2.4. Void Volume Fraction Analyses

X-ray computed tomography (CT) was used to quantify the void content. Therefore, scans were carried out on the Nanotom 180 NF (GE phoenix X-ray, Wunstorf, Germany) laboratory CT device. A molybdenum target and a tube voltage of 60 kV was used for the data acquisition. Similar to Plank et al. [24] and Senck et al. [25], the grey value-based 'ISO X' threshold procedure was used for the quantitative evaluation of the void content. This method was already applied to additive manufactured samples [12,26]. The CT scans at a voxel edge length of 7 µm and additional region of interest scans at a voxel edge length of 2 µm were carried out. The latter high-resolution scans were performed on S-PLA-CFC and N-PA6-CFC. Due to the different densities of these materials and different void shapes, two different ISO values were determined. The analyses volume of these high-resolution scans is $3.6 \times 3.6 \times 3.6$ mm^3 and was used as a reference for the definition of the appropriate ISO value for void analysis at a voxel edge length of 7 µm. The scans at a voxel edge length of 7 µm provide an analysis volume of the total cross-section (Y-Z section) and allow an approximate length of 13 mm (X-direction) for void analysis.

The void analysis was performed with VGStudio MAX 3.4 (Volume Graphics GmbH, Heidelberg, Germany) software. Moreover, a multistep segmentation of the high-resolution scan with a voxel edge length of 2 µm voxel size was carried out and based on these segmentations the "ISO X" threshold was estimated. For the higher density materials (PLA) an ISO value of 62.75 and for the lower density materials (PA6) an ISO value of 71.4 was defined.

2.5. Mechanical Characterisation

For all mechanical testing, a Zwick–Roell servo-hydraulic test rig with a cylinder rated at 25 kN force was used. The test rig was operated by Cubus software in displacement mode.

A Zwick–Roell force transducer and the internal displacement sensor of the cylinder were used for measurement. Additionally, a correlated solutions 3D digital image correlation (DIC) system recorded the surface displacement and calculated the surface strains of the specimens during the tests. Post-processing of the DIC data was performed in the software Vic-3D 8.

For tensile testing, MTS 647 hydraulic jaws were used to grip the specimens. Glass fibre composite end tabs were adhesively bonded to the specimens with 3M DP490 epoxy adhesive. Before testing, a speckle pattern was applied to one face of the specimens for DIC measurement. The test procedure followed ASTM standard D3039 [21] for tensile testing of composite laminates with a machine head speed of 2 mm·min^{-1}. The test was stopped at the break of the specimens.

For bending testing, 4-point bending testing according to ASTM D6272 [22] was chosen with the load span half of the support span. By this method, the central span of the specimen is in bending only, and no transverse force component exists. This central span is monitored by DIC side on, enabling monitoring of maximum deflection and strains over the thickness. The machine head speed was set at 2 mm·min^{-1} and the maximum displacement was chosen to exceed the strength of the material. For both test cases, stiffness and strength values were calculated according to the respective standards

3. Results and Discussion

3.1. Void Volume Fraction

The specimen's void volume fraction is listed in Table 6, and the reference MEX specimen is given along with CFC and post-processed CFC specimens. Voids in the injection-moulded specimens were not detected, as expected, due to high densification, and the void volume fraction in CCF filament was negligible (<0.2 vol.%). In Figure 3, CT-scanned images along with scalar void volume scale of N-PLA specimens are represented with the top midsection view and corresponding insets with the side midsection view. The black circles within N-PLA-CFC and N-PLA-CFC-PC represent voids in the TP matrix, and red arrows indicate voids within pre-cured TS based CCF filament. However, the defects in unprocessed CCF filament were minimal, and, therefore, the voids detected in CFC-processed specimens may be the attributes of epoxy-based TS. According to the dynamic mechanical analysis (DMA) of CCF filament by Adumitroaie et al. [5], these voids correlated to the inhomogeneity in the TS composition and inhomogeneous heat distribution between the CCF's constituents during AM process as seen in CT images. The high amount of blue-coloured voids seen in N-PLA-CFC, with void volume in the range of 0.01 mm^3 to 17.0 mm^3, are distributed throughout the whole sample. The post-processed N-PLA-CFC-PC CT image showed a high reduction in void volume fraction within TP binding matrix as well as CFC processed CCF filament. The total void volume fraction of N-PLA-CFC is 14.0 vol.% and, post-processed N-PLC-CFC-PC is 1.0 vol.%, a 92% reduction in void volume fraction. Also, post-processed S-PLA-CFC-PC and L-PA6-CFC-PC showed a reduction in void volume fraction between 35% to 50% depending on the weighted average fibre length [17].

Table 6. Influence of rapid consolidation on void reduction in CFC specimen. For reference, the void volume fraction of the corresponding MEX specimen with a raster angle of 0° is reported.

Material	Void (Vol. %)		
	MEX-0°	CFC	CFC-PC
N-PLA	12.2 *	14.0	1.0
S-PLA	15.9 *	16.3	8.6
N-PA6	14.3 *	16.8 *	8.3
L-PA6	27.0 *	29.2	18.2

* Void volume fraction was taken from previously published research work [14,17,20].

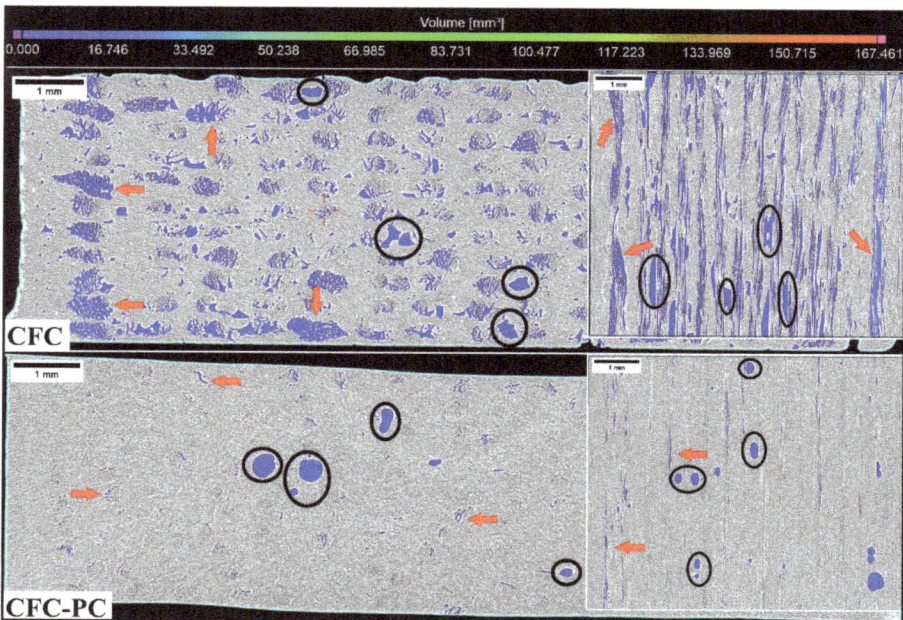

Figure 3. Top midsection CT images (X-Y) of the N-PLA with corresponding side midsection inset views (X-Z). A black circle is indicative of voids due to the thermoplastic matrix and red arrows showed voids within the CFC processed CCF.

3.2. Tensile Properties

All tensile-tested CFC specimens demonstrated longitudinal splitting mode in the defective region shown in Figure 4a. The defective region (red box) in Figure 4b, the area without binding matrix is due to the CFC print head design and pre-set instruction for CFC composite fabrication. The CFC print head flaws are contributing to defective processing and increased unreliability in the established processing settings, specifically in fibre reinforced binding matrix [27]. Furthermore, L-PA6 binding matrix has the highest void composition compared to the short fibre-filled TP and neat TP binding matrix (Table 6). Comparable results were observed earlier [12,17] showing a correlation between increased fibre lengths and increasing void volume fraction in the MEX specimens.

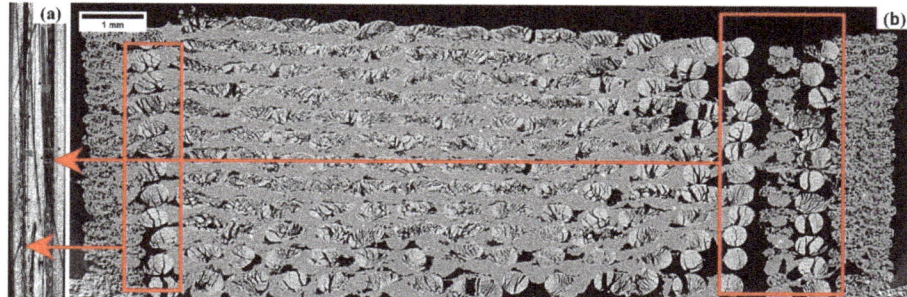

Figure 4. Tensile tested fractured image (**a**) and CT image (X-Y) (**b**) of L-PA6-CFC specimen. The longitudinal failure observed in the tensile tested image (**a**) is due to the defects observed in the CT image is indicated by arrows connected to the defective region (red box) in CT-scanned image (**b**).

Figure 5a,b shows comparative tensile test results of N-PLA-CFC and S-PLA-CFC along with injection-moulded and MEX AM specimens as reference for binding matrix. Similarly, Figure 6a,b showed tensile test results of N-PA6-CFC and L-PA6-CFC, respectively. An increase in tensile strength is observed for S-PLA-CFC (497 ± 16 MPa) compared to N-PLA-CFC (394 ± 64 MPa) and a 30% increase in S-PLA-CFC (49,850 ± 505 MPa) modulus. Also, a large standard deviation in CFC processed specimens is indicative of previously discussed defects associated with CFC AM processing.

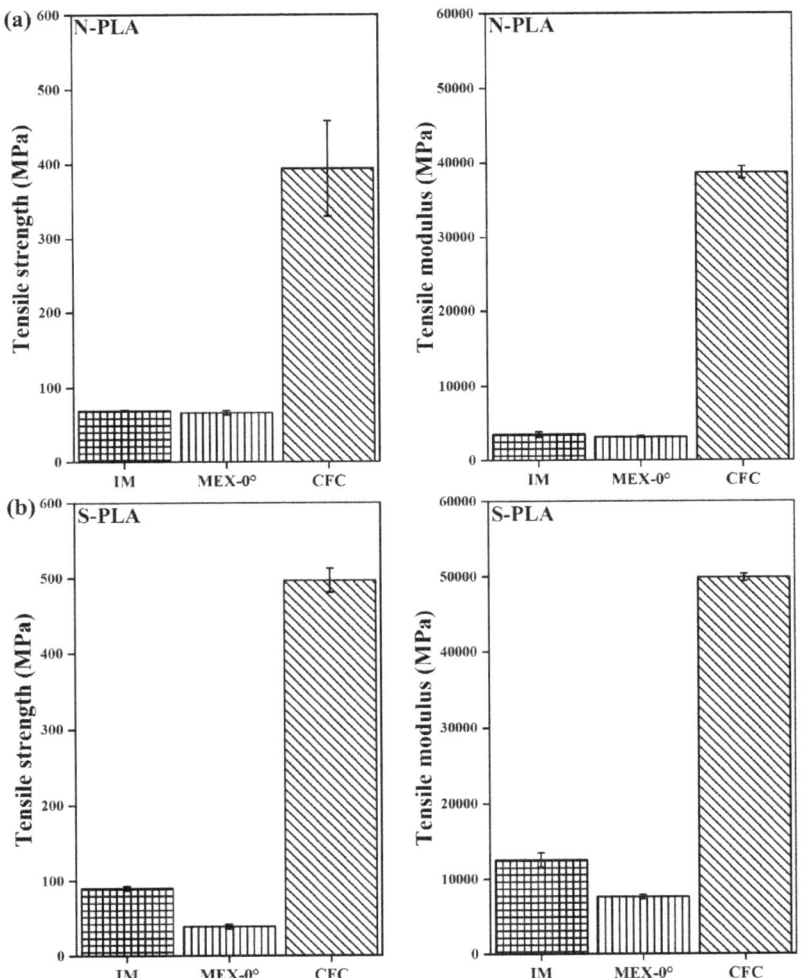

Figure 5. Tensile test data of PLA with a corresponding error bar. Where, N-PLA is neat PLA and S-PLA is short carbon fibre filled PLA. The corresponding IM, MEX-0°, and CFC are injection moulded, 3D printed and composite filament co-extruded 3D printed tensile tested specimens, respectively.

Figure 6a shows tensile test results of N-PA6-CFC along with tensile test results of L-PA6-CFC in Figure 6b. A 53% increase in L-PA6-CFC tensile strength (TS) (488 ± 27 MPa) compared to N-PA6-CFC (320 ± 17 MPa) and a 76% increase in tensile modulus (TM) in CFC processed L-PA6 (45240 ± 2740 MPa). The increased weighted average length contributes positively to the mechanical performance of CFC processed L-PA6. However, the total fibre volume fraction in CCF decreased drastically and contributed to the loss in tensile properties, approximately 67% and 80% in tensile modulus and strength, respectively. Defects detected by CT such as CCF splitting-spreading, and therefore, CCF is no longer consolidated and contributed to early failure as shown in Figure 4a.

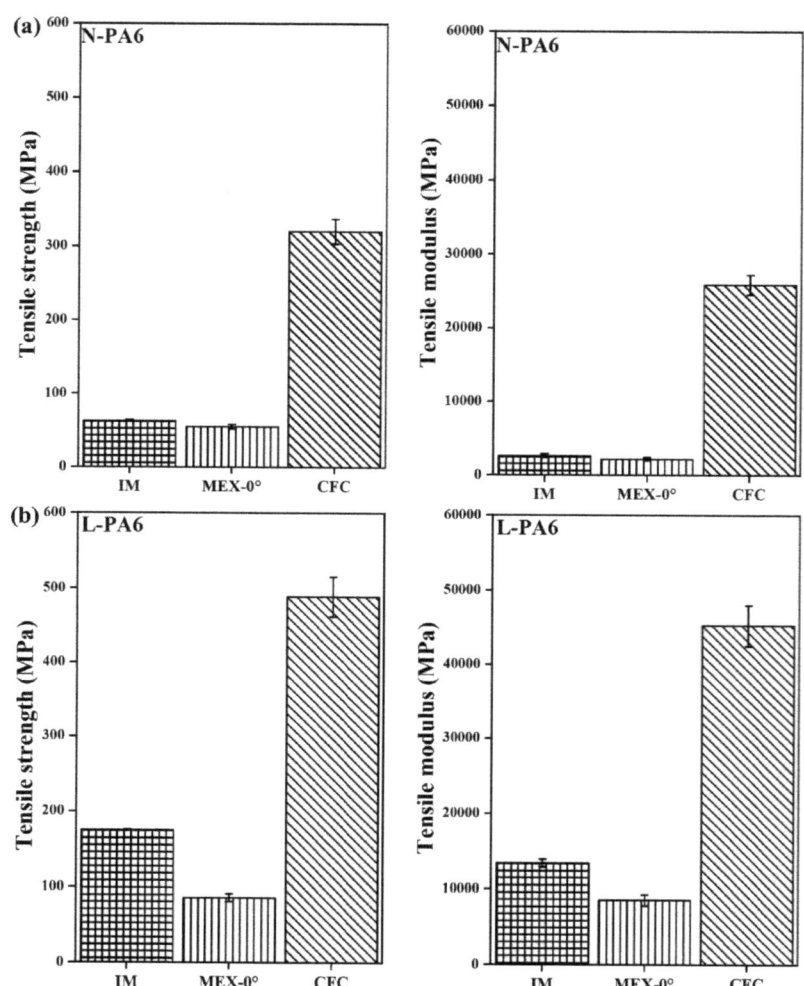

Figure 6. Tensile test data of PA6 with a corresponding error bar. Where, N-PA6 is neat PA6 and L PA6 is long carbon fibre reinforced PA6. The corresponding IM, MEX-0°, and CFC are injection moulded, 3D printed and composite filament co-extruded 3D printed tensile tested specimens, respectively.

3.3. Flexural Properties

Flexural strength is purely based on the TP binding matrix and, therefore, it is essential to perform the rapid consolidation study on the flexural standard specimens. The 4-point flexural specimens showed elastic response followed by the plastic collapse in one or more sections in the central span. Some influences of the load application rollers were visible, potentially from high contact stresses and subsequent indentation of the material. In situ detection of fracture modes by the DIC system is difficult because of the limited number of measurements on the thin specimen side.

In Figure 7a, the post-processed N-PLA-CFC-PC (488 ± 15 MPa) specimen showed increased flexural strength (FS) by approximately 50% compared to N-PLA-CFC (244 ± 16 MPa). Furthermore, there was an 11% increase in flexural modulus (FM) after post-processing for N-PLA-CFC-PC. On the contrary, the influence of short fibre filling on PLA as a binding matrix had minimal effect on the FM but an approximately 37% increase in FS after post-processing. Furthermore, N-PLA-CFC-PC coupons with the highest reduction in void volume fraction (Table 6) showed similar FS and FM (45180 ± 1390 MPa) compared to S-PLA-CFC (FS= 255 ± 5.5 MPa and FM= 46850 ± 2280 MPa) and S-PLA-CFC-PC (FS= 346 ± 11.5 MPa and FM= 47580 ± 1640 MPa).

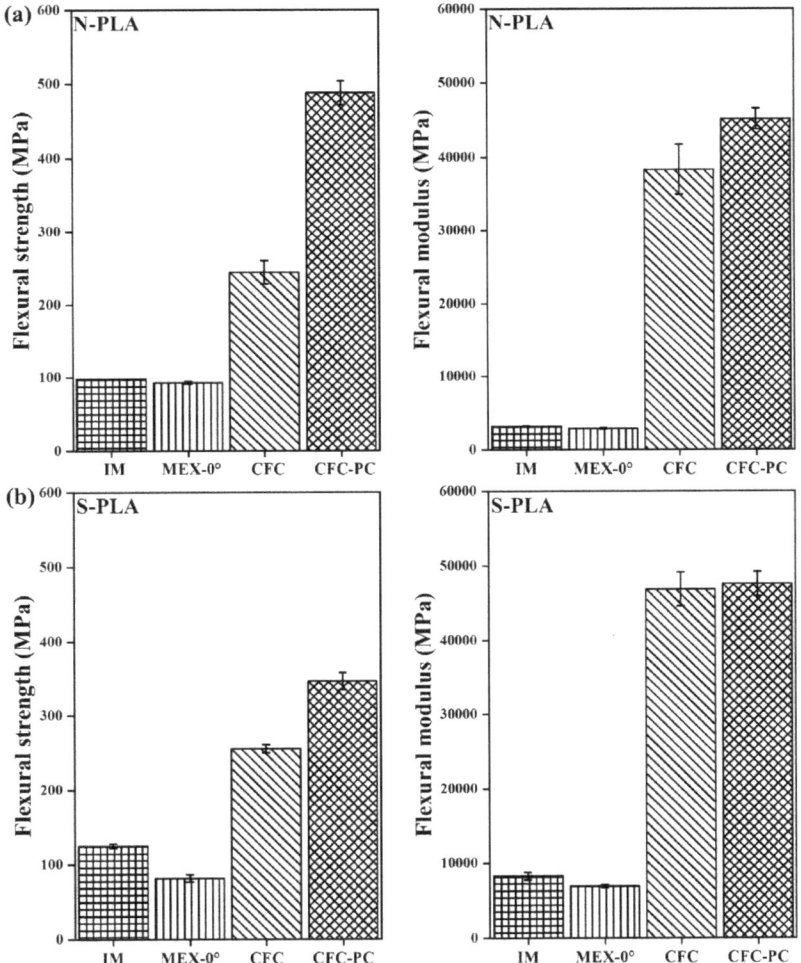

Figure 7. Flexural test data of PLA with a corresponding error bar. Where, N-PLA is neat PLA and S-PLA is short carbon fibre filled PLA. The corresponding IM, MEX-0°, CFC, and CFC-PC are injection moulded, 3D printed, composite filament co-extruded, and post-processed composite filament co-extruded 3D printed 4-pointing beding tested specimens, respectively.

In Figure 8a, post-processed N-PA6-CFC-PC (FS= 320 ± 11.5 MPa and FM = 41,600 ± 2200 MPa) demonstrates flexural properties increase by two-fold compared to N-PA6-CFC. In contrast, L-PA6-CFC showed FS lower than referred injection-moulded (L-PA-IM) binding TP matrix (Figure 8b). The cause for flexural underperformance of CCF reinforced L-PA6-CFC is discussed in the void fraction analysis section as it can be correlated to the large number of defects observed in Figure 4, and similarly tabulated in Table 6. Moreover, after a reduction in void volume fraction by approximately 38%, the FS and FM of post-processed L-PA6-CFC-PC increased.

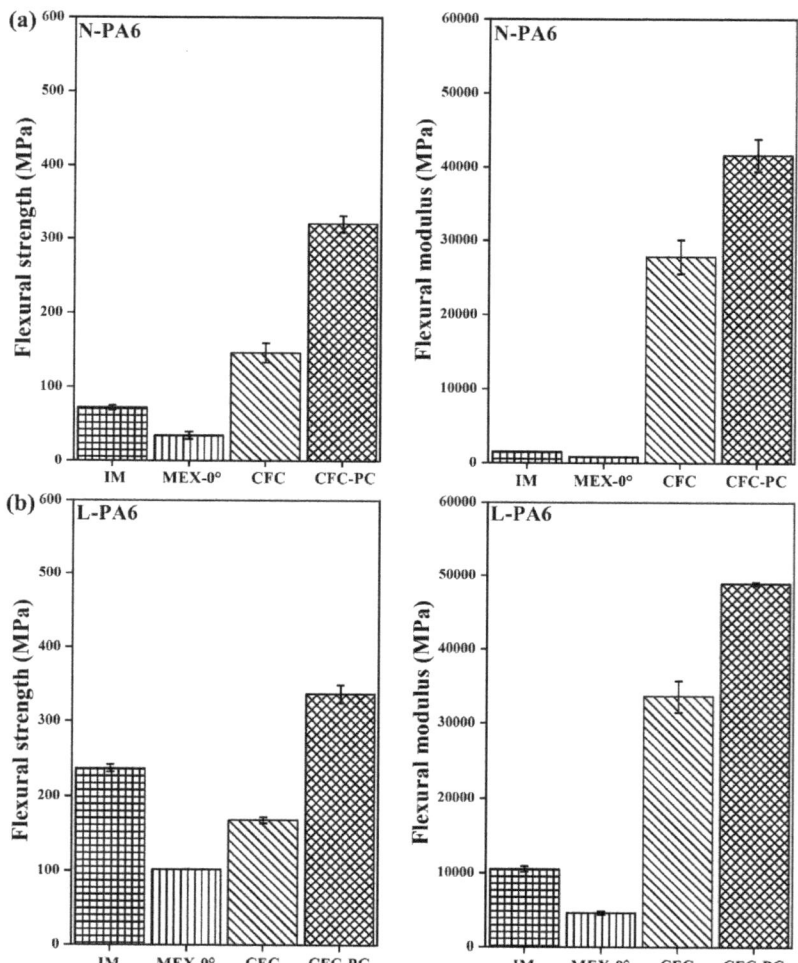

Figure 8. Flexural test data of PA6 with a corresponding error bar. Where, N-PA6 is neat PA6 and L PA6 is long carbon fibre reinforced PA6. The corresponding IM, MEX-0°, CFC, and CFC-PC are injection moulded, 3D printed, composite filament co-extruded, and post-processed composite filament co-extruded 3D printed 4-pointing beding tested specimens, respectively.

Figure 9 shows N-PA6 specimens as-printed and post-consolidated after failure. The as-printed specimens showed delamination, whereas the post-consolidation eliminates the tendency to delaminate. In contrast, N-PLA-CFC specimens exhibited plastic deformation without visible large-scale delamination. For S-PLA-CFC and L-PA6-CFC, there was plastic deformation (plastic knee) without visible delamination. Because of the large deflections, buckling of the fibres in the printed layers was visible for most specimens in the top layer where the largest compressive strains were present. Furthermore, in a few instances, the contact of the top supports led to peeling of the top layer under large deformations, however, these damage modes occurred after passing the bending strength.

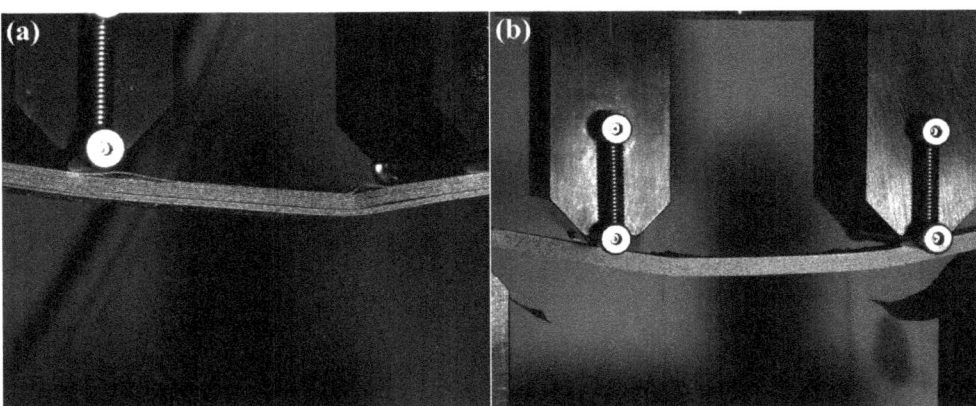

Figure 9. Specimen Nr. 3 of N-PA6-CFC showing delamination (**a**) and specimen Nr. 2 of N-PA6-CFC-PC showing buckling with little delamination at failure (**b**).

The S-PLA based binding matrix showed better tensile and flexural properties compared to L-PA6 as a binding matrix, even though the weight average fibre length of carbon fibre in L-PA6 were longer. This can be attributed to the process-induced annealing effect in the PLA matrix due to high bed temperature, maintained above the glass transition temperature [20,28].

4. Conclusions

In summary, the results show that rapid consolidation as a post-processing technique is highly beneficial, given the amount of defects detected in the CFC processed specimens. Void volumetric fraction reduction between 50% and 90% determined by CT scans was achieved, resulting in a 10% to 80% increase in flexural properties. The next valid step is to reduce the voids by applying more compaction pressure through the nozzle on the discharged fibre strands. This would be achieved by slightly reducing the layer height and thus increasing the fibre volume fraction in composite parts. This can further improve the mechanical properties of the rapid consolidated CFC parts. The compliance of upscaling the composite CFC AM to serial production is difficult due to inherent process-induced voids, and, therefore resulting ininconsistent part's quality. Hence, in this article, the researchers have shown the advantages of rapid consolidation as a post-processing methodology to overcome the perceived disadvantage in composite CFC AM.

Author Contributions: C.S. conceptualised the research work, conducted experimental design, data analysis, data curation and wrote the original draft. J.M. and B.P. conducted computed tomography experimental design, data analysis. S.S. conducted experimental design, data curation, data analysis, and proofread the final draft. G.S. supervised, proofread the final draft, and provided comments for improving the manuscript. J.S. supervised, proofread the final draft and provided comments for improving the manuscript. All authors have read and agreed to the published version of the manuscript.

Funding: This research work has been performed as part of the 3D CFRP (grant Nr. 859832), BeyondInspection (grant Nr. 874540) and Pore3D (grant Nr. 868735). Open Access Funding by the University of Linz.

Data Availability Statement: Further data is available on request from the authors.

Acknowledgments: The Researchers acknowledge the financial support by the EU funded network M-Era.Net, European Regional Development Fund (ERDF) through IWB 2014-2020, the BMVIT (Austrian Ministry for Transport, Innovation, and Technology); the FFG (Austrian Research Promotion Agency), and the federal state of Upper Austria. Special thanks to the team Wood K plus for the operational assistance and fruitful discussions.

Conflicts of Interest: The authors declare no conflict of interest. The funders had no role in the design of the study; in the collection, analyses, or interpretation of data; in the writing of the manuscript, or in the decision to publish the results.

References

1. Azarov, A.V.; Antonov, F.K.; Vasil'ev, V.V.; Golubev, M.V.; Krasovskii, D.S.; Razin, A.F.; Salov, V.A.; Stupnikov, V.V.; Khaziev, A.R. Development of a two-matrix composite material fabricated by 3D printing. *Polym. Sci. Ser. D* **2017**, *10*, 87–90. [CrossRef]
2. Azarov, A.V.; Antonov, F.K.; Golubev, M.V.; Khaziev, A.R.; Ushanov, S.A. Composite 3D printing for the small size unmanned aerial vehicle structure. *Compos. Part B Eng.* **2019**, *169*, 157–163. [CrossRef]
3. Wang, Y.; Zhang, G.; Ren, H.; Liu, G.; Xiong, Y. Fabrication strategy for joints in 3D printed continuous fiber reinforced composite lattice structures. *Compos. Commun.* **2022**, *30*, 101080. [CrossRef]
4. Struzziero, G.; Barbezat, M.; Skordos, A.A. Consolidation of continuous fibre reinforced composites in additive processes: A review. *Addit. Manuf.* **2021**, *48*, 102458. [CrossRef]
5. Adumitroaie, A.; Antonov, F.; Khaziev, A.; Azarov, A.; Golubev, M.; Vasiliev, V.V. Novel Continuous Fiber Bi-Matrix Composite 3-D Printing Technology. *Materials* **2019**, *12*, 3011. [CrossRef] [PubMed]
6. Chabaud, G.; Castro, M.; Denoual, C.; Le Duigou, A. Hygromechanical properties of 3D printed continuous carbon and glass fibre reinforced polyamide composite for outdoor structural applications. *Addit. Manuf.* **2019**, *26*, 94–105. [CrossRef]
7. Becker, C.; Oberlercher, H.; Heim, R.B.; Wuzella, G.; Faller, L.-M.; Riemelmoser, F.O.; Nicolay, P.; Druesne, F. Experimental Quantification of the Variability of Mechanical Properties in 3D Printed Continuous Fiber Composites. *Appl. Sci.* **2021**, *11*, 11315. [CrossRef]
8. Savandaiah, C.; Sieberer, S.; Steinbichler, G. Additively Manufactured Composite Lug with Continuous Carbon Fibre Steering Based on Finite Element Analysis. *Materials* **2022**, *15*, 1820. [CrossRef] [PubMed]
9. Fedulov, B.; Fedorenko, A.; Khaziev, A.; Antonov, F. Optimization of parts manufactured using continuous fiber three-dimensional printing technology. *Compos. Part B Eng.* **2021**, *227*, 109406. [CrossRef]
10. Spoerk, M.; Savandaiah, C.; Arbeiter, F.; Sapkota, J.; Holzer, C. Optimization of mechanical properties of glass-spheres-filled polypropylene composites for extrusion-based additive manufacturing. *Polym. Compos.* **2019**, *40*, 638–651. [CrossRef]
11. Spoerk, M.; Savandaiah, C.; Arbeiter, F.; Traxler, G.; Cardon, L.; Holzer, C.; Sapkota, J. Anisotropic properties of oriented short carbon fibre filled polypropylene parts fabricated by extrusion-based additive manufacturing. *Compos. Part A Appl. Sci. Manuf.* **2018**, *113*, 95–104. [CrossRef]
12. Savandaiah, C.; Maurer, J.; Gall, M.; Haider, A.; Steinbichler, G.; Sapkota, J. Impact of processing conditions and sizing on the thermomechanical and morphological properties of polypropylene/carbon fiber composites fabricated by material extrusion additive manufacturing. *J. Appl. Polym. Sci.* **2021**, *138*, 50243. [CrossRef]
13. ISO/ASTM 52900:2015. Additive Manufacturing—General Principles—Terminology. Available online: https://www.iso.org/obp/ui/#iso:std:iso-astm:52900:ed-1:v1:en (accessed on 25 March 2022).
14. Savandaiah, C.; Maurer, J.; Lesslhumer, J.; Haider, A.; Steinbichler, G. Enhancement of binding matrix stiffness in composite filament co-extrusion additive manufacturing. In Proceedings of the SPE—ANTEC®: Virtual Edition—2020, Plastic Technology Conference, Online, 10 March–5 May 2020.
15. Van de Werken, N.; Koirala, P.; Ghorbani, J.; Doyle, D.; Tehrani, M. Investigating the hot isostatic pressing of an additively manufactured continuous carbon fiber reinforced PEEK composite. *Addit. Manuf.* **2021**, *37*, 101634. [CrossRef]
16. Van de Werken, N.; Koirala, P.; Ghorbani, J.; ABEL, M.; Tehrani, M. Improving Properties of Additively Manufactured Carbon Fiber Composites via Post Pressing. In Proceedings of the American Society for Composites, Thirty-Fourth Technical Conference, Atlanta, GA, USA, 23–25 September 2019.
17. Savandaiah, C.; Maurer, J.; Plank, B.; Steinbichler, G.; Sapkota, J. Rapid Consolidation of 3D Printed Composite Parts Using Compression Moulding for Improved Thermo-Mechanical Properties. *RPJ* **2022**. accepted.
18. Total Corbion. Luminy®L175. Available online: https://www.totalenergies-corbion.com/media/eushodia/pds-luminy-l175-190507.pdf (accessed on 25 March 2022).
19. DSM. Akulon®F132-E1. Available online: https://plasticsfinder.com/api/document/tech/Akulon%C2%AE%20F132-E1/EAuBundva/en (accessed on 25 March 2022).
20. Savandaiah, C.; Plank, B.; Maurer, J.; Lesslhumer, J.; Steinbichler, G. Comparative Study of Filled and Unfilled Polylactic Acid Produced via Injection Molding and 3D Printing. In Proceedings of the SPE ANTEC®Classic: Plastic Technology Conference, Online, 10–21 May 2021.
21. *ASTM D3039-17*; Standard Test Method for Tensile Properties of Polymer Matrix Composite Materials. ASTM International: West Conshohocken, PA, USA, 2017. Available online: https://www.astm.org/d3039_d3039m-17.html (accessed on 25 March 2022).

22. *ASTM D6272-17*; Standard Test Method for Flexural Properties of Unreinforced and Reinforced Plastics and Electrical Insulating Materials by Four-Point Bending. ASTM International: West Conshohocken, PA, USA, 2020. Available online: https://www.astm.org/d6272-17e01.html (accessed on 25 March 2022).
23. *ASTM D618-21*; Standard Practice for Conditioning Plastics for Testing. ASTM International: West Conshohocken, PA, USA, 2021. Available online: https://www.astm.org/d0618-21.html (accessed on 25 March 2022).
24. Plank, B.; Rao, G.; Kastner, J. Evaluation of CFRP-Reference Samples for Porosity made by Drilling and Comparison with Industrial Porosity Samples by Means of Quantitative X-ray Computed Tomography. In Proceedings of the 7th International Symposium on NDT in Aerospace, Bremen, Germany, 16–18 November 2015; p. 10.
25. Senck, S.; Scheerer, M.; Revol, V.; Dobes, K.; Plank, B.; Kastner, J. Non-Destructive Evaluation of Defects in Polymer Matrix Composites for Aerospace Applications Using X-ray Talbot-Lau Interferometry and Micro CT. In Proceedings of the 58th AIAA/ASCE/AHS/ASC Structures, Structural Dynamics, and Materials Conference, Grapevine, TX, USA, 9–13 January 2017.
26. Khudiakova, A.; Berer, M.; Niedermair, S.; Plank, B.; Truszkiewicz, E.; Meier, G.; Stepanovsky, H.; Wolfahrt, M.; Pinter, G.; Lackner, J. Systematic analysis of the mechanical anisotropy of fibre-reinforced polymer specimens produced by laser sintering. *Addit. Manuf.* **2020**, *36*, 101671. [CrossRef]
27. Albrecht, H.; Savandaiah, C.; Lepschi, A.; Loew-Baselli, B.; Haider, A. Parametric study in co-extrusion-based additive manufacturing of continuous fiber-reinforced plastic composites. In Proceedings of the II International Conference on Simulation for Additive Manufacturing, Sim-AM—II, Pavia, Italy, 11–13 September 2019.
28. Bhandari, S.; Lopez-Anido, R.A.; Gardner, D.J. Enhancing the interlayer tensile strength of 3D printed short carbon fiber reinforced PETG and PLA composites via annealing. *Addit. Manuf.* **2019**, *30*, 100922. [CrossRef]

Development of 3D Printed Biodegradable Mesh with Antimicrobial Properties for Pelvic Organ Prolapse

Jiongyu Ren [1,2], Rebecca Murray [3,4,5], Cynthia S. Wong [6], Jilong Qin [7], Michael Chen [1,2,3], Makrina Totsika [7], Andrew D. Riddell [5,8], Andrea Warwick [5], Nicholas Rukin [3,5] and Maria A. Woodruff [1,2,*]

1. Centre for Biomedical Technologies, Queensland University of Technology (QUT), Brisbane, QLD 4000, Australia; edward.ren@qut.edu.au (J.R.); m66.chen@hdr.qut.edu.au (M.C.)
2. School of Mechanical, Medical and Process Engineering, Queensland University of Technology (QUT), Brisbane, QLD 4000, Australia
3. Herston Biofabrication Institute, Metro North Health, Brisbane, QLD 4029, Australia; becca.murray@health.qld.gov.au (R.M.); nicholas.rukin@health.qld.gov.au (N.R.)
4. Australian Institute for Bioengineering and Nanotechnology, The University of Queensland, Brisbane, QLD 4072, Australia
5. Redcliffe Hospital, Metro North Health, Redcliffe, QLD 4020, Australia; adriddell@hotmail.com (A.D.R.); mda99amw@yahoo.com (A.W.)
6. Aikenhead Centre for Medical Discovery (ACMD), St Vincent's Hospital, Melbourne, VIC 3065, Australia; c_wong32@hotmail.com
7. Centre for Immunology and Infection Control, School of Biomedical Sciences, Queensland University of Technology, Brisbane, QLD 4000, Australia; jilong.qin@qut.edu.au (J.Q.); makrina.totsika@qut.edu.au (M.T.)
8. Northside Clinical Unit, School of Clinical Medicine, The University of Queensland, Brisbane, QLD 4072, Australia
* Correspondence: mia.woodruff@qut.edu.au

Abstract: To address the increasing demand for safe and effective treatment options for pelvic organ prolapse (POP) due to the worldwide ban of the traditional polypropylene meshes, this study introduced degradable polycaprolactone (PCL)/polyethylene glycol (PEG) composite meshes fabricated with melt-electrowriting (MEW). Two PCL/PEG mesh groups: 90:10 and 75:25 (PCL:PEG, wt%) were fabricated and characterized for their degradation rate and mechanical properties, with PCL meshes used as a control. The PCL/PEG composites showed controllable degradation rates by adjusting the PEG content and produced mechanical properties, such as maximal forces, that were higher than PCL alone. The antibacterial properties of the meshes were elicited by coating them with a commonly used antibiotic: azithromycin. Two dosage levels were used for the coating: 0.5 mg and 1 mg per mesh, and both dosage levels were found to be effective in suppressing the growth of *S. aureus* bacteria. The biocompatibility of the meshes was assessed using human immortalized adipose derived mesenchymal stem cells (hMSC). In vitro assays were used to assess the cell viability (LIVE/DEAD assay), cell metabolic activity (alamarBlue assay) and cell morphology on the meshes (fluorescent and electron microscopy). The cell attachment was found to decrease with increased PEG content. The freshly drug-coated meshes showed signs of cytotoxicity during the cell study process. However, when pre-released for 14 days in phosphate buffered saline, the initial delay in cell attachment on the drug-coated mesh groups showed full recovery at the 14-day cell culture time point. These results indicated that the PCL/PEG meshes with antibiotics coating will be an effective anti-infectious device when first implanted into the patients, and, after about 2 weeks of drug release, the mesh will be supporting cell attachment and proliferation. These meshes demonstrated a potential effective treatment option for POP that may circumvent the issues related to the traditional polypropylene meshes.

Keywords: pelvic organ prolapse; controllable degradation rate; polycaprolactone; polyethylene glycol; antibacterial; biocompatible

1. Introduction

Pelvic organ prolapse (POP) and incontinence are common and significant problems for women. It has been estimated that half of all women will have either symptoms or signs of prolapse after the menopause. Studies suggest that 10–15% of women in developed countries will undergo surgery for prolapse during their lifetime [1]. Patients with symptoms of urinary and faecal incontinence and POP, were commonly treated via implantation of polypropylene (PP) pelvic meshes [2,3]. The pelvic mesh was expected to reinforce the pelvic organ, as well as prevent recurrence of the symptoms. Data from the Therapeutic Goods Administration (TGA) showed that 151,000 meshes have been implanted in Australia since 1998 [4] and 3.7 million world-wide between 2005 and 2013 [5]. However, complications of pelvic mesh implantation such as erosion into vagina, infection, pain and discomfort were occurring, with some patients requiring further surgery [5]. The high rates of complication prompted the Food and Drug Administration (FDA) to issue 2 warnings against the use of certain pelvic meshes [6], leading to a worldwide withdrawal of a number of products, such as Gynecare Prolift®, Prolift+M™, Prosima™, and Anterior Pinnacle™ kits [7]. New Zealand was the first country in the world to ban the use of transvaginal POP mesh products in 2017 and followed by UK in 2018 [8]. According to TGA data, in 2018 Australia cancelled the approval of all mesh devices placed through the vagina for POP and required all mesh devices to be subject to a comprehensive review before being supplied. The worldwide bans have created a significant unmet clinical need of providing women with good treatment options and viable therapeutics.

Meshes used in the past were predominantly made from polypropylene, a non-degradable polymer. The risk of erosion was in part due to the trans-vaginal placement of mesh and the difference in mechanical properties of the mesh compared to natural tissue. Apart from erosion, another major issue with the implanted meshes is that they can become infected. Studies have shown high rates of infection in meshes whereby two-thirds of these patients often develop infection 2–4 years following implantation [9]. Bacteria that are commonly found in the infected meshes are *Escherichia coli* and *Staphylococcus aureus* and 31% of patients swabbed presented with multibacterial infections consisting of *P. mirabilis*, *E. coli*, *Staphylococcus*, *Streptococcus* and *Enterococcus* [10].

Prior to the worldwide ban, the commercially available pelvic meshes were generally created via knitting or weaving techniques [11,12]. These polypropylene meshes were designed to exhibit high tensile strength to support the pelvis. The stiffness of commercial meshes ranges from 11 N/mm for SmartMesh to 28 N/mm for Gynemesh [11]. However, the stiffness of the meshes also contributed to the failure due to their mismatch in mechanical properties compared to the vaginal tissue [11].

Additive manufacturing techniques are widely used in tissue engineering and regenerative medicine to fabricate three-dimensional (3D) printed scaffolds or meshes. One technique that provides a high degree of control in scaffold fabrication is melt-electrowriting (MEW), which is well established in our lab [13,14]. In this study, biodegradable polymers, polycaprolactone (PCL) and polyethylene glycol (PEG), both FDA approved for medical devices, were used to fabricate composite meshes.

In this project, we aim to fabricate resorbable mesh with tailorable degradation rate and antibacterial properties as a potential solution to the clinically used polypropylene meshes. It is anticipated that creating a tissue-substitute that is anti-bacterial and imparts compatible biological and mechanical properties such as tensile strength and stiffness can address current limitations of the pelvic mesh.

2. Materials and Methods

Testing was performed to investigate the effect of mesh material (PCL, PCL/PEG 90:10, PCL/PEG 75:25) mesh geometry via cross hatch spacing (1 mm vs. 1.5 mm) and mesh state (control, degraded, mock loaded, drug loaded (0.5 and 1 mg/mL), drug released). Various groupings were tested mechanically via tensile testing for antibiotic loading and release, antimicrobial test and biocompatibility (Figure 1).

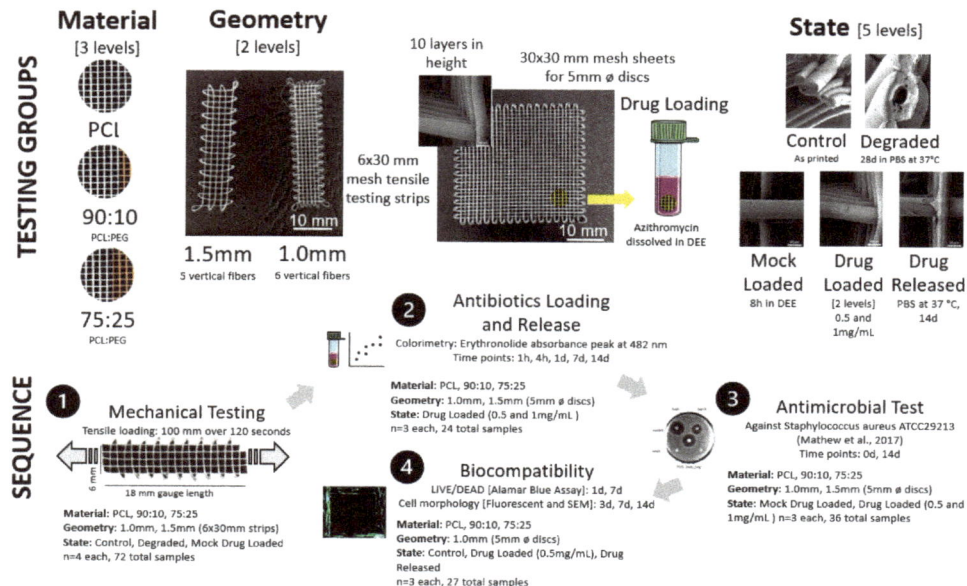

Figure 1. The testing groups showing details on the sample material (PCL, 90:10, 75:25), geometry in terms of spacing and sample size, and state (**top**). The testing sequence is also shown with details of testing type, sample groupings and sample size (**bottom**).

2.1. PCL/PEG Composite Preparation

Medical grade polycaprolactone (mPCL) (Purasorb® PC 12, Corbion Purac Biomaterials, The Netherlands) and PEG (Mw 20,000, Sigma-Aldrich, Australia) were dissolved in chloroform in weight ratios of 90:10 and 75:25. The polymer composites were mixed for 8 h and left in the fume hood until complete evaporation of solvents. The PCL/PEG 90:10 and PCL/PEG 75:25 were used for mesh printing with an in-house built melt-electrowriting (MEW) device.

2.2. Melt-Electrowriting of PCL and PCL/PEG Meshes

The MEW device produces the meshes using an applied pressure to extrude a molten polymer through a positively charged nozzle onto a grounded motorized collector plate according to the parameters detailed in Table 1 below. The collector plate translates in x and y directions controlled by a Gcode using computer programming. The details of the MEW process can be found in our previous study [13]. PCL and PCL/PEG meshes were produced with a 90° cross-hatched fibre deposition in 10-layer high sheets (30 × 6 mm and 30 × 30 mm). The meshes were fabricated with a fibre spacing of 1 mm and 1.5 mm with the groups of meshes used in this study summarized in Figure 1 and the MEW printing parameters used are shown in Table 1.

Table 1. MEW parameters of PCL and PCL/PEG composite meshes.

Mesh Type	Voltage (kV)	Temperature (°C)	Tip to Collector Distance (mm)	Needle Gauge	Air Pressure (MPa)	Plate Speed
PCL/PEG 90:10	4.5	95	5	21	0.08	300
PCL/PEG 75:25	4.5	95	5	21	0.08	300
PCL (Control)	6	90	5	21	0.05	600

2.3. Mesh Degradation

2.3.1. Physiological Condition Degradation

Three mesh samples from each group were weighed and immersed in 10 mL phosphate-buffered saline (PBS). The samples in PBS solution were placed on a rotator in an oven and the temperature was kept at 37 °C for 28 days. The meshes were assessed for mass loss at 1 day, 7 d, 14 d, 21 d and 28 d time points.

2.3.2. Accelerated Degradation

Three mesh samples of each group were weighed and immersed in 10 mL 5 M NaOH at 37 °C until the samples became irretrievable as reported previously [15]. The meshes were assessed for mass loss at 1 h, 3 h, 6 h, 10 h, 1 d, 3 d and 7 d.

2.4. Mechanical Testing

The 30 × 6 mm mesh strips were used for the tensile testing with a Tytron 250 Microforce Testing System (MTS Systems Corp., Minnesota, MN, USA) using a load cell (Model 661.11B-02, 2.5 mN resolution). Test samples consisted of two mesh geometries (1.0 mm and 1.5 mm) and three materials (PCL, 90:10, 75:25), resulting in 18 groups (n = 4 each). All samples were tested using an 18 mm gauge length to stretch for 100 mm over 120 s. Force-displacement data were obtained from the tests and the stress-strain curves were obtained by calculating the stress (σ) from the force divided by the average cross-sectional area of the meshes (Equation (1)) and strain (ε) by normalising the initial displacement (Equation (2)). The test cross sectional area was approximated as the sum of the circular longitudinal fibres, where the 1.0 mm meshes had 60 fibres (6 across and 10 layers) and the 1.5 mm mesh had 50 fibres (5 across × 10 layers), and the fibre diameter was taken as the average of 5 measurements from SEM images for each material (PCL = 55.6 µm, 90:10 = 66.4 µm, 75:25 = 51.4 µm). Elastic modulus and yield strength were calculated from a 0.2% offset linear best fit line between 30–70% strain. Ultimate tensile strength (UTS) is the maximum stress the mesh endured, and maximum force is defined as the maximum force (N) per 1 cm cross-sectional width (enables comparison to other mesh types).

$$\sigma = \frac{F}{A} = \frac{F}{\pi \left(\frac{\emptyset}{2}\right)^2 l n} \qquad (1)$$

where F is the measured force, A is the mesh cross sectional area, $\emptyset$ is the diameter of fibres, l is the number of mesh layers and n is the number of vertical fibres (5 for 1.5 mm meshes and 6 for 1.0 mm meshes).

$$\varepsilon = \frac{\Delta L}{L} = \frac{d - d_o}{L} \qquad (2)$$

where ΔL is the change in sample length, L is the gauge length of samples (18 mm), d is the measured displacement and d_o is the initial displacement value.

2.5. Antibiotics Loading and Release Profile

Azithromycin, a broad-spectrum antibiotic, was coated onto the meshes and assessed for its antibacterial potency.

2.5.1. Antibiotics Loading

Azithromycin was loaded onto the meshes using a method adopted from a previous study [16]. Briefly, the 30 × 30 mm mesh sheets were cut into disks of 5 mm in diameter and placed in 1 mL flat bottom centrifuge tubes with screwable caps. The drug loading solutions were prepared by dissolving 0.5 mg and 1 mg of azithromycin in 100 µL of diethyl ether (DEE), and the mesh disks were incubated with the loading solutions for 8 h at room temperature with mild agitation. Following the incubation, the meshes were air dried in a fume hood for complete DEE evaporation and the drug loaded meshes were kept at −20 °C for further analysis.

2.5.2. Antibiotics Loading Efficiency

Azithromycin concentration was measured by colorimetry by mixing the azithromycin with 43% sulfuric acid solution [16]. Erythronolide, the hydrolysis degrative product of azithromycin, exhibits an absorbance peak at 482 nm, and this was used to quantify the concentration of azithromycin based on the absorbance intensity [17]. The azithromycin standard solutions were prepared by incubating the drug powder at concentrations of 1, 5, 10, 15, 20 and 25 mg mL^{-1} in 43% sulfuric acid for 30 min, and the absorbance was read at 482 nm. The standard curve was plotted based on the absorbance intensity and drug concentration and a linear equation was obtained from the standard curve. The drug-loaded meshes were placed in fresh 2 mL centrifuge tubes and 500 µL of 43% sulfuric acid was added. After 30 min of incubation, the absorbance was read at 482 nm in triplicate. The azithromycin concentration was calculated using the standard curve.

2.5.3. Antibiotics Release Profile

The drug-loaded meshes were incubated in 500 µL of PBS at 37 °C with rotation, and 250 µL of PBS solution was taken for drug concentration measurement at predetermined time points: 1 h, 4 h, 1 d, 7 d and 14 d. 250 µL of fresh PBS was added to the mesh incubation tubes at each time point. Standard solutions of azithromycin (1, 5, 10, 15, 20, 25 mg/mL) were prepared by mixing drug powder in 250 µL PBS. The standard solutions were mixed with 250 µL sulfuric acid solution (43%) for 30 min. Absorbance was read at 482 nm to obtain the measurements of standard solutions and mesh samples. The standard curve was obtained using the standard solutions, and concentration of azithromycin released from the mesh samples was calculated based on the standard curve.

2.6. Antimicrobial Test

The antimicrobial capacity of the mesh samples was conducted against *S. aureus* ATCC 25923 by a disk diffusion method [16]. Briefly, the bacterial strain was inoculated onto a brain heart infusion agar (Oxoid). After 24 h incubation at 37 °C, bacterial colonies were isolated and suspended in sterile saline until the turbidity was compatible with 0.5 Mac Farland. *S. aureus* suspension (100 µL) was spread onto a Mueller–Hinton agar (Oxoid) plate. The PCL and PCL/PEG meshes (5 mm) with two doses of azithromycin (1 and 0.5 mg, n = 3 for each dose) after 0 and 14 d of release in PBS at 37 °C were sterilized for 30 min under UV and pasted onto the agar plate and incubated for 18 h at 37 °C. Azithromycin antimicrobial susceptibility disks (15 µg, Oxoid) were used as the positive control. Unloaded meshes and mock treated meshes in DEE were used as the negative controls. The bacterial growth on the plate was visualized directly after incubation of the plates at 37 °C for 18 h, and the diameter of the inhibition zone was measured according to clinical and laboratory standards institute (CLSI M02-A10) recommendations.

2.7. Biocompatibility Test

The biocompatibility of the samples was tested using human immortalized adipose derived mesenchymal stem cells (hMSC) (ATCS CRC4000, ATCC). Nine MEW pelvic mesh groups were selected to test their biocompatibility: PCL_non-loaded control, PCL_drug loaded (0.5 mg azithromycin), PCL_drug released (drug-loaded meshes released for 14 days in PBS as described in 2.4.3), 90:10_non-loaded control, 90:10_ drug loaded, 90:10_ drug released; 75:25_non-loaded control, 75:25_drug loaded, 75:25_ drug released. All mesh samples were cut into disks of 5 mm in diameter and sterilized for 30 s with 70% ethanol. The mesh samples were air dried overnight in a biosafety cabinet and further sterilized with 20 min UV radiation on each side prior to cell seeding.

2.7.1. Cell Seeding

The hMSC cells (passage 5) were seeded onto the mesh sample disks at a density of 1×10^4 cells/disk and cultured at 37 °C, 5% CO_2 in ATCC Mesenchymal Stem Cell Basal Medium (ATCPCS500030) supplemented with ATCC Mesenchymal Stem Cell

Growth Kit (ATCPCS500040) and 0.2 mg/mL Geneticin selective antibiotics (G418 Sulphate, Thermo Fisher Scientific, Brisbane, Australia). The cells on the mesh samples were assessed for their viability and morphology with the following in vitro assays: alamarBlue assay, LIVE/DEAD assay, fluorescent microscopy and scanning electron microscopy (SEM) imaging.

2.7.2. Cell Viability Assessment: LIVE/DEAD and alamarBlue Assays

The alamarBlue assay (Thermo Fisher Scientific, Brisbane, Australia) was used to quantitatively assess the cell metabolism following the manufacturer protocol. Briefly, at 1 d, 7 d and 14 d timepoints, the MSC medium was removed from 4 samples of each mesh group, and the samples were rinsed with PBS and transferred to a fresh 48-well plate. The samples were then incubated with 330 µL of fresh culture medium containing 10% (v/v) of alamarBlue solution for 4 h at 37 °C with 5% CO_2. After the incubation, 3 aliquots of 100 µL from each sample medium were transferred to black-wall 96-well plates. The fluorescence was read at 545/590 nm (excitation/emission) with a CLARIOstar microplate reader (BMG Labtech, Mornington, Australia).

LIVE/DEAD assay was used to show the distribution of live and dead cells attached on the mesh samples. Briefly, after 1 d and 7 d of cell seeding, the mesh samples were moved to a fresh 48-well plate and washed twice with PBS solution. The disk samples were incubated for 30 min in 300 µL of LIVE/DEAD staining solution (Thermo Fisher Scientific, Brisbane, Australia) containing 2 µM calcein and 4 µM ethidium. The stained samples were imaged with a fluorescent microscope (Zeiss Axio Observer 7, Carl Zeiss, Oberkochen, Germany) immediately after the staining.

2.7.3. Cell Morphology: Fluorescent Microscopy and SEM

The cell morphology stain by immunofluorescence and microscope imaging were performed as described previously [18]. Briefly, at 3 d, 7 d and 14 d timepoints, cell culture medium was removed and the mesh samples were transferred into a fresh 48-well plate. The samples were then washed in PBS and fixed in 4% paraformaldehyde (Sigma, Melbourne, Australia) for 30 min at room temperature. Following a rinse in PBS and permeabilization in 0.2% Triton X-100 solution (Sigma-Aldrich, Australia), the samples were incubated with 0.5% bovine serum albumin for 10 min. The samples were then immersed for 45 min in staining solution containing 0.8 U/mL Alexa Fluor® 488 Phalloidin (Thermo Fisher Scientific, Brisbane, Australia) and 5 µg/mL 4′,6-diamino-2-phenylindole (DAPI; Thermo Fisher Scientific, Brisbane, Australia). The samples were imaged with a fluorescent microscope (Axio Observer 7, Carl Zeiss, Oberkochen, Germany).

SEM sample preparation and imaging were performed as described previously [19]. Briefly, at 3 d, 7 d and 14 d timepoints, the mesh samples were fixed in 2.5% glutaraldehyde (ProSciTech, Townsville, Australia) immediately after cell culture. The samples were washed in PBS buffer (Electron Microscopy Sciences, Hatfield, PA, USA) and dehydrated in graded ethanol solutions and dried with Hexamethyldisilazane (Sigma Aldrich, Melbourne, Australia). The gold sputter-coated samples were imaged using a TESCAN MIRA3 SEM (Tescan, Brno-Kohoutovice, Czech Republic).

2.8. Statistical Analysis

The statistical tests were performed by two-way ANOVA with GraphPad Prism 9 software (GraphPad Software Inc., San Diego, CA, USA). For the mechanical testing data, there was repeated violation of the assumption of normal distribution; therefore, the non-parametric Kruskal–Wallis test with post hoc analysis using Bonferroni corrected Wilcoxon signed-rank test was used. A $p < 0.05$ was considered a significant result.

3. Results

3.1. Degradation Test

The physiological and accelerated degradation curves are shown in Figure 2, with the composite meshes shown to degrade faster than the PCL. In physiological conditions, the mass loss % increased with increasing PEG content in the mesh fibres. In the accelerated degradation test, both the 90:10 and 75:25 mesh groups showed significantly greater mass loss % compared to that of the PCL mesh group (control). The 90:10 and 75:25 groups became unretrievable after 10 h of degradation, while the PCL meshes were retrievable after 3 days in NaOH.

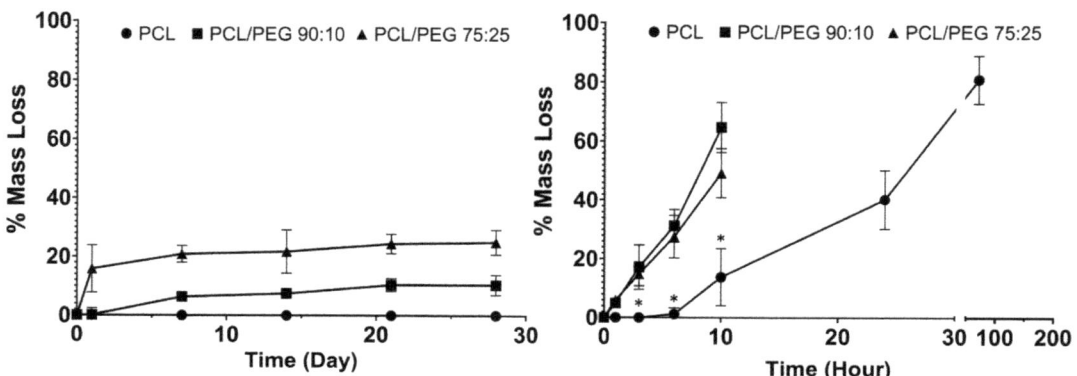

Figure 2. Physiological degradation (10 mL PBS at 37 °C) of PCL and PCL/PEG meshes (**left**). Accelerated degradation (10 mL 5 M NaOH at 37 °C) of PCL and PCL/PEG meshes (**right**). * shows significant less mass loss % in the PCL meshes compared to both PCL/PEG groups ($n = 3$, $p < 0.05$).

3.2. Mechanical Testing

The material composition was found to substantially affect the tensile properties of the meshes, with the composite materials (90:10 and 75:25) having consistently higher strength (yield strength UTS and maximum force) than PCL (Figure 3). The stiffness was also found to increase with increasing PEG content, with 75:25 typically having a higher Young's modulus than 90:10, and both higher than PCL in the non-degraded state (Figure 3; a 276% increase for 90:10 over PCL and a 615% increase for 75:25 over PCL for 1.0 mm controls). However, for the 75:25 composite, the degradation in 28 days of PBS reduced the stiffness by −46% in comparison to the 1 mm 75:25 control.

The mesh geometry with the smaller mesh spacing (1.0 mm), and, therefore, more cross fibres to bear a load, was found to have a non-significant increase in tensile strength (UTS and maximum force) when compared to 1.5 mm spacing for the same material and state ($p > 0.99$).

The degradation of the meshes in 28 days of PBS did not result in any significant reduction in the mesh UTS or yield strength for the as printed samples for any material (degraded 1.0 mm vs. control 1.0 mm, $p > 0.97$). Further, the test setup effect of drug loading via immersion in DEE for 8 h did not result in any significant changes to the mesh tensile properties for any material (mock loaded 1.0 mm vs. control 1.0 mm, $p > 0.99$).

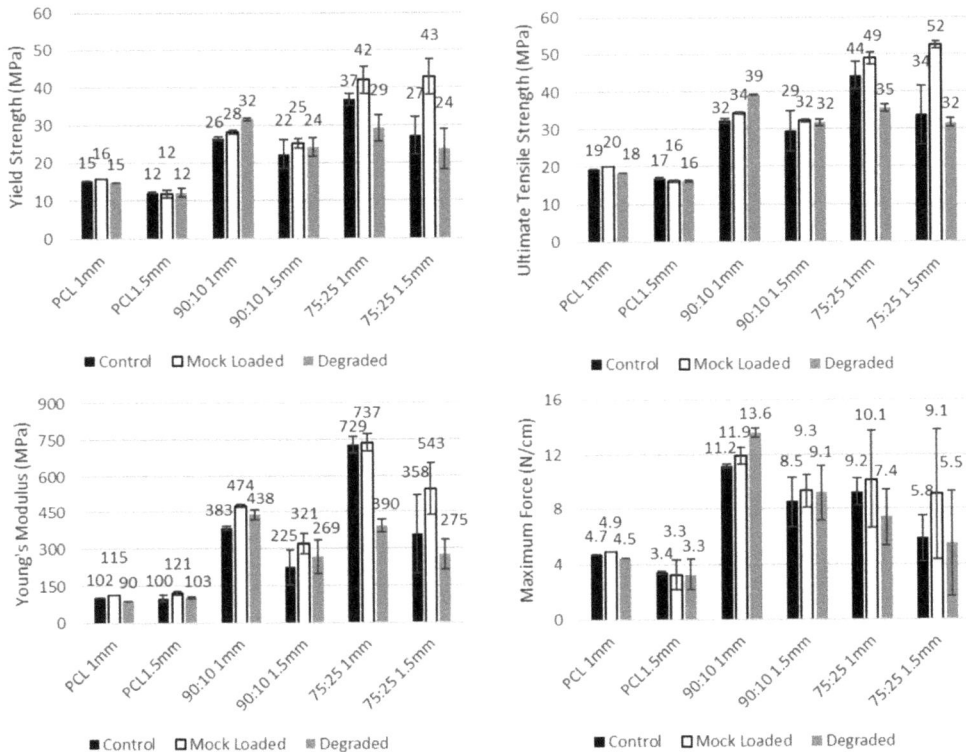

Figure 3. Tensile testing results, including strength (yield strength (**top left**), ultimate tensile strength (**top right**) and max force per centimetre (**bottom right**)) and stiffness (Young's modulus (**bottom left**)) measures for each grouping. Results are given in mean with error bar denoting standard deviation.

3.3. Antibiotics Release

Across all the materials, the mesh geometry affected the drug release, with the smaller mesh spacing (1 mm) having significantly higher cumulative antibiotics release than the larger 1.5 mm spacing (Figure 4). The 1 mg drug loaded samples had higher cumulative release than the 0.5 mg samples for all the materials (Figure 4).

3.4. Antimicrobial Test

Clear inhibition zones were observed for all the material groups (PCL, 90:10, 75:25) for both spacings (1.5 mm and 1.0 mm) at day 0 and day 14 (Figure 5). The 14 d measurements resulted in a significantly smaller inhibition zone compared with 0 d.

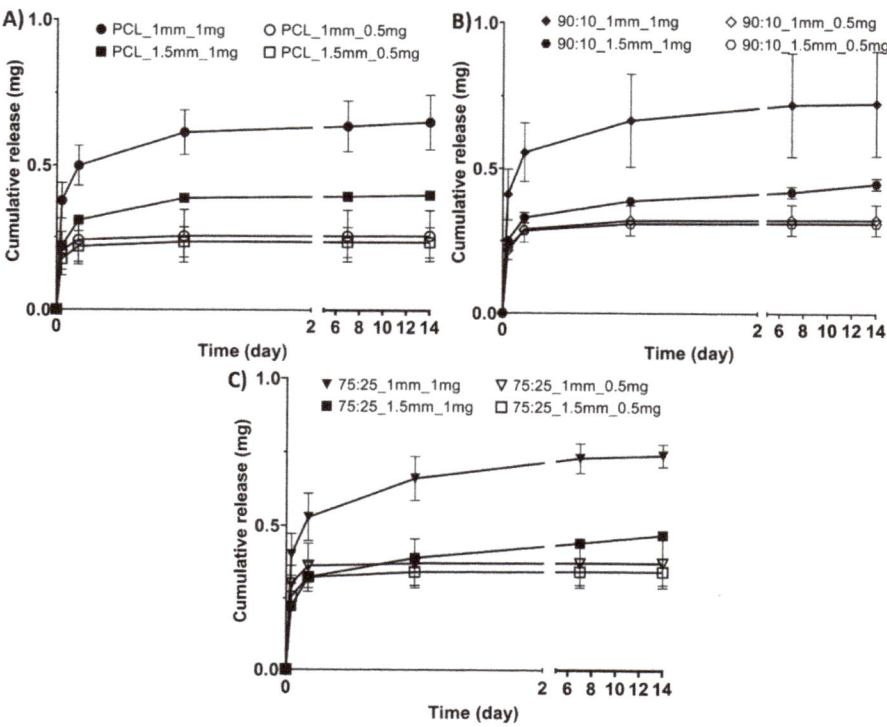

Figure 4. Cumulative antibiotic release profiles for the sample groups: (**A**) PCL groups, (**B**) 90:10 groups, (**C**) 75:25 groups.

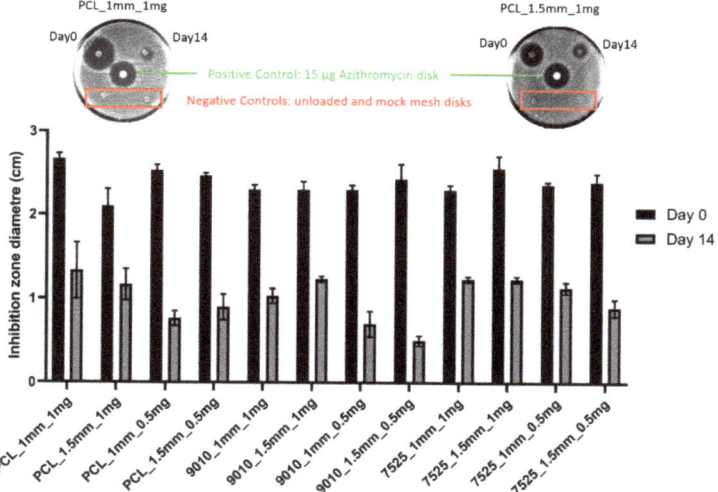

Figure 5. Inhibition zone measurements for day 0 and day 14 for all testing groups (**bottom**). A representative image is shown of the testing plates (**top**), where clear inhibition zones can be seen for the fabricated meshes at day 0 and day 14 and for the positive control, whereas the mock loaded and unloaded negative controls did not exhibit the inhibition zone.

3.5. Biocompatibility Test

3.5.1. LIVE/DEAD and alamarBlue Assay

The LIVE/DEAD results showed the viable cells in green fluorophore and dead cells in red at 1-day and 7-day time points (Figure 6). As shown in Figure 6A, the cells were viable in the PCL non-loaded control and drug-released mesh groups; few live cells were observed in the drug-loaded group. In Figure 6C, a small number of live cells were found in the 90:10 non-loaded control group and a number of viable cells were high in the drug-released group; the cells on the drug-loaded meshes were mostly dead. In Figure 6E, a few viable cells were found on the 75:25 non-loaded control mesh, and the drug-loaded mesh samples showed almost no cell attachment. The drug-released group showed higher cell viability than the freshly drug-coated and control groups. The overall trend in the LIVE/DEAD assay showed higher numbers of viable cells at the 7-day time point, especially in the drug-released mesh groups.

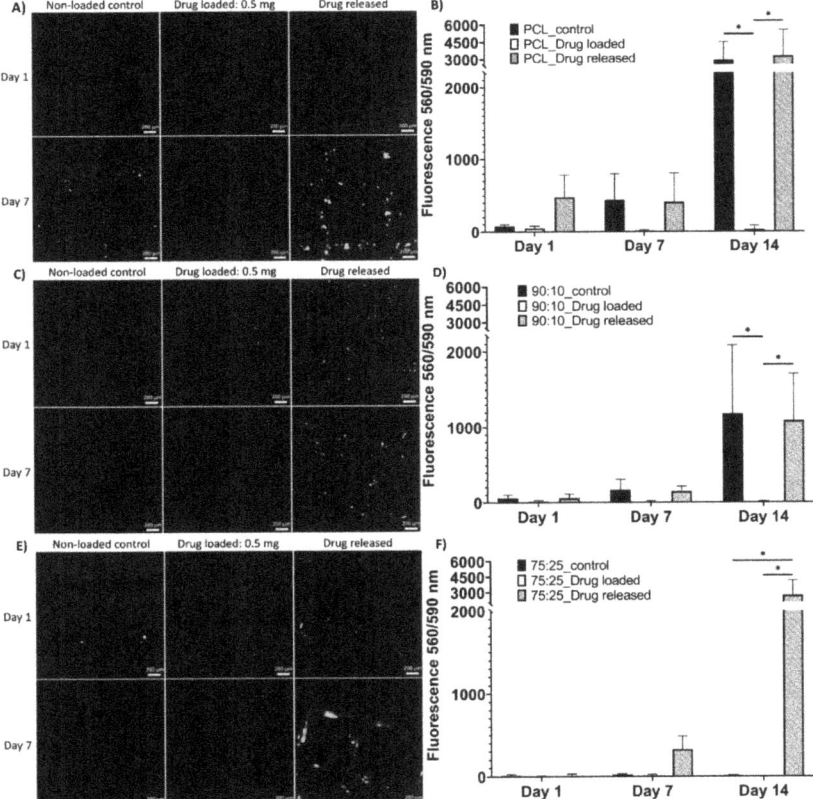

Figure 6. LIVE/DEAD assay and alamarBlue assay of all sample groups at all assessment time points. LIVE/DEAD assay results ((**A**) for PCL meshes, (**C**) for 90:10 meshes, (**E**) for 75:25 meshes) showing viable cells in green fluorescence. AlamarBlue results ((**B**) for PCL meshes, (**D**) for 90:10 meshes, (**F**) for 75:25 meshes) show mean± standard error mean (scale = 200 μm). In graph B and graph D, both non-loaded control group and drug released group showed significantly higher cell metabolic activity at the 14-day time point compared to the drug-loaded group. In graph F, significantly higher cell metabolic activity was found at the 14-day time point in the drug-released group compared to both the drug-loaded and non-loaded control groups. * shows statistical significance $p < 0.05$.

The AlamarBlue assays showed cell metabolic activity that indicated cell viability and proliferation on the mesh samples at the 1-day, 7-day and 14-day time points (Figure 6). Noticeably, at the 14-day time point, a statistical difference was found between the drug-released group and drug-loaded group in all mesh materials (Figure 6B,D,F), between the control group and drug-loaded group in PCL and 90:10 materials (Figure 6B,D) and between the drug-released group and non-loaded control group in 75:25 material (Figure 6F).

3.5.2. Cell Morphology by Fluorescent Microscopy

The cell morphology on the mesh structures was examined with a fluorescent microscope, and the nuclei of the cells were shown in blue fluorophore and cell skeletons were shown in green (Figures 7–9).

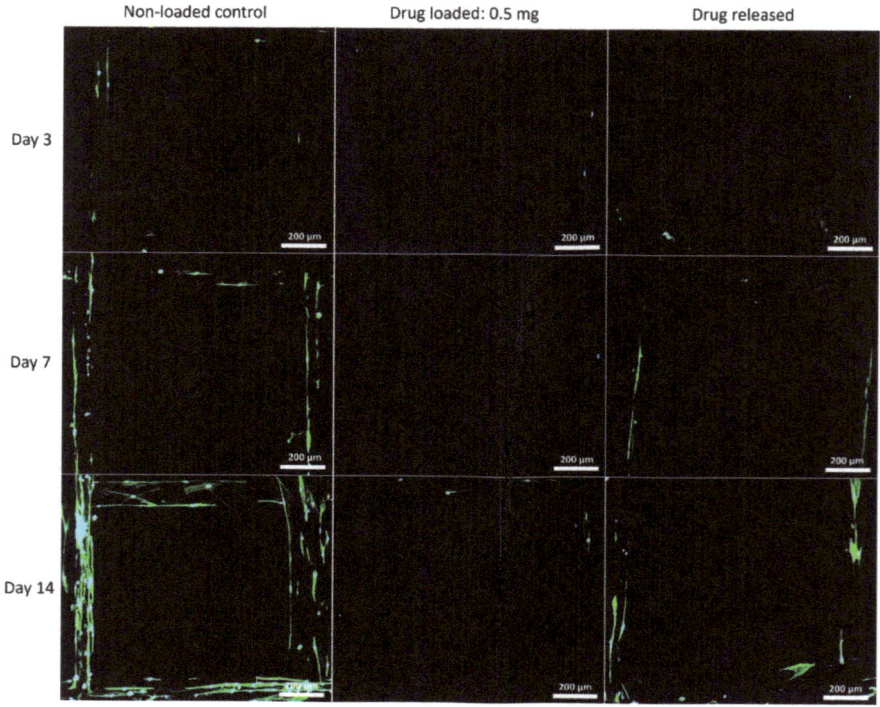

Figure 7. MSC cell morphology on PCL mesh groups at all time points. Cell nuclei are shown in blue and cytoskeleton shown in green.

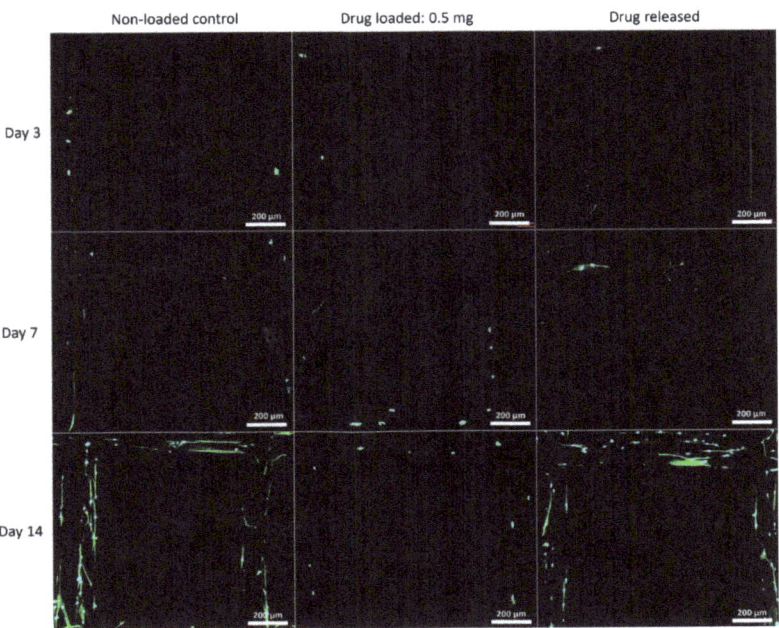

Figure 8. MSC cell morphology on 90:10 mesh groups at all time points. Cell nuclei are shown in blue and cytoskeleton shown in green.

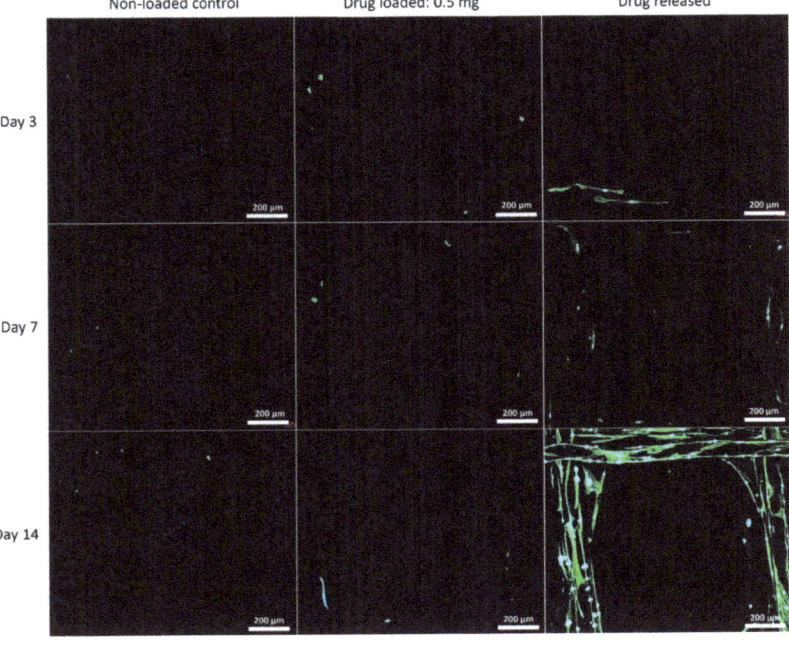

Figure 9. MSC cell morphology on 75:25 mesh groups at all time points. Cell nuclei are shown in blue and cytoskeleton shown in green.

3.5.3. Cell Morphology by SEM

The cell morphology was assessed with SEM to show the interaction of MSC cells and the mesh surface (Figures 10–12).

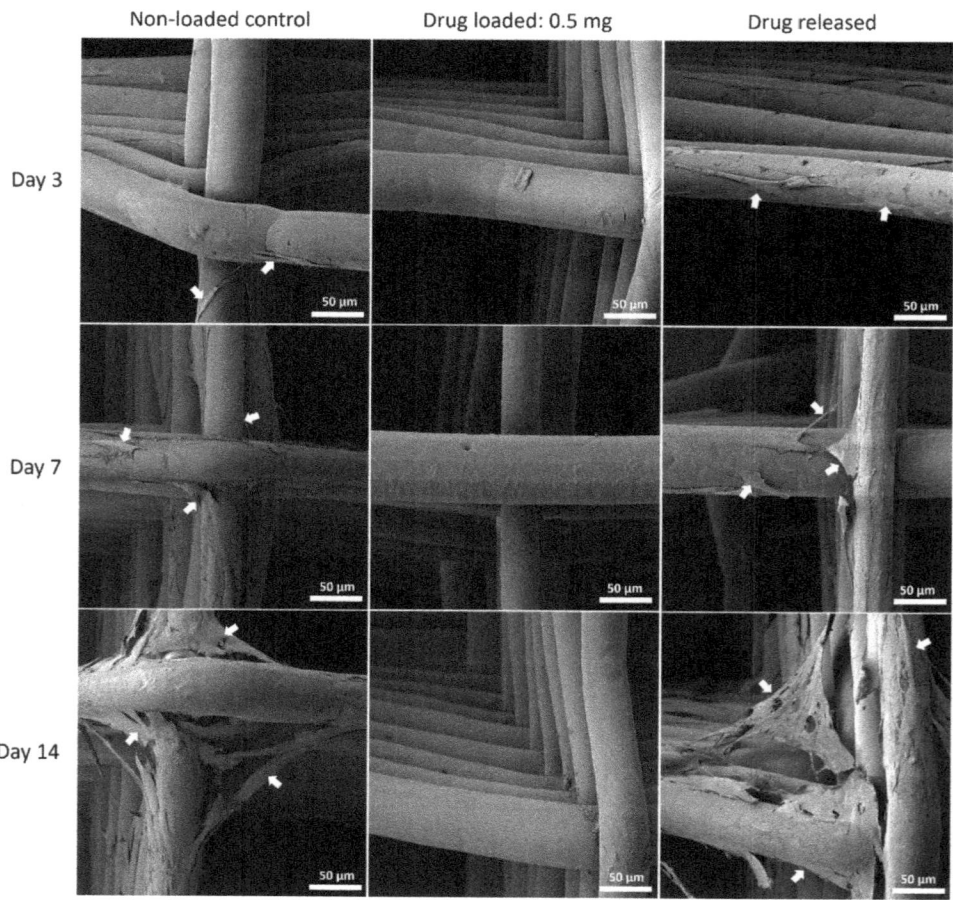

Figure 10. MSC cell morphology by SEM on PCL mesh groups at all time points. Arrows pointing at representative cells or cell sheets attached on mesh fibres.

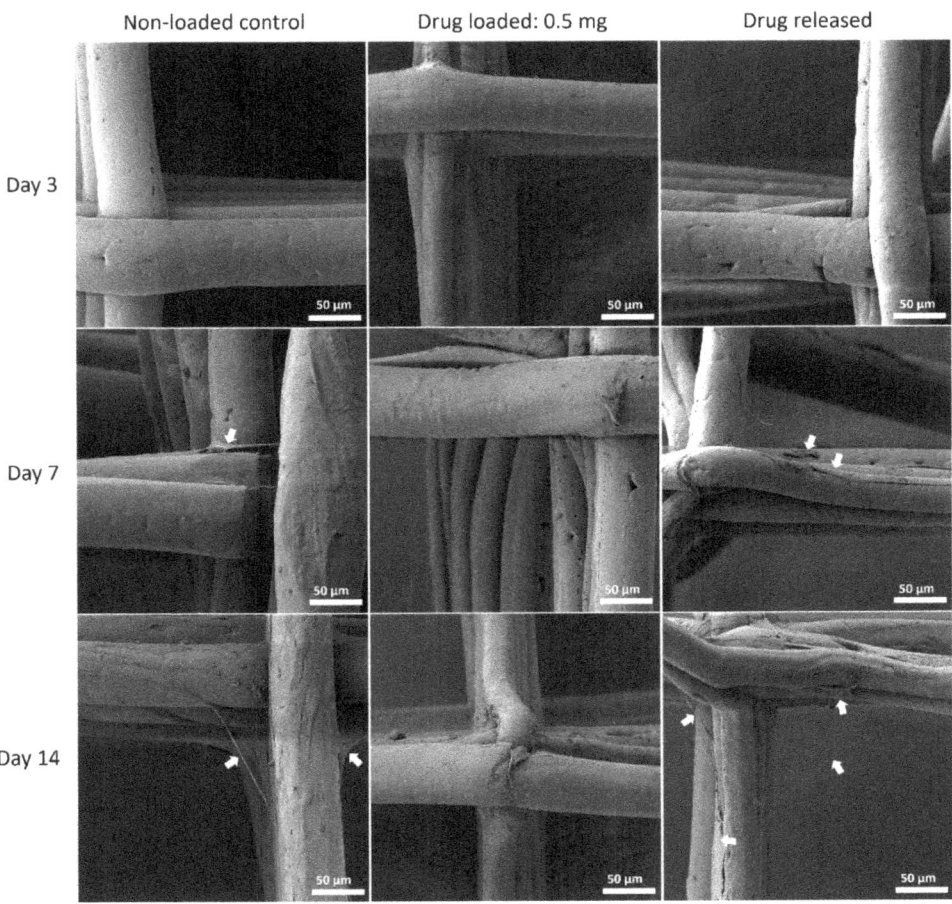

Figure 11. MSC cell morphology on PCL/PEG 90:10 mesh groups at all time points by SEM. Arrows pointing at representative cells or cell sheets attached on mesh fibres.

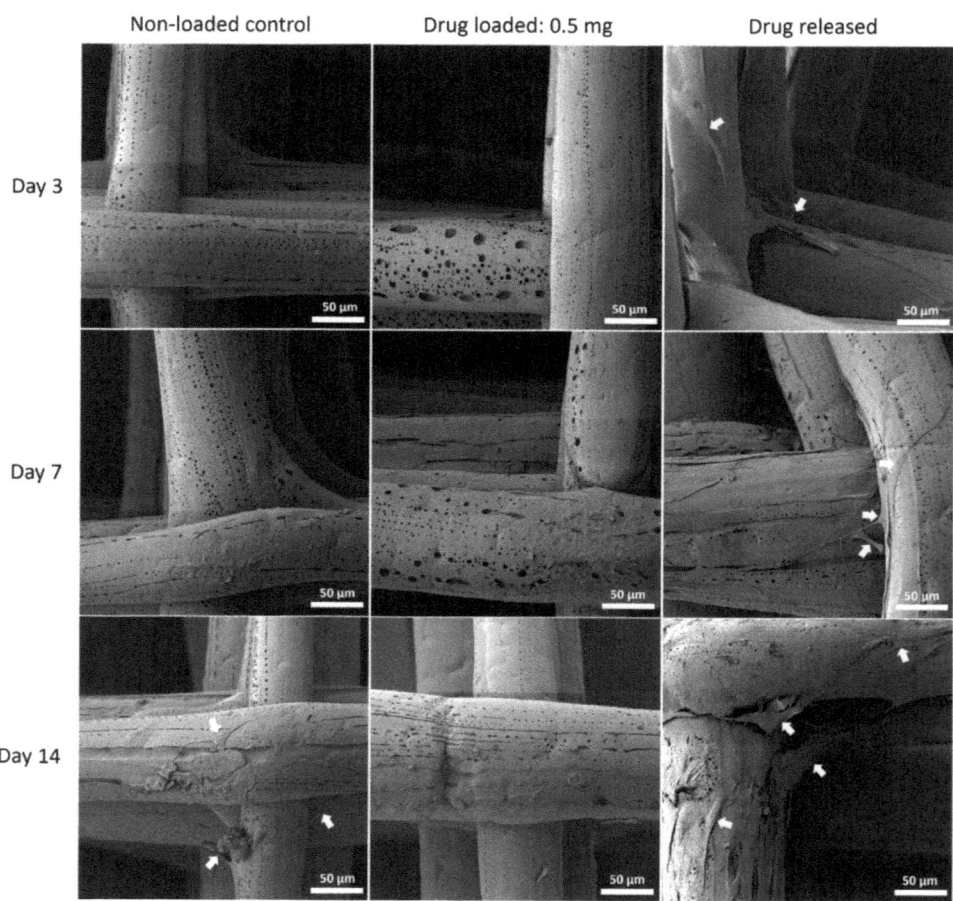

Figure 12. MSC cell morphology on PCL/PEG 75:25 mesh groups at all time points by SEM. Arrows pointing at representative cells or cell sheets attached on mesh fibres.

4. Discussion

The ban on pelvic mesh implants generated an urgent need for an alternate mesh product to treat POP. One of the main causes of mesh failure is the non-compliance of the polypropylene mesh. Studies have investigated methods to modify the polypropylene meshes, for example, producing different knitted mesh structures [20] or coating the mesh with electrospun polymer composite of polylactic acid (PLA) and PCL [21].

There has also been a growing interest in creating the pelvic meshes with biodegradable polymers such as PLA [22], PLGA [23] and PCL [24]. The use of non-degradable polymer such as polypropylene for implants tend to lead to scarring and other foreign body responses. The advantage of using biodegradable polymer such as PCL, PLA and PLGA is that they are FDA approved and widely used as biomaterials for tissue engineering purposes including pelvic meshes [25,26]. PCL, although commonly used as biomaterials, is known for its hydrophobicity which impedes cell adhesion [27]. Incorporating PEG, which is a soft, hydrophilic polymer, to create a hybrid scaffold is often a strategy employed to enhance surface hydrophilicity [28–30]. PEG is widely used in the field of drug delivery to produce a controlled sustainable delivery system by blending with polyurethane [31] and PCL [32,33]. PLA-PEG scaffold has also been electrospun for neural tissue engineering [30].

Electrospinning is a manufacturing technique that has gained increasing interest for fabricating pelvic meshes due to its ability to create microstructures mimicking the extracellular matrix, which has been shown to enhance tissue integration [24,26]. This present study fabricated the meshes using melt-electrowriting (MEW), a technique similar to electrospinning with the exception that MEW enables the meshes to be produced with a high degree of control and precision and does not include toxic solvents [34]. Additionally, the fibre thickness, porosity and scaffold height can be customised to produce scaffolds with optimal properties for pelvic meshes.

In this study, composites of PCL and PEG meshes with two different ratios (90:10 and 75:25 PCL:PEG) were fabricated using MEW. PEG was a good choice for the composites as it has a similar melting point to that of PCL ($\Delta Tm \sim 3\ °C$) [35]. As medical grade PCL has a slow degradation rate (~4 years), the hydrophilic nature of PEG was utilised to alter the degradation rate of the composites. The degradation rate of PCL/PEG mesh is controllable by adjusting the content of PEG in the composite to match the rate of tissue regeneration in the pelvic floor. The degradation profile at physiological conditions (in PBS at 37 °C) showed an average of 10% mass loss in the 90:10 mesh group and ~25% mass loss in the 75:25 groups. No weight changes were observed in the PCL group (Figure 2). These results indicate that the hydrophilic PEG component was dissolved by the aqueous PBS solution and washed off from the mesh fibres, while the PCL showed no sign of degradation. Interestingly, the degraded 75:25 samples formed hollow structures, indicating the PEG component of the composite might have been more central when manufactured via MEW (Figure S1). Other PCL composites, such as PCL/PLA, showed a weight loss of less than 5% in PBS at 37 °C after 30 days, whereas PCL/PGA exhibited a loss of 55%, which would be too fast for a POP application [36]. Polypropylene, despite being a non-degradable polymer in vitro, when implanted as a mesh, underwent superficial degradation in 33% of the patients [37]. After 3 months of implantation, peeling of the fibre surface, cracks and flaking of the polymer were observed.

To examine the long-term degradation behaviour of PCL mesh and PCL/PEG composite meshes, we used 5 M NaOH for an accelerated degradation test method. The hydrophobic PCL meshes lost 80% of their mass within 3 days, while both PCL/PEG mesh groups lost their structure and became unretrievable after 10 h. At the 10-h time point, the PCL/PEG 90:10 group showed almost 5-fold increase in degradation rate and the PCL/PEG 75:25 group showed a 3.7-fold increase compared to PCL alone. Although both composite groups had a significantly faster degradation rate than PCL, interestingly, the 90:10 group degraded faster than the 75:25 group in accelerated conditions. This could be due to the change of crystallization of PCL with the increased presence of PEG when the fibres cooled down to room temperature after MEW printing [35,38]. The increased crystallinity of PCL alters its physical properties, such as increased melting temperature, degradation rate, and increased the mechanical strength with higher stiffness.

The addition of PEG to PCL increased the tensile strength of the meshes compared to the PCL control group. Increasing the PEG content to 25% exhibited an average increase in ultimate tensile strength of 99% to 34 MPa (1.5 mm) and 127% (44 MPa; 1.0 mm) in the 75:25 group). Polypropylene meshes, such as Gynemesh, exhibited tensile strength of 2.59 MPa, which is markedly lower than the PCL/PEG composites [39]. Electrospun meshes fabricated using other polymers also exhibited similar tensile strength as Gynemesh, such as PLA (3.5 MPa) and PLGA/PCL (3.6 MPa). Interestingly, PLA fibres, when aligned, produce meshes with tensile strength that increased to 22 MPa, which is similar to our PCL/PEG composites, which comprised aligned fibres [39]. Additionally, the increase in PEG also corresponded to a significant increase in stiffness (358 MPa (1.5 mm, $p = 0.79$) and 729 MPa (1 mm, $p = 0.02$) in Young's modulus) compared to PCL alone. This increased stiffness is also likely associated with the increased crystallinity of the PCL with increased proportion of PEG. Stiffer polypropylene-based meshes, especially Gynemesh with a Young's modulus of 9 MPa, have been shown to disrupt ECM remodelling and produce protein responses similar to vaginal degeneration [39,40]. Additionally, the stiffness of meshes can influence the rate of mesh-related complications [41,42], resulting in increasing risk of mesh exposure [40].

It is noted that there are very limited studies in the literature that utilise MEW to create meshes for POP application. Most of the studies fabricated meshes via electrospinning, producing scaffolds with lower mechanical strength than MEW scaffolds. For example, the tensile strengths of our PCL/PEG composites (~30 MPa for 90:10 group) were higher than those of other PCL composites, as shown by researchers such as Vashaghian et al. [26], whereby electrospun PCL/PLGA and PCL/Gelatin exhibited tensile strengths of 12.4 ± 1.6 MPa and 3.5 ± 0.9 MPa, respectively. The stiffness of the PVDF electrospun scaffolds ranged from 13.1 to 25.8 MPa [43] and was 10 to 20 times lower than the PCL/PEG composites. Irrespective of manufacturing techniques, the stiffness of these scaffolds was still too high when compared to premenopausal healthy vaginal tissues, which measured at 0.79 MPa [25]. On the other hand, while it is desirable to have lower mechanical properties, electrospun scaffolds have non-uniform small pore sizes, which hinders cellular infiltration and tissue integration. The ability of MEW to better control fibre thickness and pore size has the advantage of tailoring the scaffold's parameters to obtain the desirable properties.

Although parameters such as Young's modulus and ultimate tensile strength are commonly used to assess the mechanical characteristics of meshes, they can be difficult to compare when the mesh structures and sizes vary. Pott et al. proposed an alternative approach for mesh strength comparison by measuring maximal force that the mesh sustained over 1 cm mesh width (N/cm) [44]. As shown in Figure 3, the maximal force values of the PCL meshes with 1 and 1.5 mm spacing were 4.7 and 3.4 N/cm, respectively. Such maximal force is not adequate as the clinically relevant force for hernia repair was noted as 32 N/cm (lateral) and 22 N/cm (cranial/caudal) [44]. The PCL–PEG composite meshes improved the maximal force, whereby a 10% PEG addition exhibited an increase of 136% for the 1 mm-spaced mesh and 148% for the 1.5 mm-spaced meshes. Changes in scaffold architecture through features such as interwoven fibres, varying fibre orientation, and altering the composite composition may enable the mesh strength to approach clinically relevant levels. The pore size and shape are important to take into consideration when designing the meshes. These parameters have been shown to influence mechanical strength, in particular, the strength of tissue ingrowth. The pore sizes of commercially available meshes were wide-ranging, with 1.1 mm measured in Novasilk mesh to Ultrapro with 4 mm pore size [25]. Large pore sizes (> 1 mm) have been shown to integrate better with tissue and exhibited more tissue ingrowth in pigs [45,46]. The mechanical strength of tissue ingrowth was enhanced as the pores increased from 1 mm to 5 mm [46]. In addition, the meshes with hexagonal pores encouraged the strongest tissue ingrowth in pigs after 90 days of implantation [46]. The pore sizes of 1 and 1.5 mm of our meshes were chosen based on the literature and aimed at facilitating good integration with the host tissue.

Since the meshes aim to be ultimately implanted in humans, the biocompatibility of the various types of meshes was assessed. MSCs are commonly used in tissue engineering for cell-based therapy [47] owing to their ability to differentiate into various types of cells, such as smooth muscle cells [48] and endothelial cells [49]. MSCs also produce various types of growth factors, including VEGF, which will assist with production of blood vessels and tissue integration [50]. Additionally, MSCs are immunomodulatory, producing cytokines that regulate immune cells, such as T cells, to influence the activation of the cells involved in wound healing and tissue repair [51]. Seeding of MSCs on the PCL/PEG scaffolds has the potential of creating meshes that encourage healing and tissue integration, ultimately improving the outcome of the POP treatment. Another advantage of using MSCs for the biocompatibility test in this study is that MSCs are found to be superior to other cell types for in vitro cytotoxicity tests because they are a more accurate modelling of in vivo conditions [52].

Although PEG increased the mechanical properties of the PCL meshes, it appeared to produce a less conducive surface for cell attachment. As shown in Figures 9 and 12, there were fewer MSCs observed after 14 days of culture when PEG was increased from

10% to 25%. The addition of PEG to PCL increased the hydrophilicity of the meshes, but PEG is also known to have an anti-cell attachment due to decreased initial random motility coefficient, which reduces cell-substrate adhesion in hMSC [53]. This effect is also increased with a higher content of PEG in the substrate. Endothelial cells cultured on PU/PEG composite also demonstrated initial lower cell proliferation when compared to PU scaffolds [54]. Other studies have shown that the relationship between surface wettability and cell adhesion displays a bell shape distribution rather than a linear relationship, with the ideal hydrophilicity being cell-dependent [55,56]. Polypropylene meshes cultured with endothelial cells in vitro showed a reduction in viability of almost 50% after 3 days [57], while, in vivo, they induced a proinflammation response in 27 patients, demonstrated by an increase in M1 macrophages [58].

Another complication faced by women with mesh implants is infection of the mesh. An anti-bacterial property was incorporated into the composite meshes fabricated in this study. Azithromycin was selected as the antibiotic for this study due to its broad antimicrobial coverage and previously demonstrated effectiveness in PCL fibres loaded with azithromycin [16]. Mathew et al. demonstrated that electrospun PCL fibres loaded with azithromycin via a solvent evaporation technique inhibited the growth of *Staphylococcus aureus* for 14 days [16]. Azithromycin is a widely used macrolide antibiotic that inhibits bacterial protein synthesis, which provides coverage of many Gram-positive bacteria and most Gram-negative bacteria, including 'atypical' bacteria, such as mycoplasma and mycobacteria species [59]. Its clinical uses include treatment of community-acquired pneumonia, otitis media, pharyngitis, cervical infections, pelvic inflammatory disease, skin infections as an alternative to beta-lactam antibiotics and as prophylaxis to vulnerable patients with advanced acquired immunodeficiency syndrome [60,61]. In this study, the antibacterial property was maintained across all three types of meshes, with the 75:25 PCL:PEG 0.5 mg azithromycin group still producing an inhibition zone of 10 mm after 14 days (Figure 5).

Although azithromycin is an effective bactericide, our study indicated that it may be cytotoxic above certain concentrations. Azithromycin has been shown to be cytotoxic to fibroblasts at a concentration of 0.05% while, at 0.025%, it was biocompatible [62]. As shown in our study by the LIVE/DEAD assay and alamarBlue assay, there were very few viable cells in the azithromycin-loaded groups across all three types of meshes throughout the 14 days of culture (Figure 6). However, in the drug-released groups, increased cell proliferation was observed across the experimental period, especially the 75:25 group (Figure 6F). As the drug released, the concentration of azithromycin on the meshes decreased gradually, thereby enabling cells to proliferate with time. This suggested that, even though there was a delay of cell attachment and proliferation when the drug-coated meshes were first incubated, the cytotoxicity decreased over time, enabling cells to eventually proliferate on the meshes. It is anticipated that the design and composition of these meshes will provide initial mechanical support to the pelvic organs and encourage tissue integration and infection-free regeneration after implantation, as demonstrated by the biocompatible and antibacterial properties of the mesh. When the regenerated connective tissues have sufficient strength to take over the mechanical load, the meshes are expected to fully degrade and be replaced by natural tissues.

Confined by the scope of the study and current understanding of biodegradable mesh usage for POP, there are limitations of this study: the laydown patterns of the MEW meshes were limited to the 90° cross hatch, only 2 PCL:PEG composites were studied and two antibiotic dosage levels were chosen. These mesh designs are substantially different from the woven, knitted and braided implants typically made of polyester, polytetrafluoroethylene, polypropylene, polyethylene and nylon [63]. Future studies will need to investigate the effect of different pore shapes to obtain mechanical properties comparable to the vaginal tissue. With better understanding of the regenerative rates of the connective tissues in the pelvic floor, PCL/PEG composite meshes can be prepared for a degradation rate that matches the tissue regeneration. Future experiments will include further optimizing the dosage of azithromycin to acquire a balance between being bactericidal and biocompatible. In vivo implantation studies in small and large animal models will also be needed to demonstrate the preclinical performance of these meshes.

In this study, altering the mesh material composition led to the greatest effect on the mesh material properties, providing evidence that mesh material properties may be tailorable via polymer composites fabricated via MEW. Careful consideration should be given to the stiffness of the designed mesh as higher stiffness meshes have influenced: the rate of mesh-related complications [41,42], tissue remodelling response through stress shielding [64,65] and breakdown of collagen and elastin [66,67], resulting in an increasing risk of mesh exposure [40].

5. Conclusions

This study has demonstrated that melt-electrowritten (MEW) composite meshes comprising PCL and PEG showed a controllable degradation rate by adjusting the PEG content and produced mechanical properties, such as maximal forces, that are higher than PCL alone and move towards the forces observed clinically. Antibacterial properties with slow releasing capabilities were successfully incorporated into the meshes, albeit the concentration used warrants further adjustment. A biodegradable mesh that is compliant and antibacterial appears possible to manufacture using a version of 3D printing (MEW) and would provide a much needed and urgent treatment for women with POP.

Supplementary Materials: The following supporting information can be downloaded at: https://www.mdpi.com/article/10.3390/polym14040763/s1, Figure S1. Cross sectional SEM image of a representative sample from the 75:25 degraded group (after 28 days immersion in PBS), showing the formation of a hollow structure after the PEG content was dissolved in PBS solution, indicating the PEG component of the composite might have been more central when manufactured via MEW.

Author Contributions: Conceptualization, J.R., C.S.W., M.C., M.T., A.D.R., A.W., N.R. and M.A.W.; Data curation, J.R., R.M., C.S.W., J.Q. and M.T.; Formal analysis, J.R., R.M. and J.Q.; Funding acquisition, M.A.W., M.T. and A.W.; Investigation, M.A.W.; Methodology, M.A.W., J.R. and C.S.W.; Project administration, J.R., R.M., M.T. and A.W.; Resources, N.R. and M.A.W.; Software, M.A.W.; Supervision, M.A.W.; Writing–original draft, J.R., R.M. and C.S.W.; Writing–review & editing, J.R., R.M., C.S.W. and M.A.W. All authors have read and agreed to the published version of the manuscript.

Funding: This research and the APC were funded by Metro North Hospital and Health Services-Queensland University of Technology Collaborative Research Grants, grant number CRG172-2019.

Institutional Review Board Statement: Not applicable.

Informed Consent Statement: Not applicable.

Data Availability Statement: The raw data required to reproduce these findings also form part of an ongoing study, but they are available to download on request.

Acknowledgments: The authors would like to thank Yanan Xu for her technical assistance. The data reported were obtained using the resources of the Central Analytical Research Facility, Research Infrastructure, Queensland University of Technology, with funding from the Faculty of Engineering.

Conflicts of Interest: The authors declare no conflict of interest.

References

1. Khron, A.; Llewellyn, G. *Sterilisation*; Community Affairs References Committee: Canberra, Australia, 2013.
2. Gibson, W.; Wagg, A. Are older women more likely to receive surgical treatment for stress urinary incontinence since the introduction of the mid-urethral sling? An examination of Hospital Episode Statistics data. *BJOG Int. J. Obstet. Gynaecol.* **2015**, *123*, 1386–1392. [CrossRef] [PubMed]
3. Funk, M.J.; Edenfield, A.L.; Pate, V.; Visco, A.G.; Weidner, A.C.; Wu, J.M. Trends in use of surgical mesh for pelvic organ prolapse. *Am. J. Obstet. Gynecol.* **2012**, *208*, 79.e1–79.e7. [CrossRef] [PubMed]
4. The Senate Community Affairs Reference Committee. Number of Women in Australia Who Have Had Transvaginal Mesh Implants and Related Matters. 2018. Available online: https://www.health.gov.au/resources/publications/the-number-of-women-in-australia-who-have-had-transvaginal-mesh-implants-and-related-matters (accessed on 10 January 2022).

5. Keltie, K.; Elneil, S.; Monga, A.; Patrick, H.; Powell, J.; Campbell, B.; Sims, A.J. Complications following vaginal mesh procedures for stress urinary incontinence: An 8 year study of 92,246 women. *Sci. Rep.* **2017**, *7*, 12015. [CrossRef] [PubMed]
6. Sedrakyan, A.; Chughtai, B.; Mao, J. Regulatory warnings and use of surgical mesh in pelvic organ prolapse. *JAMA Intern. Med.* **2016**, *176*, 275–277. [CrossRef] [PubMed]
7. Edwards, S.L.; Werkmeister, J.A.; Rosamilia, A.; Ramshaw, J.A.; White, J.F.; Gargett, C.E. Characterisation of clinical and newly fabricated meshes for pelvic organ prolapse repair. *J. Mech. Behav. Biomed. Mater.* **2013**, *23*, 53–61. [CrossRef]
8. Mangir, N.; Dikici, B.A.; Chapple, C.R.; MacNeil, S. Landmarks in vaginal mesh development: Polypropylene mesh for treatment of SUI and POP. *Nat. Rev. Urol.* **2019**, *16*, 675–689. [CrossRef] [PubMed]
9. Marcus-Braun, N.; von Theobald, P. Mesh removal following transvaginal mesh placement: A case series of 104 operations. *Int. Urogynecol. J.* **2010**, *21*, 423–430. [CrossRef]
10. De Tayrac, R.; Letouzey, V. Basic science and clinical aspects of mesh infection in pelvic floor reconstructive surgery. *Int. Urogynecol. J.* **2011**, *22*, 775–780. [CrossRef]
11. Feola, A.; Barone, W.; Moalli, P.; Abramowitch, S. Characterizing the ex vivo textile and structural properties of synthetic prolapse mesh products. *Int. Urogynecol. J.* **2013**, *24*, 559–564. [CrossRef]
12. De Maria, C.; Santoro, V.; Vozzi, G. Biomechanical, Topological and Chemical Features That Influence the Implant Success of an Urogynecological Mesh: A Review. *Biomed Res. Int.* **2016**, *2016*, 1267521. [CrossRef]
13. Paxton, N.C.; Ren, J.; Ainsworth, M.; Solanki, A.K.; Jones, J.R.; Allenby, M.C.; Stevens, M.M.; Woodruff, M.A. Rheological Characterization of Biomaterials Directs Additive Manufacturing of Strontium-Substituted Bioactive Glass/Polycaprolactone Microfibers. *Macromol. Rapid Commun.* **2019**, *40*, 1900019. [CrossRef] [PubMed]
14. Pang, L.; Paxton, N.C.; Ren, J.; Liu, F.; Zhan, H.; Woodruff, M.A.; Bo, A.; Gu, Y. Development of Mechanically Enhanced Polycaprolactone Composites by a Functionalized Titanate Nanofiller for Melt Electrowriting in 3D Printing. *ACS Appl. Mater. Interfaces* **2020**, *12*, 47993–48006. [CrossRef] [PubMed]
15. Lam, C.X.F.; Savalani, M.M.; Teoh, S.-H.; Hutmacher, D.W. Dynamics of in vitro polymer degradation of polycaprolactone-based scaffolds: Accelerated versus simulated physiological conditions. *Biomed. Mater.* **2008**, *3*, 034108. [CrossRef] [PubMed]
16. Mathew, A.; Vaquette, C.; Hashimi, S.; Rathnayake, I.; Huygens, F.; Hutmacher, D.W.; Ivanovski, S. Antimicrobial and Immunomodulatory Surface-Functionalized Electrospun Membranes for Bone Regeneration. *Adv. Health Mater.* **2017**, *6*, 1601345. [CrossRef]
17. Sultana, N.; Arayne, M.S.; Hussain, F.; Fatima, A. Degradation studies of azithromycin and its spectrophotometric determination in pharmaceutical dosage forms. *Pak. J. Pharm. Sci.* **2006**, *19*, 98–103.
18. Gräfe, D.; Gernhardt, M.; Ren, J.; Blasco, E.; Wegener, M.; Woodruff, M.A.; Barner-Kowollik, C. Enzyme-Degradable 3D Multi-Material Microstructures. *Adv. Funct. Mater.* **2021**, *31*, 2006998. [CrossRef]
19. Ren, J.; Blackwood, K.A.; Doustgani, A.; Poh, P.P.; Steck, R.; Stevens, M.M.; Woodruff, M.A. Melt-electrospun polycaprolactone strontium-substituted bioactive glass scaffolds for bone regeneration. *J. Biomed. Mater. Res. Part A* **2013**, *102*, 3140–3153. [CrossRef]
20. Mirjavan, M.; Asayesh, A.; Jeddi, A.A.A. The effect of fabric structure on the mechanical properties of warp knitted surgical mesh for hernia repair. *J. Mech. Behav. Biomed. Mater.* **2017**, *66*, 77–86. [CrossRef]
21. Lu, Y.; Fu, S.; Zhou, S.; Chen, G.; Zhu, C.; Li, N.; Ma, Y. Preparation and biocompatibility evaluation of polypropylene mesh coated with electrospinning membrane for pelvic defects repair. *J. Mech. Behav. Biomed. Mater.* **2018**, *81*, 142–148. [CrossRef]
22. Mangır, N.; Hillary, C.J.; Chapple, C.R.; MacNeil, S. Oestradiol-releasing Biodegradable Mesh Stimulates Collagen Production and Angiogenesis: An Approach to Improving Biomaterial Integration in Pelvic Floor Repair. *Eur. Urol. Focus* **2017**, *5*, 280–289. [CrossRef]
23. Vashaghian, M.; Zandieh-Doulabi, B.; Roovers, J.-P.; Smit, T.H. Electrospun Matrices for Pelvic Floor Repair: Effect of Fiber Diameter on Mechanical Properties and Cell Behavior. *Tissue Eng. Part A* **2016**, *22*, 1305–1316. [CrossRef] [PubMed]
24. Paul, K.; Darzi, S.; McPhee, G.; Del Borgo, M.P.; Werkmeister, J.A.; Gargett, C.E.; Mukherjee, S. 3D bioprinted endometrial stem cells on melt electrospun poly ε-caprolactone mesh for pelvic floor application promote anti-inflammatory responses in mice. *Acta Biomater.* **2019**, *97*, 162–176. [CrossRef] [PubMed]
25. Mancuso, E.; Downey, C.; Doxford-Hook, E.; Bryant, M.G.; Culmer, P. The use of polymeric meshes for pelvic organ prolapse: Current concepts, challenges, and future perspectives. *J. Biomed. Mater. Res. Part B Appl. Biomater.* **2019**, *108*, 771–789. [CrossRef] [PubMed]
26. Vashaghian, M.; Ruiz-Zapata, A.; Kerkhof, M.H.; Doulabi, B.Z.; Werner, A.; Roovers, J.P.; Smit, T.H. Toward a new generation of pelvic floor implants with electrospun nanofibrous matrices: A feasibility study. *Neurourol. Urodynamics* **2016**, *36*, 565–573. [CrossRef] [PubMed]
27. Woodruff, M.; Hutmacher, D.W. The return of a forgotten polymer—Polycaprolactone in the 21st century. *Prog. Polym. Sci.* **2010**, *35*, 1217–1256. [CrossRef]
28. Guazzelli, E.; Galli, G.; Martinelli, E. The Effect of Poly(ethylene glycol) (PEG) Length on the Wettability and Surface Chemistry of PEG-Fluoroalkyl-Modified Polystyrene Diblock Copolymers and Their Two-Layer Films with Elastomer Matrix. *Polymers* **2020**, *12*, 1236. [CrossRef] [PubMed]
29. Shi, L.; Zhang, J.; Zhao, M.; Tang, S.; Cheng, X.; Zhang, W.; Li, W.; Liu, X.; Peng, H.; Wang, Q. Effects of polyethylene glycol on the surface of nanoparticles for targeted drug delivery. *Nanoscale* **2021**, *13*, 10748–10764. [CrossRef] [PubMed]

30. Lins, L.C.; Wianny, F.; Livi, S.; Hidalgo, I.A.; Dehay, C.; Duchet-Rumeau, J.; Gérard, J.-F. Development of Bioresorbable Hydrophilic–Hydrophobic Electrospun Scaffolds for Neural Tissue Engineering. *Biomacromolecules* **2016**, *17*, 3172–3187. [CrossRef] [PubMed]
31. Salimi, S.; Wu, Y.; Barreiros, M.I.E.; Natfji, A.A.; Khaled, S.; Wildman, R.; Hart, L.R.; Greco, F.; Clark, E.A.; Roberts, C.; et al. A 3D printed drug delivery implant formed from a dynamic supramolecular polyurethane formulation. *Polym. Chem.* **2020**, *11*, 3453–3464. [CrossRef]
32. Ranganathan, S.I.; Kohama, C.; Mercurio, T.; Salvatore, A.; Benmassaoud, M.M.; Kim, T.W.B. Effect of temperature and ultraviolet light on the bacterial kill effectiveness of antibiotic-infused 3D printed implants. *Biomed. Microdevices* **2020**, *22*, 1–14. [CrossRef]
33. Stewart, S.A.; Domínguez-Robles, J.; McIlorum, V.J.; Gonzalez, Z.; Utomo, E.; Mancuso, E.; Lamprou, D.A.; Donnelly, R.F.; Larrañeta, E. Poly(caprolactone)-Based Coatings on 3D-Printed Biodegradable Implants: A Novel Strategy to Prolong Delivery of Hydrophilic Drugs. *Mol. Pharm.* **2020**, *17*, 3487–3500. [CrossRef] [PubMed]
34. Lanaro, M.; Luu, A.; Lightbody-Gee, A.; Hedger, D.; Powell, S.K.; Holmes, D.W.; Woodruff, M.A. Systematic design of an advanced open-source 3D bioprinter for extrusion and electrohydrodynamic-based processes. *Int. J. Adv. Manuf. Technol.* **2021**, *113*, 2539–2554. [CrossRef]
35. Nguyen Tri, P.; Prud'Homme, R.E. Crystallization and Segregation Behavior at the Submicrometer Scale of PCL/PEG Blends. *Macromolecules* **2018**, *51*, 7266–7273. [CrossRef]
36. Vieira, A.; Guedes, R.; Marques, A. Degradation and Viscoelastic Properties of PLA-PCL, PGA-PCL, PDO and PGA Fibres. *Mater. Sci. Forum* **2010**, *636–637*, 825–832. [CrossRef]
37. Clavé, A.; Yahi, H.; Hammou, J.-C.; Montanari, S.; Gounon, P.; Clavé, H. Polypropylene as a reinforcement in pelvic surgery is not inert: Comparative analysis of 100 explants. *Int. Urogynecol. J.* **2010**, *21*, 261–270. [CrossRef]
38. Douglas, P.; Albadarin, A.; Sajjia, M.; Mangwandi, C.; Kuhs, M.; Collins, M.N.; Walker, G. Effect of poly ethylene glycol on the mechanical and thermal properties of bioactive poly(ε-caprolactone) melt extrudates for pharmaceutical applications. *Int. J. Pharm.* **2016**, *500*, 179–186. [CrossRef]
39. Powers, S.; Burleson, L.K.; Hannan, J.L. Managing female pelvic floor disorders: A medical device review and appraisal. *Interface Focus* **2019**, *9*, 20190014. [CrossRef]
40. Liang, R.; Abramowitch, S.; Knight, K.M.; Palcsey, S.; Nolfi, A.; Feola, A.; Stein, S.; Moalli, P.A. Vaginal degeneration following implantation of synthetic mesh with increased stiffness. *BJOG: Int. J. Obstet. Gynaecol.* **2012**, *120*, 233–243. [CrossRef]
41. Mistrangelo, E.; Mancuso, S.; Nadalini, C.; Lijoi, D.; Costantini, S. Rising use of synthetic mesh in transvaginal pelvic reconstructive surgery: A review of the risk of vaginal erosion. *J. Minim. Invasive Gynecol.* **2007**, *14*, 564–569. [CrossRef]
42. Visco, A.G.; Weidner, A.C.; Barber, M.D.; Myers, E.R.; Cundiff, G.W.; Bump, R.C.; Addison, W. Vaginal mesh erosion after abdominal sacral colpopexy. *Am. J. Obstet. Gynecol.* **2001**, *184*, 297–302. [CrossRef]
43. Ding, J.; Deng, M.; Song, X.-C.; Chen, C.; Lai, K.-L.; Wang, G.-S.; Yuan, Y.-Y.; Xu, T.; Zhu, L. Nanofibrous biomimetic mesh can be used for pelvic reconstructive surgery: A randomized study. *J. Mech. Behav. Biomed. Mater.* **2016**, *61*, 26–35. [CrossRef] [PubMed]
44. Pott, P.P.; Schwarz, M.L.R.; Gundling, R.; Nowak, K.; Hohenberger, P.; Roessner, E.D. Mechanical Properties of Mesh Materials Used for Hernia Repair and Soft Tissue Augmentation. *PLoS ONE* **2012**, *7*, e46978. [CrossRef] [PubMed]
45. Weyhe, D.; Cobb, W.; Lecuivre, J.; Alves, A.; Ladet, S.; Lomanto, D.; Bayon, Y. Large pore size and controlled mesh elongation are relevant predictors for mesh integration quality and low shrinkage—Systematic analysis of key parameters of meshes in a novel minipig hernia model. *Int. J. Surg.* **2015**, *22*, 46–53. [CrossRef]
46. Lake, S.; Ray, S.; Zihni, A.M.; Thompson, D.M.; Gluckstein, J.; Deeken, C.R. Pore size and pore shape—But not mesh density—Alter the mechanical strength of tissue ingrowth and host tissue response to synthetic mesh materials in a porcine model of ventral hernia repair. *J. Mech. Behav. Biomed. Mater.* **2015**, *42*, 186–197. [CrossRef]
47. Solovieva, A.; Miroshnichenko, S.; Kovalskii, A.; Permyakova, E.; Popov, Z.; Dvořáková, E.; Kiryukhantsev-Korneev, P.; Obrosov, A.; Polčák, J.; Zajickova, L.; et al. Immobilization of Platelet-Rich Plasma onto COOH Plasma-Coated PCL Nanofibers Boost Viability and Proliferation of Human Mesenchymal Stem Cells. *Polymers* **2017**, *9*, 736. [CrossRef] [PubMed]
48. Gu, W.; Hong, X.; Le Bras, A.; Nowak, W.; Bhaloo, S.I.; Deng, J.; Xie, Y.; Hu, Y.; Ruan, X.Z.; Xu, Q. Smooth muscle cells differentiated from mesenchymal stem cells are regulated by microRNAs and suitable for vascular tissue grafts. *J. Biol. Chem.* **2018**, *293*, 8089–8102. [CrossRef]
49. Khaki, M.; Salmanian, A.H.; Abtahi, H.; Ganji, A.; Mosayebi, G. Mesenchymal Stem Cells Differentiate to Endothelial Cells Using Recombinant Vascular Endothelial Growth Factor -A. *Rep. Biochem. Mol. Biol.* **2018**, *6*, 144–150.
50. Beckermann, B.M.; Kallifatidis, G.; Groth, A.; Frommhold, D.; Apel, A.; Mattern, J.; Salnikov, A.V.; Moldenhauer, G.; Wagner, W.; Diehlmann, A.; et al. VEGF expression by mesenchymal stem cells contributes to angiogenesis in pancreatic carcinoma. *Br. J. Cancer* **2008**, *99*, 622–631. [CrossRef]
51. Pittenger, M.F.; Discher, D.E.; Péault, B.M.; Phinney, D.G.; Hare, J.M.; Caplan, A.I. Mesenchymal stem cell perspective: Cell biology to clinical progress. *NPJ Regen. Med.* **2019**, *4*, 22. [CrossRef]
52. Scanu, M.; Mancuso, L.; Cao, G. Evaluation of the use of human Mesenchymal Stem Cells for acute toxicity tests. *Toxicol. Vitr.* **2011**, *25*, 1989–1995. [CrossRef]
53. Briggs, T.; Treiser, M.D.; Holmes, P.F.; Kohn, J.; Moghe, P.V.; Arinzeh, T.L. Osteogenic differentiation of human mesenchymal stem cells on poly(ethylene glycol)-variant biomaterials. *J. Biomed. Mater. Res. Part A* **2008**, *91A*, 975–984. [CrossRef] [PubMed]

54. Wang, H.; Feng, Y.; Fang, Z.; Yuan, W.; Khan, M. Co-electrospun blends of PU and PEG as potential biocompatible scaffolds for small-diameter vascular tissue engineering. *Mater. Sci. Eng. C* **2012**, *32*, 2306–2315. [CrossRef]
55. Dowling, D.; Miller, I.; Ardhaoui, M.; Gallagher, W. Effect of Surface Wettability and Topography on the Adhesion of Osteosarcoma Cells on Plasma-modified Polystyrene. *J. Biomater. Appl.* **2010**, *26*, 327–347. [CrossRef] [PubMed]
56. Hao, L.; Yang, H.; Du, C.; Fu, X.; Zhao, N.; Xu, S.; Cui, F.; Mao, C.; Wang, Y. Directing the fate of human and mouse mesenchymal stem cells by hydroxyl–methyl mixed self-assembled monolayers with varying wettability. *J. Mater. Chem. B* **2014**, *2*, 4794–4801. [CrossRef] [PubMed]
57. Sezer, U.A.; Sanko, V.; Gulmez, M.; Aru, B.; Sayman, E.; Aktekin, A.; Aker, F.V.; Demirel, G.Y.; Sezer, S. Polypropylene composite hernia mesh with anti-adhesion layer composed of polycaprolactone and oxidized regenerated cellulose. *Mater. Sci. Eng. C* **2019**, *99*, 1141–1152. [CrossRef]
58. Nolfi, A.L.; Brown, B.N.; Liang, R.; Palcsey, S.L.; Bonidie, M.J.; Abramowitch, S.D.; Moalli, P.A. Host response to synthetic mesh in women with mesh complications. *Am. J. Obstet. Gynecol.* **2016**, *215*, 206.e1–206.e8. [CrossRef]
59. Bakheit, A.H.; Al-Hadiya, B.M.; Abd-Elgalil, A.A. Azithromycin. *Profiles Drug Subst. Excip. Relat. Methodol.* **2014**, *39*, 1–40. [CrossRef]
60. Parnham, M.J.; Haber, V.E.; Giamarellos-Bourboulis, E.J.; Perletti, G.; Verleden, G.M.; Vos, R. Azithromycin: Mechanisms of action and their relevance for clinical applications. *Pharmacol. Ther.* **2014**, *143*, 225–245. [CrossRef]
61. Rowland, K.; Ewigman, B. Azithromycin for PID beats doxycycline on all counts. *J. Fam. Pract.* **2007**, *56*, 1006–1009.
62. Kurnia, S. Biocompatibility of Azitromicyn on Connective Tissue. *Indones. J. Trop. Infect. Dis.* **2015**, *2*, 42–45. [CrossRef]
63. Cosson, M.; Debodinance, P.; Boukerrou, M.; Chauvet, M.P.; Lobry, P.; Ego, A. Mechanical properties of synthetic implants used in the repair of prolapse and urinary incontinence in women: Which is the ideal material? *Int. Urogynecol. J.* **2003**, *14*, 169–178. [CrossRef] [PubMed]
64. Goel, V.K.; Lim, T.-H.; Gwon, J.; Chen, J.-Y.; Winterbottom, J.M.; Park, J.B.; Weinstein, J.N.; Ahn, J.-Y. Effects of Rigidity of an Internal Fixation Device A Comprehensive Biomechanical Investigation. *Spine* **1991**, *16*, S155–S161. [CrossRef] [PubMed]
65. Huiskes, H.R.; Weinans, H.; Grootenboer, H.J.; Dalstra, M.M.; Fudala, B.; Slooff, T.T. Adaptive bone-remodeling theory applied to prosthetic-design analysis. *J. Biomech.* **1987**, *20*, 1135–1150. [CrossRef]
66. Rumian, A.P.; Draper, E.R.C.; Wallace, A.L.; Goodship, A.E. The influence of the mechanical environment on remodelling of the patellar tendon. *J. Bone Jt. Surg. Br. Vol.* **2009**, *91*, 557–564. [CrossRef] [PubMed]
67. Yamamoto, E.; Hayashi, K.; Yamamoto, N. Effects of stress shielding on the transverse mechanical properties of rabbit patellar tendons. *J. Biomech. Eng.* **2000**, *122*, 608–614. [CrossRef] [PubMed]

Article

Additive Manufacturing and Characterization of Metal Particulate Reinforced Polylactic Acid (PLA) Polymer Composites

Ved S. Vakharia [1], Lily Kuentz [2], Anton Salem [3], Michael C. Halbig [4,*], Jonathan A. Salem [4] and Mrityunjay Singh [5,*]

1. NASA Pathway Intern, Department of Mechanical and Aerospace Engineering, University of California, San Diego, CA 92092, USA; ved.vakharia@gmail.com
2. NASA Intern Currently at Department of Geology, University of Oregon, Eugene, OR 97403, USA; lilykuentz@gmail.com
3. NASA Intern Currently at VulcanForms, Inc., Burlington, MA 01803, USA; antonsalem1@gmail.com
4. NASA Glenn Research Center, Cleveland, OH 44135, USA; jonathan.a.salem@nasa.gov
5. Ohio Aerospace Institute, Cleveland, OH 44142, USA
* Correspondence: michael.c.halbig@nasa.gov (M.C.H.); mrityunjaysingh@oai.org (M.S.)

Citation: Vakharia, V.S.; Kuentz, L.; Salem, A.; Halbig, M.C.; Salem, J.A.; Singh, M. Additive Manufacturing and Characterization of Metal Particulate Reinforced Polylactic Acid (PLA) Polymer Composites. *Polymers* **2021**, *13*, 3545. https://doi.org/10.3390/polym13203545

Academic Editor: Matthew Blacklock

Received: 21 September 2021
Accepted: 7 October 2021
Published: 14 October 2021

Publisher's Note: MDPI stays neutral with regard to jurisdictional claims in published maps and institutional affiliations.

Copyright: © 2021 by the authors. Licensee MDPI, Basel, Switzerland. This article is an open access article distributed under the terms and conditions of the Creative Commons Attribution (CC BY) license (https://creativecommons.org/licenses/by/4.0/).

Abstract: Affordable commercial desktop 3-D printers and filaments have introduced additive manufacturing to all disciplines of science and engineering. With rapid innovations in 3-D printing technology and new filament materials, material vendors are offering specialty multifunctional metal-reinforced polymers with unique properties. Studies are necessary to understand the effects of filament composition, metal reinforcements, and print parameters on microstructure and mechanical behavior. In this study, densities, metal vol%, metal cross-sectional area %, and microstructure of various metal-reinforced Polylactic Acid (PLA) filaments were characterized by multiple methods. Comparisons are made between polymer microstructures before and after printing, and the effect of printing on the metal-polymer interface adhesion has been demonstrated. Tensile response and fracture toughness as a function of metal vol% and print height was determined. Tensile and fracture toughness tests show that PLA filaments containing approximately 36 vol% of bronze or copper particles significantly reduce mechanical properties. The mechanical response of PLA with 12 and 18 vol% of magnetic iron and stainless steel particles, respectively, is similar to that of pure PLA with a slight decrease in ultimate tensile strength and fracture toughness. These results show the potential for tailoring the concentration of metal reinforcements to provide multi-functionality without sacrificing mechanical properties.

Keywords: Polylactic Acid (PLA); 3-D printing; polymer composites; multifunctionality; fused filament fabrication; metal-reinforced PLA; mechanical properties

1. Introduction

3-D printing is a fast-evolving technology due to its ease of access and usefulness in various avenues of science and engineering. In contrast to traditional subtractive manufacturing techniques, additive manufacturing creates objects in an additive process through a layer-by-layer joining method. This method presents the benefit of materials savings, as well as new design approaches. The introduction of 3-D printing has pushed industries, such as automotive, aerospace, energy, and medical, to new heights [1,2]. Recent technological advancements center on the implementation of multifunctional materials [3–5]. With the development of new lightweight multifunctional components and systems, there is a need for new 3-D printable materials and robust manufacturing processes.

In just a few years, there have been significant advances in the development of polymer 3-D printing systems ranging from commercially available high-end fused deposition modeling (FDM), also referred to as fused filament fabrication (FFF) machines to desktop 3-D printers. 3-D printing can be advantageous to traditional manufacturing processes depending on manufacturing constraints. 3-D printing technologies allow for far more complex geometries, eliminate the need for tooling, and their versatility introduces printing with varying materials and parameters. However, many low-cost printers can only fabricate materials from acrylonitrile butadiene styrene (ABS) and polylactic acid (PLA) systems. As a result, these are the most studied 3-D printed materials [6–8]; however, low-cost manufacturing could benefit from multifunctional ABS and PLA material production capability.

There is literature available on the PLA material system for 3-D printing. Tymrak et al. [9] measured the tensile strength and elastic modulus of printed components made from low-cost 3-D printers. They found an average tensile strength and elastic modulus for PLA to be 56.6 MPa and 3368 MPa, respectively. A life cycle analysis performed by Wittbrodt et al. and Kreiger et al. [10,11] suggests that 3-D printing as a manufacturing method has a lower environmental impact than that of conventional, subtractive manufacturing for a variety of products. Additive manufacturing can reuse reclaimed materials as a filament in a variety of manufacturing methods [12]. However, there is an informational void for modified PLA systems, particularly those with metallic particle reinforcements. Other reinforcement materials such as synthetic or natural fibers, ceramics, and metals have been explored [13]. Namiki et al. impregnated carbon fiber into PLA filament to take advantage of continuous fiber reinforcement [14]. Several other studies have similarly investigated carbon fiber, and modified carbon fiber reinforced PLA filaments [15–22]. A previous study characterized metal-reinforced PLA filaments, but Fafenrot et al. only explored bronze and magnetic iron reinforced PLA [23]. A very recent study successfully implemented continuous metal fibers (copper and steel wire) into PLA filaments and enhanced the composite's mechanical properties [24]. Liu et al. characterized the mechanical response of metal-based PLA filaments; specifically, copper and aluminum powder. They found that enhanced strength is also possibly dependent on printing properties such as the raster angle [25].

Traditionally, PLA is one of the most commonly used polymers for 3-D printing. Recently, composite filament materials have emerged, such as particulate reinforced PLA. The additions to the PLA aim to improve the multi-functionality by affecting such properties as strength, thermal conductivity, electrical conductivity, magnetic attraction, and radiation shielding. Metal-oxide semiconductors have been manufactured from sintered 3-D printed materials [26], and filaments with TiO_2 and MoO_3 were explored for their antimicrobial and antifungal properties [27]. Multiple groups pursued multifunctionality to try and introduce electroconductive properties to 3-D printed materials [28–30]. However, the specifics of these materials are highly variable. Cost can range from expensive industrial grade to inexpensive pellets (approximately $1/lb) that can be used to create filament. Density, and perhaps most significantly, metal volume percent, can vary from manufacturer to manufacturer. These material details play a vital role in determining the mechanical, thermal, and electrical properties of PLA-based systems and, therefore, govern a product's design.

In this study, PLA was printed using desktop 3-D printers. The microstructure, density, tensile strength, elastic modulus, and fracture toughness of these materials were determined as a function of their print height and metallic composition. This research aims to determine the role that starting filaments and print height play in determining the mechanical properties of 3-D printed materials. Through the thorough characterization and better understanding of the commercially available filaments and the resultant printed material properties, these composite materials can be better tailored to improve specific properties to be used in multifunctional applications.

2. Materials and Methods

This study used commercially available Makerbot Replicator 2X (TYPE, MakerBot, Brooklyn, New York, NY, USA) and Orion Delta 3-D printers (TYPE, SeeMeCNC, Ligonier, IND, USA). These machines were used to manufacture ASTM D 638 [31] tensile coupons (Figure 1), as well as fracture toughness specimens (Figure 2). These specimens were printed with 0.1 mm, 0.2 mm, 0.3 mm, and 0.4 mm layer heights, with an overall thickness between 9.6 and 11.2 mm. In addition to varying the layer height of the printed specimens, the print material was changed. Typical extruder temperatures were 200–220 °C and bed temperatures were 45–60 °C depending on the filaments. Different materials used in this study were pure PLA (Makerbot), bronze-reinforced PLA (colorFabb), copper-reinforced PLA (colorFabb), magnetic iron-reinforced PLA (Proto-pasta), and stainless steel-reinforced PLA (Proto-pasta). These materials will hereby be referred to as PLA, Br-PLA, Cu-PLA, MI-PLA, and SS-PLA, respectively. Three tensile coupons and three fracture toughness specimens of each layer height and material type were printed and tested.

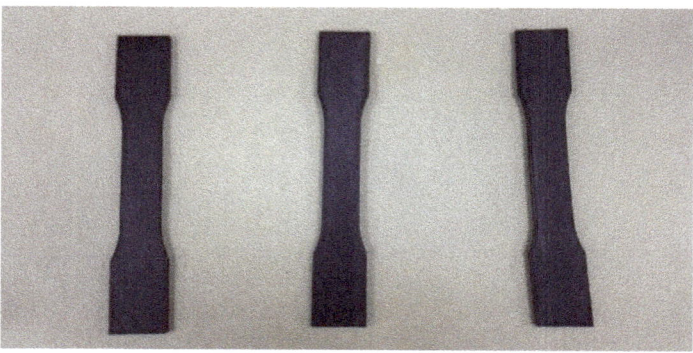

Figure 1. ASTM D 638 Tensile Specimens (6 inches long).

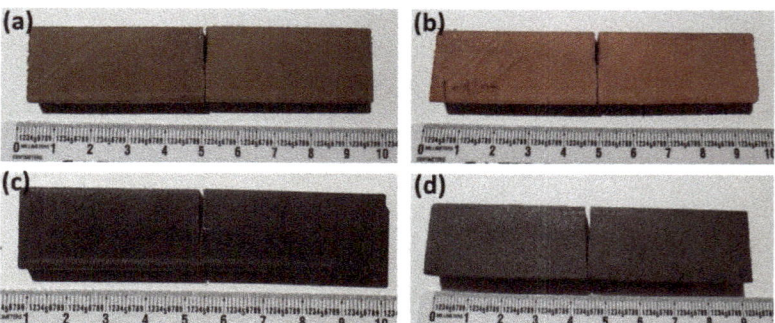

Figure 2. (**a**) Br-PLA, (**b**) Cu-PLA, (**c**), MI-PLA, (**d**) and SS-PLA specimens after undergoing fracture toughness testing.

Prior to testing, the bulk density of each material was measured along with that from the Archimedes buoyancy method. Through thermogravimetric analysis (TGA) of bulk filament pieces, metallic particle weight percentage was obtained, from which metallic volume percentage was calculated. Samples from printed coupons and as-supplied filament were mounted in epoxy and polished to allow for microstructural analysis of the cross-sections using an optical microscope. Measurements of metallic particle area percentage and porosity with the PLA were conducted on the optical micrographs using ImageJ (National Institutes of Health, Bethesda, MD, USA)).

For tensile testing, the strains were determined by using photogrammetry. The elastic moduli of materials were determined in two different manners: dynamically and mechanically. The dynamic testing consisted of impulse excitation in nominal conformance with ASTM C1259 [32]. Mechanical modulus was determined from the tensile data. Generally, dynamic and mechanical results were similar, and the dynamic testing ceased. Fracture toughness was determined according to D5045 [33]. The thickness, width, and crack length of each specimen were approximately 10, 20, and 9 mm, respectively. Some of the thickness or plasticity criteria were not fulfilled; however, additional specimens of some materials tested with a greater thickness (10 mm) resulted in similar values, implying reasonable trends in the data and conclusions.

3. Results and Discussions
3.1. Material Characterization

Figure 3 compares density measurements for all materials used in this study. There is a discrepancy between "in-house" water immersion density measurements and those provided by the filament producer, particularly the 0.26 g/cc and 0.43 g/cc difference present in the Br-PLA and Cu-PLA materials, respectively. The difference between filament producer-provided measurements and those carried out in this study is at least 0.1 g/cc for every material. It is critical to have an accurate density of source materials before designing a 3-D printed structure, and therefore the values need to be assessed and confirmed. This information can be found by simply characterizing material before use and then designing structures accordingly.

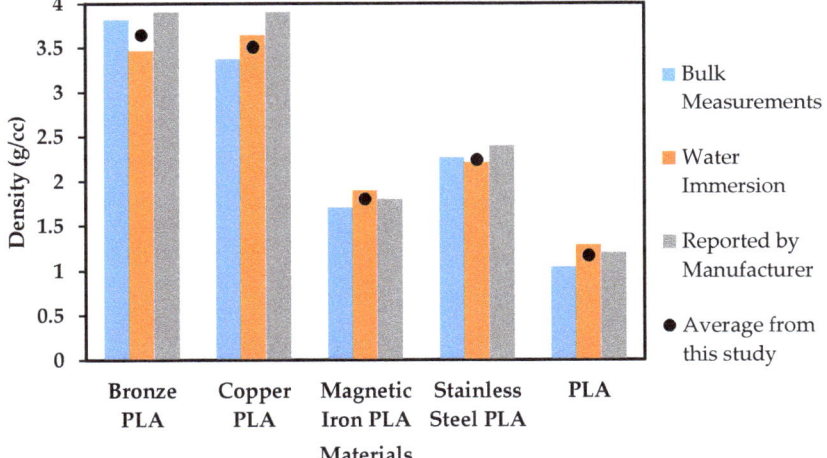

Figure 3. Filament densities obtained from multiple methods and reported by filament manufacturers.

Figure 4 shows the TGA results, performed in air, of the materials used in this study. By approximately 400 °C, the PLA has burned away, and the remaining weight can be associated with the metallic particle reinforcements. For example, approximately 80 wt% of bronze and copper are remaining once the PLA is removed from each sample at 400 °C. The metal wt% and associated vol% are displayed in Table 1.

Table 1. Metal wt% and vol% of materials obtained from TGA of as-received filament.

Material	Metal wt%	Metal vol%
Br-PLA	80.35%	36.02%
Cu-PLA	80.57%	36.41%
MI-PLA	48.33%	11.05%
SS-PLA	58.87%	18.09%
PLA	0.00%	0.00%

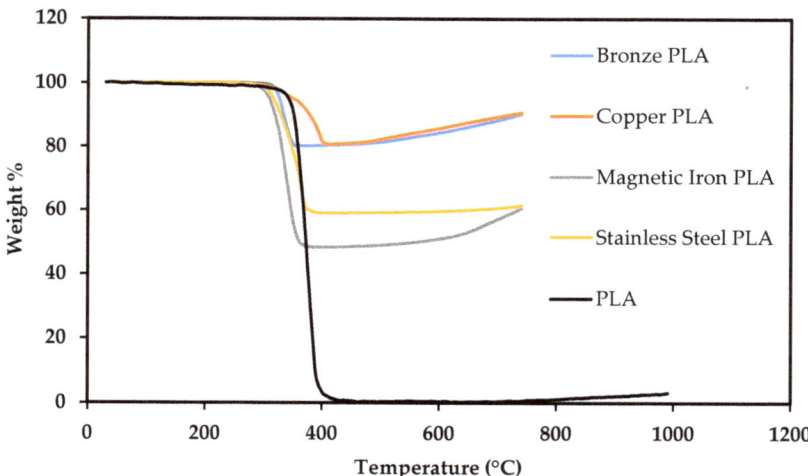

Figure 4. TGA results of materials conducted in air.

Fafenrot et al. conducted a similar study on Br-PLA and MI-PLA filaments that were obtained from the same sources as in this study. They observed very similar metal vol% of 36 and 12, respectively. Their calculations were also conducted via TGA measurements [23].

Br-PLA and Cu-PLA exhibit a much higher metal vol% than MI-PLA or SS-PLA. Direct comparisons of mechanical strength between these materials cannot be conducted from these measurements, in part due to density difference and varying reinforcement vol% values. It can be rationally predicted that material with 36 vol% copper is far less likely to behave as pure PLA than a material with only 11% magnetic iron. The volume of metal reinforcement should significantly affect mechanical properties in comparison to pure PLA. The multifunctionality of material is significantly dependent on its composition and, in this case, the amount of metal reinforcement in PLA. An objective of this research is to determine if multifunctionality may be tailored to a specific purpose if this metal vol% can be controlled.

Figure 5a shows micrographs for Br-PLA, Cu-PLA, MI-PLA, and SS-PLA filament (before printing), and Figure 5b shows micrographs of the same materials after printing. The bright, white regions in the micrographs are the metallic particles, with the surrounding material being PLA. In all the micrographs in Figure 5a,b, except perhaps the Br-PLA filament prior to printing (Figure 5a), there are dark spots that have shapes resembling those of the metallic particles. Due to the size and shape similarities between them, we hypothesize these dark spots are created from metallic particles pulling out of the PLA rather than the porosity created during printing. This most likely occurred during the polishing of the mounted samples. The abrasive surface of the polishing pads caught onto the edges of the hard metal particles sticking out above the soft PLA, and pulled them out. Therefore, area percentages of metallic particles were obtained by combining the approximate amounts of bright and dark spots as shown in Figure 6. The area percentages determined from the micrographs from Figure 5a,b can be seen in Figure 7 and Table 2.

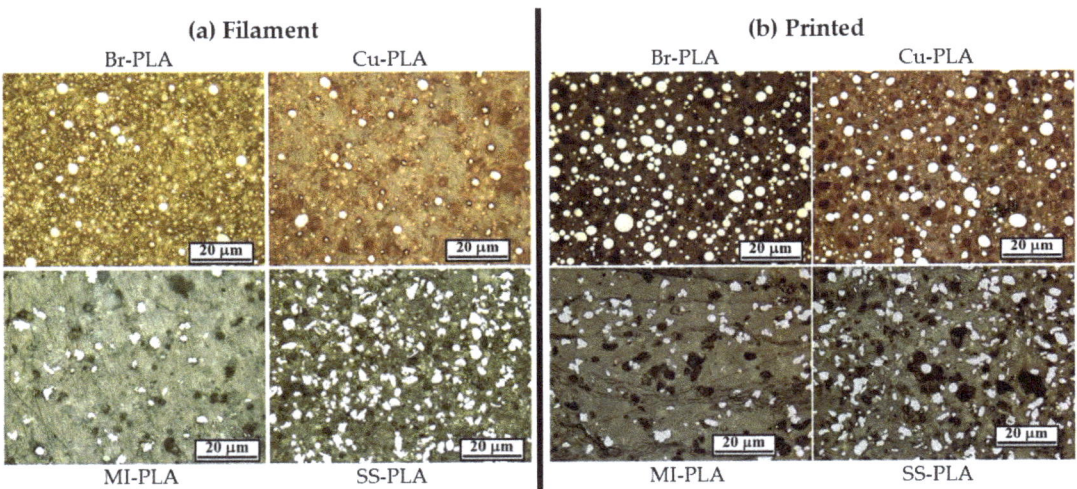

Figure 5. Micrographs of Br-PLA, Cu-PLA, MI-PLA, and SS-PLA (**a**) filaments before printing, and (**b**) parts after printing.

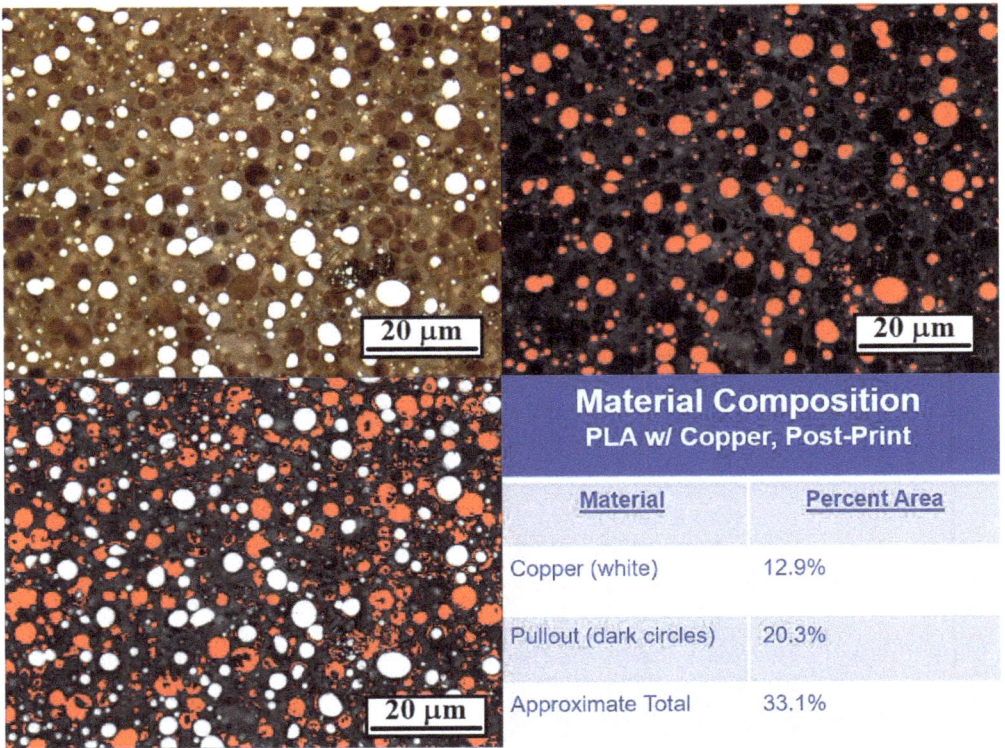

Figure 6. Area% calculation of Cu-PLA (post-print) where bright white spots (metal particles) are added with dark spots (pullout locations) to obtain total area% particles (performed with ImageJ).

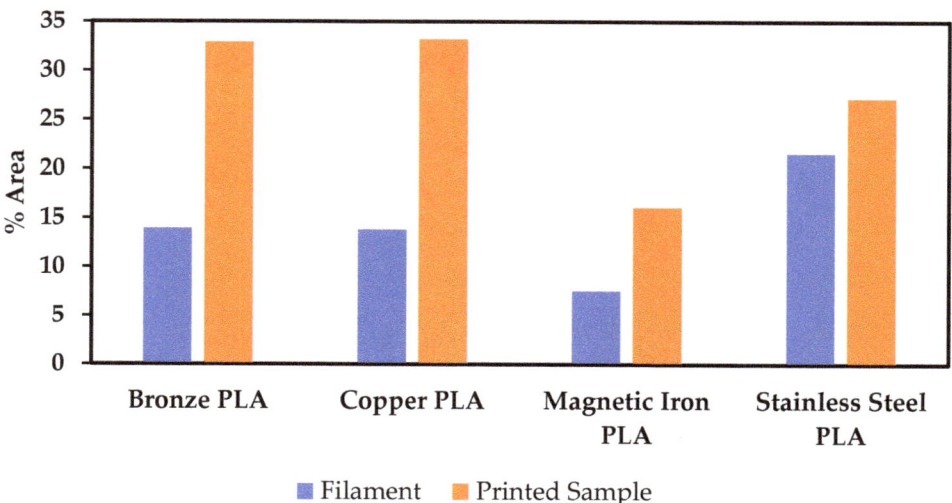

Figure 7. The measured area percent of Br-PLA, Cu-PLA, MI-PLA, and SS-PLA before and after printing.

Table 2. The measured area percent of Bronze (Br), Copper (Cu), Magnetic Iron (MI), and Stainless Steel (SS) in reinforced PLA filament and printed samples.

Material	Metal Area%	Pullout Area%	Total Area%
Br-Filament	13.9%	N/A	13.9%
Br-Printed	20.4%	12.5%	32.9%
Cu-Filament	3.2%	10.6%	13.8%
Cu-Printed	12.9%	20.3%	33.2%
MI-Filament	3.4%	4.1%	7.5%
MI-Printed	8.1%	8.0%	16.1%
SS-Filament	19.4%	2.1%	21.5%
SS-Printed	14.2%	13.0%	27.2%

The cross-sectional area % measured (using ImageJ) for particles in the printed materials in Figure 5b is similar to the vol% of metallic particles in these filaments as obtained through TGA experiments; however, that is not the case for results obtained from analyses of the filament micrographs displayed in Figure 5a. Since TGA experiments were conducted on the as-received filament, it should be expected that the metallic particle area % of the filament micrographs (when including the pullout amounts) should be similar to the vol% obtained from TGA measurements.

The possibility that gravity effects during the printing of the bulk samples could cause the discrepancies between the metal area % determined from the micrographs and the vol% calculated from the TGA experiments was considered. These samples were created with commercially available desktop 3-D printers that were not designed to print polymer filament reinforced with metallic particles. As the filament is heated at the printer head, the decreased viscosity may allow for metal particles just above the heated region to migrate down to the print head and be dispersed into the more viscous regions of the filament. In other words, the structure maintaining the position of these metal particles loses its strength, allowing for the metal to migrate in the direction forced by gravity. However, the gravity effect is most likely occurring within a printed layer height, which should only affect the location and not the amount of detectable reinforcement. All printed layers are within the polished plane in the micrograph, and therefore the metallic particles should be visible. There are also certain issues in changing the nozzle diameter due to abrasion as well as sticking of some particulates in the nozzle orifice. These factors should be evaluated carefully in future investigations.

In addition, the printing procedure may cause the plastic filament to shrink, causing the metallic particles to occupy a higher area % than they did before printing. However, PLA has proven to have minor warping or shrinking issues upon cooling, unlike ABS, which warps drastically after printing [34,35].

There seems to be significantly more metal reinforcement in printed samples. The most likely explanation of why filament cross-sections show less area % than printed parts is that the printing process has a beneficial effect on the adhesion between the metallic reinforcement and the PLA. The mounting and polishing procedure to prepare polished sections was consistent between filaments and printed samples, yet less reinforcement is detected in the filament. The PLA was exposed to a different temperature and cooling time in each print, possibly affecting the PLA interface around the particles.

Gupta et al. performed tribological studies on printed samples of the same materials used in this study and found that the presence of metallic reinforcements, regardless of the metal, decreases the wear rate and coefficient observed in PLA [36]. Studies weren't conducted to compare the tribological behavior between the filament and the printed sample of a given material. However, with the increase in pullout observed in this study, we expect the wear rate to be higher in filaments than in printed samples. The micrographs in Figure 5a are blurrier than those in Figure 5b due to the polishing finish, and this limits the ability to recognize all of the pullout locations of the metallic particles. That could also be an effect of fewer particles being present, resulting in a smoothing of the matrix. Therefore, a different method was used to estimate the amount of pullout in the polished filament samples. If it is assumed that the reinforcement amount is correctly identified in the printed samples (taking into account observed pullout) and through the TGA experiments, then we can calculate how much reinforcement was pulled out of the filaments during the polishing procedure. That data is shown in Table 3 and Figure 8. Except for SS-PLA, there was less pullout detected in the printed samples. To obtain the best accuracy in future assessments of particulate vol%, multiple images of each sample will be assessed in order to provide a larger database. In addition, performing TGA on the printed material should provide optimal information about the metallic particle content.

Table 3. Area % of total reinforcement observed in filament and printed micrographs. Pullout amounts in filament micrographs are estimated by subtracting the detected metallic particles from the total (metal + pullout) of their printed counterparts.

Material	Metal Area%	Pullout Area%	Total Area%	Metal Vol% (TGA)
Br-Filament	13.9%	19.0% *		36.02%
Br-Printed	20.4%	12.5%	32.9%	
Cu-Filament	3.2%	30.0% *		36.41%
Cu-Printed	12.9%	20.3%	33.2%	
MI-Filament	3.4%	12.7% *		11.05%
MI-Printed	8.1%	8.0%	16.1%	
SS-Filament	19.4%	7.8% *		18.09%
SS-Printed	14.2%	13.0%	27.2%	

* Calculated

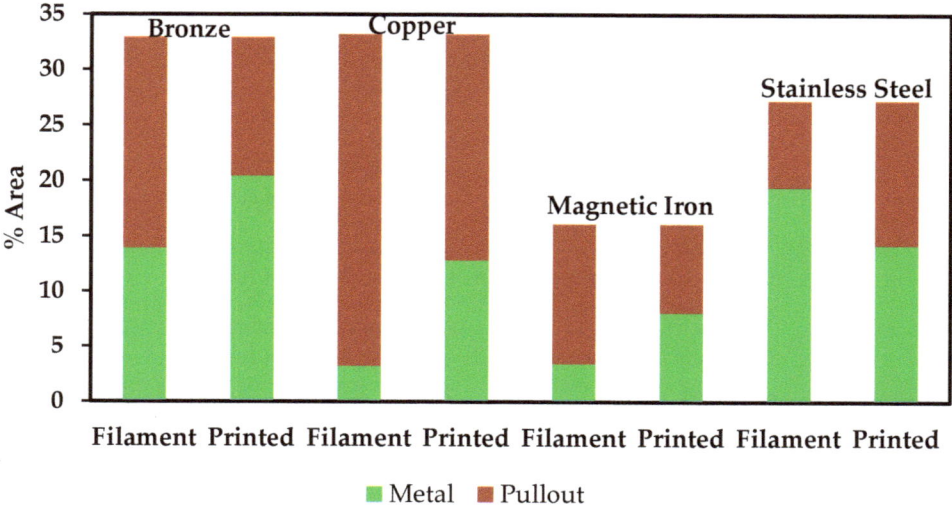

Figure 8. A visualization of the data presented in Table 3 showing the area percentage of total reinforcement observed in filament and printed micrographs.

This pullout observation supports the hypothesis that PLA-metal interface adhesion may be weak, especially before undergoing a heating and cooling procedure via printing, and should be optimized. It is disconcerting that a simple polishing procedure easily removes at least half of the metallic particles from printed samples. The weak PLA-metal interface adhesion does not provide promise for the use of these materials on exposed surfaces that may experience friction with their surroundings, hindering the design process for these materials. As the filament is heated and printed, it may be unavoidable for the metallic particles to move around in the PLA. However, the adhesion at the PLA-metal interface can be improved. Optimization of print temperature, extrusion rate, raster angle, and bed temperature may improve interface adhesion.

3.2. Mechanical Properties

Figure 9a–d show the data obtained from tensile testing for PLA samples printed with 0.1, 0.2, 0.3, and 0.4 mm layers, respectively. Three samples of each material were tested. One test of PLA printed with a print layer height of 0.2 mm failed prematurely due to a poor experimental setup. The overall results are consistent and provide confidence in the repeatability of the printing process. Strain to failure is variable; however, this is common for polymeric materials. Figure 10a shows the average tensile response of each of the four PLA print heights. The Young's modulus of each material was calculated from the slope of the linear, elastic portion of the tensile response. The highest stress withstood by the material is taken as the ultimate strength. Figure 10b shows Young's moduli and ultimate tensile strengths of the same samples. A slight decrease in modulus with print layer height is exhibited; however, little effect is seen on ultimate strength. PLA behaves consistently, regardless of print height. Therefore, metal-reinforced PLA can be compared to PLA printed at any of these four print heights. This consistency, regardless of print height, is not the case for all standard 3-D printed materials. For example, a previous study showed that ABS layer height plays a critical role in determining the 3-D printed polymer's tensile performance [37]. Once again, material choice proves to be a governing parameter when designing 3-D printed structures.

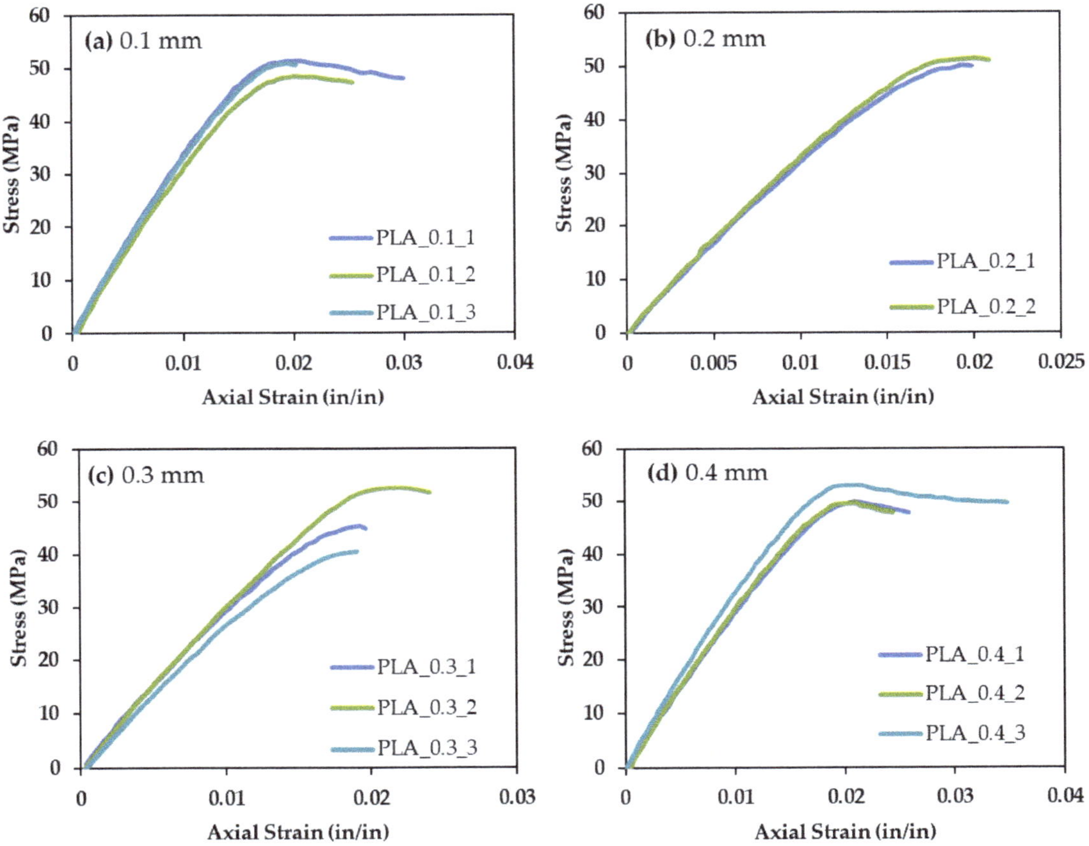

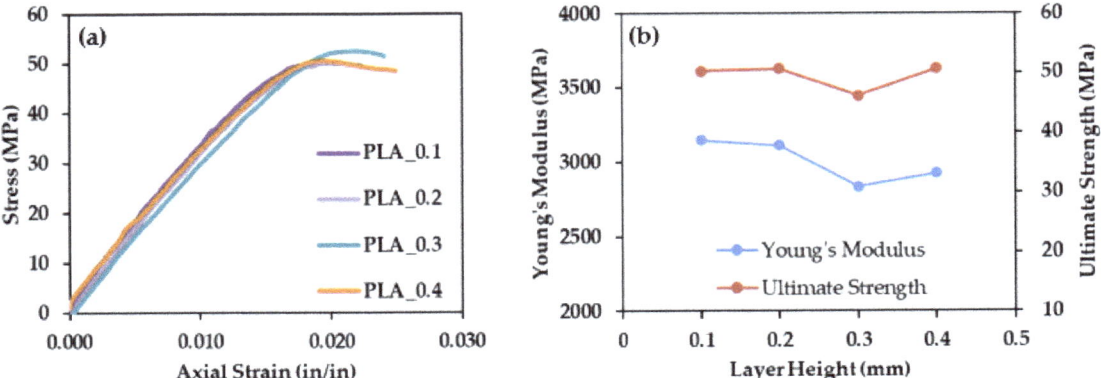

Figure 9. Tensile data for PLA with a print layer height of (a) 0.1 mm, (b) 0.2 mm, (c) 0.3 mm, and (d) 0.4 mm.

Figure 10. (a) Average tensile response and (b) Young's modulus and Ultimate Strength for PLA printed at 0.1, 0.2, 0.3, and 0.4 mm print layer heights.

Figure 11 shows the average tensile response for metal-reinforced PLA samples printed at 0.3 mm print height compared to PLA samples printed at 0.3 mm print height. All the materials exhibit significant deformation prior to failure, with the Br-PLA and Cu-PLA exhibiting the largest strains (>3%). All metal-filled materials are weaker than pure PLA.

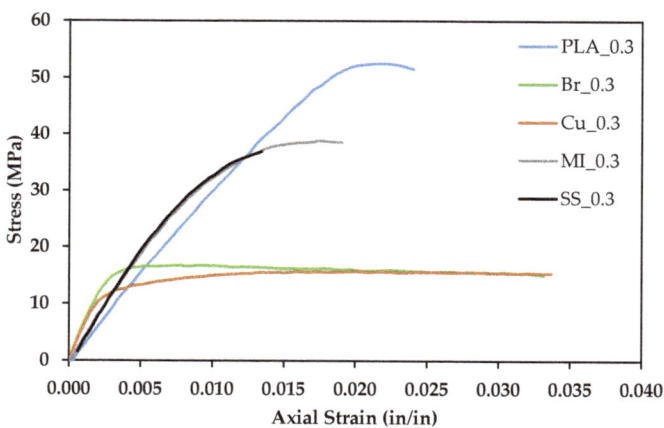

Figure 11. Average tensile response for PLA and metal reinforced PLA at 0.3 mm print layer height.

Fracture toughness was calculated with Equation (1),

$$K_Q = \left(\frac{6P_Q\sqrt{a}}{BW}\right)\frac{1.99 - \frac{a}{W}\left(1 - \frac{a}{W}\right)\left(2.15 - 3.93\frac{a}{W} + 2.7\left(\frac{a}{W}\right)^2\right)}{\left(1 + 2\frac{a}{W}\right)\left(1 - \frac{a}{W}\right)^{\frac{3}{2}}}, \quad (1)$$

where P_Q is the applied load, B is the thickness of the specimen, a is the crack length, and W is the width of the specimen. An example of the data obtained from the three-point flexural tests used to calculate fracture toughness is shown in Figure 12. Figure 13a–e show fracture surfaces for all materials printed with a 0.3 mm print layer height, with Figure 13f showing the fracture surface of Br-PLA at a higher magnification. Print layers in each material are clearly visible, but only pure PLA is free of porosity between layers. Figure 13g displays the average fracture toughness of the materials investigated in this study. All images and data shown in Figure 13 are obtained from prints utilizing a 0.3 mm print layer height.

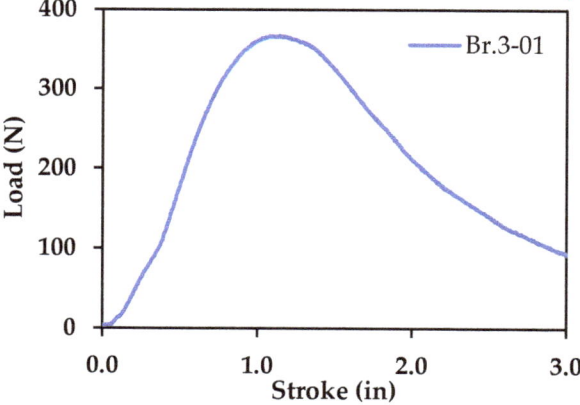

Figure 12. Data obtained from a three-point flexural test on Br-PLA.

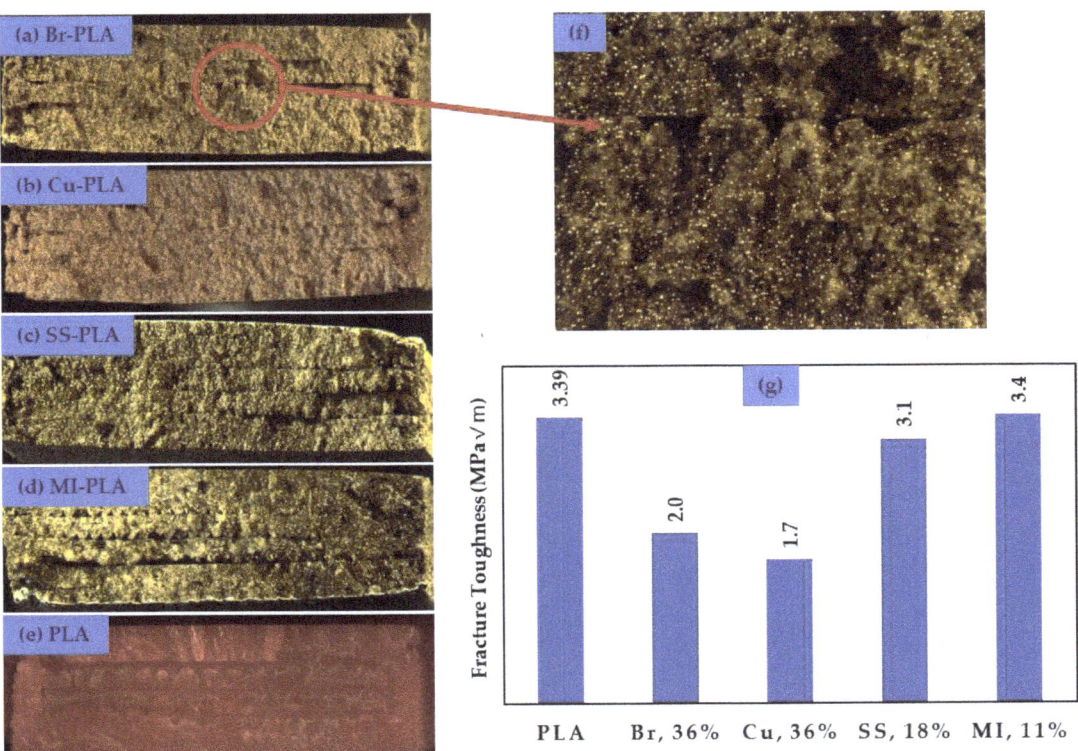

Figure 13. Fracture surfaces of (**a**) Br-PLA, (**b**) Cu-PLA, (**c**) SS-PLA, (**d**) MI-PLA, and (**e**) pure PLA, printed with a 0.3 mm print layer height. Br-PLA is shown at higher magnification in (**f**). (**g**) Corresponding fracture toughness values for materials (printed at a 0.3 mm layer height).

In all mechanical tests, SS-PLA and MI-PLA performed nearly as well as pure PLA. These three material compositions show similar tensile responses, but the presence of metallic particles causes a lower ultimate strength and earlier onset of plastic deformation. In contrast, there is a significant and noticeable change in mechanical behavior in Br-PLA and Cu-PLA. These two material compositions have much lower ultimate strength and fracture toughness values. Large additions of metal particles in the Br- and Cu-PLA may have resulted in lower fracture toughness due to the pull-out of the particles from the PLA.

The Br-PLA and Cu-PLA filaments contained approximately 36 metal vol%, while the MI-PLA and SS-PLA only contained 11 and 18 metal vol%, respectively. The high vol% addition of bronze and copper plays a vital role in causing this change in mechanical response. The significant presence of bronze and copper transforms PLA into a significantly more ductile structural material at a high cost of reduced ultimate strength and fracture toughness. They begin to deform plastically immediately after a strain is applied. These two materials show potential for novel 3-D printed structural design if the goal is for a more resilient (yet weaker) material.

However, a more beneficial scenario would be if the presence of metal additions only introduced new optical, magnetic, electrical, or thermal properties without affecting the mechanical properties of the polymer. Interestingly, the presence of magnetic iron and stainless steel do not have drastic effects on the tensile performance of PLA. There is a slight increase in the Young's modulus of MI-PLA and SS-PLA. However, since the ultimate strength decreases with metal addition, the increased stiffness is not necessarily a positive change in tensile response. The presence of magnetic iron or stainless steel does not drastically affect the tensile response of PLA due to its small concentration. Behavior that is more similar to pure PLA is beneficial because it allows the introduction of multifunctionality without sacrificing mechanical strength or toughness. Perhaps a lower vol% of bronze or copper in the PLA would result in a similar tensile response to MI-PLA and SS-PLA filaments.

Table 4 shows the Young's modulus, ultimate strength, Poisson's ratio, and fracture toughness for each material. In every case, metallic particle reinforcement increases the Young's modulus, leading to a slightly stiffer material and a decrease in ultimate strength and fracture toughness. Fafenrot et al. observed similar tensile responses in their Br-PLA and MI-PLA filaments. Their data shows Br-PLA deforming plastically almost immediately after the strain is applied, while MI-PLA behaves similar to pure PLA but with a slightly earlier onset of plastic deformation. Both trends agree with those shown in this study. They also measured ultimate tensile strength for Br-PLA, MI-reinforced, and pure PLA processed using varying temperatures and nozzle diameters. Their data did vary with changing print parameters but did not stray much from an average value. Their values for ultimate tensile strength ranged between 12 and 20 MPa for Br-PLA, 35 and 40 MPa for MI-PLA, and 45 and 50 MPa for pure PLA filaments [22]. These values and trends are similar to those found in this study as well.

Table 4. Average Young's modulus (MPa), ultimate strength (MPa), Poisson's ratio, and fracture toughness (MPa$\sqrt{m}$) for all materials used in this study.

Material	Young's Modulus (MPa)	Ultimate Strength (MPa)	Poisson's Ratio	Fracture Toughness (MPa$\sqrt{m}$)
Br-PLA (0.3 mm)	5401	17	0.29	2.0
Cu-PLA (0.3 mm)	4496	16	0.25	1.7
MI-PLA (0.3 mm)	3638	39	0.32	3.4
SS-PLA (0.3 mm)	3910	38	0.33	3.1
PLA (0.1 mm)	3145	50	0.33	N/A
PLA (0.2 mm)	3110	51	0.34	N/A
PLA (0.3 mm)	2835	46	0.33	3.39
PLA (0.4 mm)	2925	51	0.34	N/A

4. Conclusions

There is potential to use metal-reinforced PLA as a material when designing novel 3-D printed structures. Though they may exhibit a slightly lower tensile strength and fracture toughness than pure PLA, this difference can be considered an acceptable sacrifice for introducing multifunctionality to 3-D printed materials. This study shows that small amounts of metal addition in the form of particulates can introduce multifunctionality to a polymer filament without drastically changing its tensile properties.

In this study, comparisons between polymer microstructures before and after printing and the effect of printing on the metal-polymer interface adhesion have been demonstrated. A more significant number of metallic particles were observed in the printed PLA samples than in the pre-print filament. The disparity is possibly due to a weak metal-polymer interfacial adhesion allowing metal particles to be removed from the PLA substrate during polishing. Polishing can represent a situation in which the material surface experiences friction with its surroundings. It is detrimental for the printed materials to have their reinforcements so easily removed. The lack of metal to PLA adhesion introduces some variability to material properties, where designs are made assuming that most, if not all, metallic particles remain bonded to its polymer substrate. Perhaps this inconsistency can be avoided by using a more robust 3-D printer instead of the readily available desktop printers used in this study. However, a material with excellent metal-polymer interfacial adhesion would not present this problem. Further studies optimizing the print temperature, extrusion rate, and bed temperature can bring more insight into this metal-polymer interface.

PLA shows consistent tensile response, regardless of print layer height. Print heights of 0.1, 0.2, 0.3, and 0.4 mm layers all performed similarly in mechanical testing. Since there are unequal metal vol% amounts, a direct comparison between the metal-reinforced PLA materials could not be made. However, the large amount of bronze and copper additions caused the PLA to deform plastically almost immediately as the strain was introduced to the specimens. The high concentration of metallic particles also significantly lowered the fracture toughness of these samples. The low amounts of magnetic iron and stainless-steel addition to PLA prevented a drastic change from the mechanical response of pure PLA. Regardless of the metal, the presence of these additions caused the PLA to exhibit a lower ultimate strength and fracture toughness. As a result of a poor metal-polymer interface adhesion, the metallic particulates introduce significant porosity into the composite, which may be the prevailing reason for the poor mechanical performance. Further studies into composite printing parameters may reveal how to maximize the metal-polymer interface adhesion, minimize the porosity, and increase mechanical properties of pure PLA or better. In order to confirm the presence of significant multifunctionality, a wide variety of mechanical, thermal, magnetic, and electrical tests and other characterization need to be performed on these materials.

Author Contributions: Conceptualization, M.S., and M.C.H.; methodology and 3-D printing, L.K., and A.S.; mechanical testing and characterization, J.A.S., A.S., L.K., and V.S.V.; data analysis, V.S.V., M.C.H., J.A.S., and M.S.; writing—original draft preparation, V.S.V., M.C.H., and M.S.; writing—review and editing, V.S.V., J.A.S., M.S., and M.C.H.; supervision, M.C.H., J.A.S., and M.S., funding acquisition, J.A.S., and M.C.H. All authors have read and agreed to the published version of the manuscript.

Funding: This research was supported by the Lewis' Educational and Research Collaborative Internship Project (LERCIP), the NASA Pathways Program, and the Revolutionary Vertical Lift Technology (RVLT) Project.

Institutional Review Board Statement: Not applicable.

Informed Consent Statement: Not applicable.

Data Availability Statement: Not applicable.

Acknowledgments: The authors would like to thank the NASA Internship Project (MIP), previously known as the Lewis' Educational and Research Collaborative Internship Project (LERCIP), for the support provided to Lily Kuentz and Anton Salem and the NASA Pathways Program for the support provided to Ved Vakharia.

Conflicts of Interest: The authors declare no conflict of interest. The funders had no role in the design of the study; in the collection, analyses, or interpretation of data; in the writing of the manuscript, or in the decision to publish the results.

References

1. Liu, Z.; Wang, Y.; Wu, B.; Cui, C.; Guo, Y.; Yan, C. A critical review of fused deposition modeling 3-D printing technology in manufacturing polylactic acid parts. *Int. J. Adv. Manuf. Technol.* **2019**, *102*, 2877–2889. [CrossRef]
2. Mohammed, J.S. Applications of 3-D printing technologies in oceanography. *Methods Oceanogr.* **2016**, *17*, 97–117. [CrossRef]
3. Balasubramanian, K.; Tirumali, M.; Badhe, Y.; Mahajan, R.Y. Nano-enabled Multifunctional Materials for Aerospace Applications. In *Aerospace Materials and Material Technologies*; Prasad, N., Wanhill, P., Eds.; Springer: Singapore, 2016; pp. 439–453.
4. Baur, J.; Silverman, E. Challenges and opportunities in multifunctional nanocomposite structures for aerospace applications. *MRS Bull.* **2011**, *32*, 328–334. [CrossRef]
5. Gibson, R.F. A review of recent research on mechanics of multifunctional composite materials and structures. *Compos. Struct.* **2010**, *92*, 2793–2810. [CrossRef]
6. Daminabo; C.S.; Goel, S.; Grammatikos, S.A.; Nezhad, H.Y.; Thakur, V.K. Fused deposition modeling-based additive manufacturing (3-D printing): Techniques for polymer material systems. *Mater. Today Chem.* **2020**, *16*, 100248. [CrossRef]
7. Solomon, I.J.; Sevvel, P.; Gunasekaran, J. A Review on the Various Processing Parameters in FDM. *Mater. Today Proc.* **2021**, *37*, 509–514. [CrossRef]
8. Mohd Pu'ad, A.S.; Abdul Haq, R.H.; Mohd Noh, H.; Abdullah, H.Z.; Idris, M.I.; Lee, T.C. Review on the Fabrication of Fused Deposition Modelling (FDM) Composite Filament for Biomedical Applications. *Mater. Today: Proc.* **2020**, *29*, 228–232. [CrossRef]

9. Tymrak, B.M.; Kreiger, M.; Pearce, J.M. Mechanical properties of components fabricated with open-source 3-D printers under realistic environmental conditions. *Mater. Des.* **2014**, *58*, 242–246. [CrossRef]
10. Wittbrodt, B.T.; Glover, A.G.; Laureto, J.; Anzalone, G.C.; Oppliger, D.; Irwin, J.L.; Pearce, J.M. Life-cycle economic analysis of distributed manufacturing with open-source 3-D printers. *Mechatronics* **2013**, *23*, 713–726. [CrossRef]
11. Kreiger, M.; Pearce, J.M. Environmental life cycle analysis of distributed 3-D printing and conventional manufacturing of polymer products. *ACS Sustain. Chem. Eng.* **2013**, *1*, 1511–1519. [CrossRef]
12. Despeisse, M.; Baumers, M.; Brown, P.; Charnley, F.; Ford, S.J.; Garmulewicz, A.; Knowles, S.; Minshall, T.H.W.; Mortara, L.; Reed-Tsochas, F.P.; et al. Unlocking value for a circular economy through 3-D printing: A research agenda. *Technol. Forecast. Soc. Chang.* **2017**, *115*, 75–84. [CrossRef]
13. Montjovent, M.O.; Mark, S.; Mathieu, L.; Scaletta, C.; Scherberich, A.; Delabarde, C.; Zambelli, P.Y.; Bourban, P.E.; Apple-gate, L.A.; Pioletti, D.P. Human fetal bone cells associated with ceramic reinforced PLA scaffolds for tissue engineering. *Bone* **2008**, *42*, 554–564. [CrossRef] [PubMed]
14. Namiki, M.; Ueda, M.; Todoroki, A.; Hirano, Y.; Matsuzaki, R. 3-D Printing of Continuous Fiber Reinforced Plastic. In Proceedings of the SAMPE Seattle 2014 International Conference and Exhibition, Seattle, DC, USA, 6 February 2014.
15. Li, N.; Li, Y.; Liu, S. Rapid prototyping of continuous carbon fiber reinforced polylactic acid composites by 3-D printing. *J. Mater. Process. Tech.* **2016**, *238*, 218–225. [CrossRef]
16. Bettini, P.; Alitta, G.; Sala, G.; Di Landro, L. Fused deposition technique for continuous fiber reinforced thermoplastic. *J. Mater. Eng. Perform.* **2017**, *26*, 843–848. [CrossRef]
17. Dickson, A.N.; Barry, J.N.; McDonnell, K.A.; Dowling, D.P. Fabrication of continuous carbon, glass and Kevlar fibre reinforced polymer composites using additive manufacturing. *Addit. Manuf.* **2017**, *16*, 146–152. [CrossRef]
18. Tian, X.; Liu, T.; Wang, Q.; Dilmurat, A.; Li, D.; Ziegmann, G. Recycling and remanufacturing of 3-D printed continuous carbon fiber reinforced PLA composites. *J. Clean. Prod.* **2017**, *142*, 1609–1618. [CrossRef]
19. Yao, X.; Luan, C.; Zhang, D.; Lan, L.; Fu, J. Evaluation of carbon fiber-embedded 3-D printed structures for strengthening and structural-health monitoring. *Mater. Des.* **2017**, *114*, 424–432. [CrossRef]
20. Matsuzaki, R.; Ueda, M.; Namiki, M.; Jeong, T.K.; Asahara, H.; Horiguchi, K.; Nakamura, T.; Todoroki, A.; Hirano, Y. Three-dimensional printing of continuous-fiber composites by in-nozzle impregnation. *Sci. Rep.* **2016**, *6*, 23058. [CrossRef] [PubMed]
21. Melenka, G.W.; Cheung, B.K.; Schofield, J.S.; Dawson, M.R.; Carey, J.P. Evaluation and prediction of the tensile properties of continuous fiber-reinforced 3-D printed structures. *Compos. Struct.* **2016**, *153*, 866–875. [CrossRef]
22. Parandoush, P.; Tucker, L.; Zhou, C.; Lin, D. Laser assisted additive manufacturing of continuous fiber reinforced thermoplastic composites. *Mater. Des.* **2017**, *131*, 186–195. [CrossRef]
23. Fafenrot, S.; Grimmelsmann, N.; Wortmann, M.; Ehrmann, A. Three-dimensional (3-D) printing of polymer-metal hybrid materials by fused deposition modeling. *Materials* **2017**, *10*, 1199. [CrossRef]
24. Hamidi, A.; Tadesse, Y. Single step 3-D printing of bioinspired structures via metal-reinforced thermoplastic and highly stretchable elastomer. *Compos. Struct.* **2019**, *210*, 250–261. [CrossRef]
25. Liu, Z.; Lei, Q.; Xing, S. Mechanical characteristics of wood, ceramic, metal and carbon fiber-based PLA composites fabricated by FDM. *J. Mater. Res. Tech.* **2019**, *8*, 3741–3751. [CrossRef]
26. Salea, A.; Prathumwan, R.; Junpha, J.; Subannajui, K. Metal oxide semiconductor 3-D printing: Preparation of copper(ii) oxide by fused deposition modelling for multi-functional semiconducting applications. *J. Mater. Chem. C* **2017**, *5*, 4614–4620. [CrossRef]
27. Horst, D.J.; Tebcherani, S.M.; Kubaski, E.T.; de Almeida Vieira, R. Bioactive potential of 3-D-printed oleo-gum-resin disks: B. papyrifera, C. myrrha, and S. benzoin loading nanooxides—TiO_2, P25, Cu_2O, and MoO_3. Hindawi Bioinorg. *Chem. Appl.* **2017**, 6398167.
28. Ahn, B.Y.; Walker, S.B.; Slimmer, S.C.; Russo, A.; Gupta, A.; Kranz, S.; Duoss, E.B.; Malkowski, T.F.; Lewis, J.A. Planar and three-dimensional printing of conductive inks. *J. Vis. Exp.* **2011**, *58*, e3189. [CrossRef] [PubMed]
29. Russo, A.; Ahn, B.Y.; Adams, J.J.; Duoss, E.B.; Bernhard, J.T.; Lewis, J.A. Pen-on-paper flexible electronics. *Adv. Mater.* **2011**, *23*, 3426–3430. [CrossRef] [PubMed]
30. Grimmelsmann, N.; Martens, Y.; Schäl, P.; Meissner, H.; Ehrmann, A. Mechanical and electrical contacting of electronic components on textiles by 3-D printing. *Procedia Technol.* **2016**, *26*, 66–71. [CrossRef]
31. ASTM D638-14. *Standard Test Method for Tensile Properties of Plastics*; ASTM International: West Conshohocken, PA, USA, 2014.
32. ASTM C1259-15. *Standard Test Method for Dynamic Young's Modulus, Shear Modulus, and Poisson's Ratio for Advanced Ceramics by Impulse Excitation of Vibration*; ASTM International: West Conshohocken, PA, USA, 2015.
33. ASTM D5045-14. *Standard Test Methods for Plane-Strain Fracture Toughness and Strain Energy Release Rate of Plastic Materials*; ASTM International: West Conshohocken, PA, USA, 2014.
34. Kamran, M.; Saxena, A. A comprehensive study on 3-D printing technology. *MIT Inter. J. Mech. Eng.* **2016**, *6*, 63–69.
35. Spoerk, M.; Holzer, C.; Gonzalez-Gutierrez, J. Material extrusion-based additive manufacturing of polypropylene: A review on how to improve dimensional inaccuracy and warpage. *J. Appl. Polym. Sci.* **2019**, *137*, 48545. [CrossRef]
36. Gupta, S.; Halbig, M.C.; Singh, M. Particulate reinforced polymer matrix composites: Tribological behavior and 3-D printing by fused filament fabrication. *Adv. Mater. Process.* **2020**, *178*, 12–18.
37. Salem, A.; Singh, M.; Halbig, M.C. 3-D printing and characterization of polymer composites with different reinforcements. In *Advanced Processing and Manufacturing Technologies for Nanostructured and Multifunctional Materials II: A Collection of Papers Presented*

at the 39th International Conference on Advanced Ceramics and Composites; Ohji, T., Singh, M., Halbig, M., Eds.; John Wiley & Sons, Inc.: Hoboken, NJ, USA, 2015; pp. 113–122.

MDPI
St. Alban-Anlage 66
4052 Basel
Switzerland
www.mdpi.com

Polymers Editorial Office
E-mail: polymers@mdpi.com
www.mdpi.com/journal/polymers

Disclaimer/Publisher's Note: The statements, opinions and data contained in all publications are solely those of the individual author(s) and contributor(s) and not of MDPI and/or the editor(s). MDPI and/or the editor(s) disclaim responsibility for any injury to people or property resulting from any ideas, methods, instructions or products referred to in the content.

www.ingramcontent.com/pod-product-compliance
Lightning Source LLC
LaVergne TN
LVHW070728100526
838202LV00013B/1193